Student Solutions Manual
for Masterton and Hurley's
Chemistry
Principles and Reactions
Fifth Edition

Cassandra T. Eagle
Appalachian State University

THOMSON
™
BROOKS/COLE

Australia • Canada • Mexico • Singapore • Spain • United Kingdom • United States

The Thinker (detail) by Auguste Rodin; The Rodin Museum, Philadelphia: Gift of Jules E. Mastbaum

Printed in the United States of America
2 3 4 5 6 7   07 06 05 04

Printer: Globus Printing

ISBN: 0-534-40882-6

For more information about our products, contact us at:
Thomson Learning Academic Resource Center
1-800-423-0563

For permission to use material from this text, contact us by:
Phone: 1-800-730-2214
Fax: 1-800-730-2215
Web: http://www.thomsonrights.com

Brooks/Cole—Thomson Learning
10 Davis Drive
Belmont, CA 94002-3098
USA

Asia
Thomson Learning
5 Shenton Way #01-01
UIC Building
Singapore 068808

Australia/New Zealand
Thomson Learning
102 Dodds Street
Southbank, Victoria 3006
Australia

Canada
Nelson
1120 Birchmount Road
Toronto, Ontario M1K 5G4
Canada

Europe/Middle East/South Africa
Thomson Learning
High Holborn House
50/51 Bedford Row
London WC1R 4LR
United Kingdom

Latin America
Thomson Learning
Seneca, 53
Colonia Polanco
11560 Mexico D.F.
Mexico

Spain/Portugal
Paraninfo
Calle/Magallanes, 25
28015 Madrid, Spain

Dedication

I dedicate this *Student Solutions Manual* to Margaret G. Werts, who in fighting the greatest battle is teaching me about what really matters in life.

Acknowledgements

I thank Dale Wheeler for being an excellent accuracy checker for this work.

I thank the people who support and encourage me.   Thank you for being there for me, for caring for me and for showing me what love really is.

Jone Eagle
Richard Shea
Bobby Sharp
Sharon Sharp
Margaret G. Werts
Mary Beth Hege
Cathy Burleson
Al Burleson
Cama Duke
Glenda Hubbard
Mary McKinney
Marianne Romanat
Sarah Dunagan
Ed and Joyce Fletcher
Carol Hite
Kerry Mitkus
Britainy Dearman
Emily Ford
Dale Wheeler
Gary Richardson
Kira Knight
Brenda Lawhead
Stan and Ruth Leonard
Lucinda Bowers
Patty Schaller

I especially thank my daughter, Amanda, for her love and enthusiasm for life.

To the Student

Thank you for purchasing the *Student Solutions Manual*. This manual was prepared to help you increase your understanding of Introductory Chemistry. I suggest that you use this manual as a sports player would use a coach. The coach helps you understand what play to make, but YOU make the play. After reading a solution to a problem, close this book and make sure that you can do the problem yourself: apply pencil to paper and make the play! Reading this manual and saying you can do the problem is much like watching an excellent sports player and saying that you can make the same play because you watched him or her do it. Put what you learn into practice and you will begin to understand.

This manual is one of many resources you have in your Introductory Chemistry course. I strongly suggest that you take advantage of all the resources available to you in Introductory Chemistry. As a teacher of Introductory Chemistry, I especially recommend that you consult your instructor regularly to help increase your understanding of Introductory Chemistry.

There is more than one solution to some of the problems in your text. It is wise to consult your instructor when your method of doing a problem differs from that in the text.

Table of Contents

Chapter 1: Matter and Measurement

---

2. *Refer to Section 1.1.*

    a. Silver is an **element**. It is found on the periodic table.

    b. Ethyl alcohol is a **compound**. It is not an element because it is not found in the periodic table. It is not a mixture because it can not be broken down into more than one ingredient.

    c. Milk is a **mixture**. It contains water, fat, protein, calcium, and other ingredients.

    d. Aluminum is a **element**. It is found on the periodic table.

---

4. *Refer to Section 1.1.*

    a. Wine is a homogeneous mixture, the mixture is the same throughout. This is called a **solution**.

    b. Gasoline is a **solution** of a variety of organic compounds.

    c. Chocolate chip cookie batter is a **heterogeneous mixture**. Not every spoonful has the same number of chips.

---

6. *Refer to Section 1.1.*

    a. A mixture of volatile gases can be separated by **gas-liquid chromatography**.

    b. Rubbing alcohol is a homogeneous mixture of volatile liquids that can be separated by **gas-liquid chromatography**.

---

8. *Refer to Section 1.1, Table 1.1, and the Table of Atomic Masses in the front of your text.*

    a. chlorine   Cl

    b. phosphorus   P

    c. potassium   K

    d. mercury   Hg

10. *Refer to Section 1.1 and Table 1.1.*

    a. Si   silicon

    b. S   sulfur

    c. Fe   iron

    d. Zn   zinc

12. *Refer to Section 1.2.*

    a. a tape measure or a meter stick.

    b. a graduated cylinder for low accuracy; a buret for higher accuracy.

    c. a thermometer

14. *Refer to Section 1.2 and Table 1.2.*

Convert one of the numbers to the units of the other. Once the numbers are expressed in common units, they can be compared directly.

a. $37.12 \text{ g} \times \dfrac{1 \text{ kg}}{1000 \text{ g}} = 0.03712 \text{ kg}$, thus: $37.12 \text{ g} < 0.3712 \text{ kg}$.

b. $28 \text{ m}^3 \times \left(\dfrac{100 \text{ cm}}{1 \text{ m}}\right)^3 = 28 \times 10^6 \text{ cm}^3$, thus: $28 \text{ m}^3 > 28 \times 10^2 \text{ cm}^3$.

c. $525 \text{ mm} \times \dfrac{1 \text{ m}}{1000 \text{ mm}} \times \dfrac{1 \text{ nm}}{1 \times 10^{-9} \text{ m}} = 525 \times 10^6 \text{ nm}$,

    thus $525 \text{ mm} = 525 \times 10^6 \text{ nm}$

16. *Refer to Section 1.2 and Example 1.1.*

Convert from degrees Fahrenheit to Celsius, and then from Celsius to Kelvin.

$t_{°F} = 1.8 t_{°C} + 32$

$T_K = t_{°C} + 273.15$

$350°F = 1.8\ t_{°C} + 32°F$

$350°F - 32°F = 1.8\ t_{°C}$

$t_{°C} = 318°F / 1.8 = 177°C$

Note that 1.8 is an exact number, and thus does not limit the number of significant figures, resulting in three (3) significant figures in the answer.

$T_K = t_{°C} + 273.15 = 177 + 273.15 = 450\ K = 4.50 \times 10^2\ K$

Note that 177 shows the greater uncertainty (meaning that it has fewer digits past the decimal than 273.15), thus the answer must also have no digits after the decimal.

---

18. *Refer to Section 1.2 and Example 1.1.*

Convert the melting point of gallium from Celsius to degrees Fahrenheit.

$t_{°F} = 1.8 t_{°C} + 32$

$t_{°F} = 1.8(29.76°C) + 32 = 85.57°F$

Since the melting point of gallium (85.57 °F) is higher than the temperature inside the car (85.0 °F), the gallium would still be a solid when you return to your car.

---

20. *Refer to Section 1.2 and Example 1.2.*

a.	0.136	Three (3) significant figures; zeros before the decimal are not significant.
b.	0.0001050	Four (4) significant figures; zeros before the first number are not significant.
c.	2.700×10^3	Four (4) significant figures; zeros after the decimal are significant.
d.	6×10^{-4}	One (1) significant figure.
e.	56003	Five (5) significant figures; zeros between two significant figures are significant.

---

22. *Refer to Section 1.2 and Example 1.5.*

$$\text{volume} = \pi(r)^2 l = (3.1416)(2.500\ cm)^2(1.20\ cm)$$

$$= (3.1416)(6.250\ cm^2)(1.20\ cm) = 23.6\ cm^3$$

Note that since the length has the least number of significant figures, 3, the answer can only have three significant figures.

3

24. Refer to Section 1.2.

a. $x = 0.5$

The first number shows the greatest uncertainty (meaning that it has the least number of digits past the decimal point, 1), thus the answer must also have only 1 digit past the decimal point.

b. $x = 401.4$

The first number has the least number of significant figures, 4, and thus limits the number of significant figures in the answer to 4 (recall that leading zeros are not significant).

c. $\dfrac{269.4}{1003.7} = 0.2684$

The sum of the numbers is the numerator is 269.4 (four significant figures), and thus the number of significant figures in the answer would be limited to four.

d. $x = 7.8$

The least number of significant figures is two in the number 0.36, and thus limits the number of significant figures in the answer to two.

26. Refer to Section 1.2.

a. 132.5 g

b. 298.69 cm

c. 13 lb

d. 3.4×10^2 oz

28. Refer to Section 1.2.

a. 4020.6 mL = 4.0206×10^3 mL

b. 1.006 g (This is already in proper scientific notation.)

c. 100.1°C = 1.001×10^2°C

30. Refer to Section 1.2.

a. Weight is a measured quantity and is **not exact**.

b. A boiling point is a measured quantity and is **not exact**.

c. 12 is the square root of 144 by calculation and is **exact**.

32. Refer to Section 1.2, Tables 1.2 and 1.3, and Example 1.4.

a. $6743 \text{ nm} \times \dfrac{10 \text{ Å}}{1 \text{ nm}} = 67430 \text{ Å} = 6.743 \times 10^4 \text{ Å}$

b. $6743 \text{ nm} \times \dfrac{1 \times 10^{-9} \text{ m}}{1 \text{ nm}} \times \dfrac{39.37 \text{ in}}{1 \text{ m}} = 2.655 \times 10^{-4} \text{ in}$

c. $6743 \text{ nm} \times \dfrac{1 \times 10^{-9} \text{ m}}{1 \text{ nm}} \times \dfrac{1 \text{ km}}{1 \times 10^3 \text{ m}} \times \dfrac{1 \text{ mi}}{1.609 \text{ km}} = 4.191 \times 10^{-9} \text{ mi}$

34. Refer to Section 1.2, Tables 1.2 and 1.3, and Example 1.4.

a. $1 \text{ na. mi.} \times \dfrac{6076.12 \text{ ft}}{1 \text{ na. mi.}} \times \dfrac{1 \text{ mi}}{5280 \text{ ft}} = 1.15078 \text{ mi.}$

6 significant figures, the distance is given as an exact number, thus the number of s.f.'s is constrained by the conversion factors, and 5280 ft/mi. is an exact number..

b. $1 \text{ na. mi.} \times \dfrac{6076.12 \text{ ft}}{1 \text{ na. mi.}} \times \dfrac{12 \text{ in}}{1 \text{ ft}} \times \dfrac{1 \text{ m}}{39.37 \text{ in}} = 1852 \text{ m.}$

4 significant figures, the distance is given as an exact number, as is the 12 in/ft conversion factor, thus the number of s.f.'s is constrained by the other two conversion factors.

c. $22 \text{ knots} \times \dfrac{1 \text{ na. mi./hr}}{1 \text{ knot}} \times \dfrac{1.151 \text{ mi}}{1 \text{ na mi.}} = 25 \text{ mi./hr} = 25 \text{ mph}$

2 significant figures, the first conversion factor is an exact number, while the second (from part a above) has 4 s.f.'s, thus the answer is constrained by the 2 s.f.'s given in the initial number.

$$0.25 \, \text{mi} \times \frac{5.0 \, \text{min}}{\text{mi}} = 1.2 \, \text{min}$$

A runner would take 1.2 min to complete the 0.25 mi lap.

$$0.50 \, \text{km} \times \frac{\text{mi}}{1.609 \, \text{km}} \times \frac{5.0 \, \text{min}}{\text{mi}} = 1.6 \, \text{min}$$

A runner would take 1.6 min to complete the 0.50 km lap.

Convert mg to g and dL to mL.

$$\frac{185 \, \text{mg}}{1 \, \text{dL}} \times \frac{1 \, \text{g}}{1000 \, \text{mg}} \times \frac{10 \, \text{dL}}{1 \, \text{L}} \times \frac{1 \, \text{L}}{1000 \, \text{mL}} = \frac{1.85 \times 10^{-3} \, \text{g}}{\text{mL}}$$

Volume = area x height. Convert the area to units of cm^2 and the height to cm and calculate volume. Then convert from cm^3 (mL) to liters.

$$\text{area} = 3.02 \times 10^6 \, \text{mi}^2 \times \left(\frac{1.609 \, \text{km}}{1 \, \text{mi}} \right)^2 \times \left(\frac{1 \times 10^5 \, \text{cm}}{1 \, \text{km}} \right)^2 = 7.82 \times 10^{16} \, \text{cm}^2$$

$$\text{height} = 2 \, \text{in} \times \frac{2.54 \, \text{cm}}{1 \, \text{in}} = 5.08 \, \text{cm}$$

$$\text{volume} = 7.82 \times 10^{16} \, \text{cm}^2 \times 5.08 \, \text{cm} = 3.97 \times 10^{17} \, \text{cm}^3 = 3.97 \times 10^{17} \, \text{mL}$$

$$3.97 \times 10^{17} \, \text{mL} \times \frac{1 \, \text{L}}{1000 \, \text{mL}} = 3.97 \times 10^{14} \, \text{L}$$

42. *Refer to Section 1.2 and Example 1.4.*

Note that the word five is used for 5 grains in the problem, thus this number is exact and is not used to determine the number of significant figures in the answer.

$$5\text{ grains} \times \frac{\text{lb}}{5.760 \times 10^3 \text{ grains}} \times \frac{453.6\text{ g}}{\text{lb}} \times \frac{1000\text{ mg}}{\text{g}} = 3.938 \times 10^2 \text{ mg}$$

44. *Refer to Sections 1.2 and 1.3.*

$$\frac{1.20 \times 10^2 \text{ g}}{112\text{ mL}} = 1.07\text{ g/mL}$$

46. *Refer to Section 1.3.*

Calculate the volume of the object from the change in the volume of the graduated cylinder, then calculate the density.

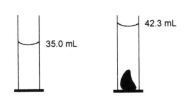

35.0 mL 42.3 mL

Volume of object = 42.3 mL – 35.0 mL = 7.3 mL

$$\text{density} = \frac{\text{mass (g)}}{\text{volume (mL)}} = \frac{11.33\text{ g}}{7.3\text{ mL}} = 1.6\text{ g/mL}$$

48. *Refer to Section 1.3, Table 1.3, and Example 1.5.*

Since the mass and the density are given, the volume (V) of the wire can be calculated.

$$10\text{ lb} \times \frac{453.6\text{ g}}{\text{lb}} \times \frac{\text{cm}^3}{2.70\text{ g}} = 1700\text{ cm}^3$$

Knowing the volume (V) of the wire, the equation can be used to determine the length (l). But first the diameter must be converted to the radius in centimeters. Remember, the diameter is equal to twice the radius.

$$\text{radius} = \frac{0.0808\text{ in}}{2} \times \frac{2.54\text{ cm}}{\text{in}} = 0.103\text{ cm}$$

$$V = \pi r^2 l$$

$$1700\text{ cm}^3 = \pi\,(0.103\text{ cm})^2\, l$$

$$l = \frac{1700 \text{ cm}^3}{\pi (1.06 \text{ cm}^2)} = 5.10 \times 10^4 \text{ cm} = 5.10 \times 10^2 \text{ m}$$

50. Refer to Section 1.3.

21% oxygen by volume = 21 L oxygen / 100 L air
Convert mass of oxygen to volume. Then convert volume of oxygen to volume of air.

$$55 \text{ kg oxygen} \times \frac{1 \text{ g oxygen}}{1 \times 10^{-3} \text{ kg oxygen}} \times \frac{1 \text{ L oxygen}}{1.31 \text{ g oxygen}} \times \frac{100 \text{ L air}}{21 \text{ L oxygen}} = 2.0 \times 10^5 \text{ L air}$$

52. Refer to Section 1.3 and Example 1.6.

Use the solubility of potassium sulfate as a conversion factor. Determine how much potassium sulfate would be dissolved in 225 g water.

$$225 \text{ g water} \times \frac{15 \text{ g potassium sulfate}}{100 \text{ g water}} = 34 \text{ g potassium sulfate}$$

The solution is **supersaturated** since a saturated solution of potassium sulfate containing 225 grams of water would only contain 34 grams of potassium sulfate.

Since 39.0 grams of potassium sulfate was used to prepare this solution, **5 grams** would be expected to crystallize out of solution.

54. Refer to Section 1.3 and Example 1.6.

Use the solubility of potassium chloride as a conversion factor.

a. at 30°C: $48.6 \text{ g water} \times \dfrac{37.0 \text{ g potassium chloride}}{100 \text{ g water}} = 18.0 \text{ g potassium chloride}$

b. at 70°C: $52.0 \text{ g potassium chloride} \times \dfrac{100 \text{ g water}}{48.3 \text{ g potassium chloride}} = 108 \text{ g water}$

c. For this part, calculate the amount of potassium chloride that will dissolve in 75 g water at each of the temperatures. Is it more than 30.0 g?

at 30°C: $75.0 \text{ g water} \times \dfrac{37.0 \text{ g potassium chloride}}{100 \text{ g water}} = 27.8 \text{ g potassium chloride}$

at 70°C: $75.0 \text{ g water} \times \dfrac{48.3 \text{ g potassium chloride}}{100 \text{ g water}} = 36.2 \text{ g potassium chloride}$

Thus, not all the potassium chloride would dissolve at 30°C, but would at 70°C.

---

56. Refer to Section 1.3.

a. **physical.** Color is observed in the absence of a chemical reaction.

b. **physical.** The state (solid) is observed in the absence of a chemical reaction.

c. **physical.** Solubility is observed in the absence of a chemical reaction.

d. **chemical.** This is a chemical reaction.

---

58. Refer to Section 1.3.

Calculate the volumes using the densities of each.

$\text{Volume of lead} = 1 \text{ g} \times \dfrac{1 \text{ cm}^3}{11.34 \text{ g}} = 8.818 \times 10^{-2} \text{ cm}^3$

$\text{Volume of oxygen} = 1 \text{ g} \times \dfrac{1 \text{ cm}^3}{1.31 \times 10^{-3} \text{ g}} = 763 \text{ cm}^3$

One gram of oxygen takes up almost 10,000 times the volume of the lead. Gases will always have a much greater volume than equal amounts of a solid.

---

60. Refer to Sections 1.2 and 1.3 and Table 1.3.

$108 \text{ carats} \times \dfrac{2.00 \times 10^2 \text{ mg}}{1 \text{ carat}} \times \dfrac{1 \text{ g}}{1000 \text{ mg}} \times \dfrac{1 \text{ lb}}{453.6 \text{ g}} = 4.76 \times 10^{-2} \text{ lbs}$

$108 \text{ carats} \times \dfrac{2.00 \times 10^2 \text{ mg}}{1 \text{ carat}} \times \dfrac{1 \text{ g}}{1000 \text{ mg}} \times \dfrac{1 \text{ cm}^3}{3.51 \text{ g}} \times \left(\dfrac{1 \text{ in}}{2.54 \text{ cm}}\right)^3 = 0.376 \text{ in}^3$

---

62. Refer to Section 1.3.

Using the mass and density, calculate the volume, and from that the cylinder's radius.

$153.2 \text{ g} \times \dfrac{1 \text{ cm}^3}{4.55 \text{ g}} = 33.7 \text{ cm}^3$

$V = \pi r^2 l$

$33.7 \text{ cm}^3 = (3.14)(r^2)(7.75 \text{ cm})$

$r^2 = 1.38 \text{ cm}^2$

$r = 1.18 \text{ cm}$

diameter $= 2r = 2.36 \text{ cm}$

---

64. *Refer to Sections 1.1 and 1.3.*

a. Chemical properties involve a chemical change in the substance, while physical properties can be observed with no change in the identity of the substance.

b. Distillation is used to separate a homogeneous mixture and involves a phase change from liquid to gas and then back to liquid. Filtration is used to separate a the liquid from the solid in a heterogeneous mixture and does not involve a phase change.

c. The solute is the substance dissolved in the solvent.

---

66. *Refer to Section 1.3.*

The less dense substance will float on top of the more dense substance. Thus, from top to bottom, the substances will be: ethyl alcohol, lead and mercury.

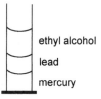

ethyl alcohol

lead

mercury

---

68. *Refer to Section 1.2*

We want: $t_{\circ F} = 2(t_{\circ C})$. Substitute this into the equation for converting Fahrenheit to Celsius.

$t_{\circ F} = 1.8(t_{\circ C}) + 32$

$2(t_{\circ C}) = 1.8(t_{\circ C}) + 32$

$0.2(t_{\circ C}) = 32$

$t_{\circ C} = 160°C$

When $t_{\circ C} = 160°C$, $t_{\circ F} = 1.8(160) + 32 = 320°F$

---

69. *Refer to Section 1.1 and Table 1.3.*

Convert gallons to km^3. Then convert nm to km and solve for area (remember: volume = area x depth).

$$31.5 \text{ gal} \times \frac{4 \text{ qt}}{1 \text{ gal}} \times \frac{57.75 \text{ in}^3}{1 \text{ qt}} \times \frac{(1 \text{ m})^3}{(39.37 \text{ in})^3} \times \frac{(1 \text{ km})^3}{(1000 \text{ m})^3} = 1.19 \times 10^{-10} \text{ km}^3$$

$$100 \text{ nm} \times \frac{1 \text{ m}}{1 \times 10^9 \text{ nm}} \times \frac{1 \text{ km}}{1000 \text{ m}} = 1.0 \times 10^{-10} \text{ km}$$

$$\frac{1.19 \times 10^{-10} \text{ km}^3}{1.0 \times 10^{-10} \text{ km}} = 1.2 \text{ km}^2$$

---

70. Refer to Section 1.3 and Table 1.3.

First calculate the volume of Al wire needed, then calculate the length. (Remember that diameter = 2r.)

$$12.0 \text{ g} \times \frac{1 \text{ cm}^3}{2.70 \text{ g}} = 4.44 \text{ cm}^3$$

$$r = \frac{0.200 \text{ in} \times 2.54 \text{ cm/in}}{2} = 0.254 \text{ cm}$$

$$V = \pi r^2 l$$

$$4.44 \text{ cm}^3 = (3.14)(0.254 \text{ cm})^2(l)$$

$$l = 21.9 \text{ cm}$$

---

71. Refer to Section 1.1 and Table 1.3.

Calculate the amount of air one breathes in a year and the amount of lead this corresponds to. Then determine the amount retained in one's lungs.

$$\frac{8.50 \times 10^3 \text{ L}}{1 \text{ day}} \times \frac{365 \text{ days}}{1 \text{ year}} = 3.10 \times 10^6 \text{ L air / year}$$

$$\frac{7.0 \times 10^{-6} \text{ g Pb}}{1 \text{ m}^3} \times \frac{1 \text{ m}^3}{1 \times 10^3 \text{ L}} \times \frac{3.10 \times 10^6 \text{ L}}{1 \text{ year}} = 2.17 \times 10^{-2} \text{ g Pb / year}$$

2.17×10^{-2} g / year $\times$ 0.75 $\times$ 0.5 = 8.1×10^{-3} g Pb / year   (retained by lungs)

Chapter 2: Atoms, Molecules, and Ions

2. Refer to Section 2.1 and "Chemistry, the Human Side."

The law of constant composition states that the ratio of the masses of elements in a compound is constant. For example, different samples of water will have the same mass ratio of hydrogen to oxygen.

4. Refer to Section 2.1 and "Chemistry, the Human Side."

a. **None** of the laws are illustrated here.

b. If the mass of the reactants (unbroken seal) equals the mass of the products (after seal is broken), one is demonstrating the **law of conservation of mass**.

c. **Neither** of the laws are illustrated here.

6. Refer to Section 2.2 and Figure 2.4.

Rutherford discovered the nucleus. His research group removed the electrons from helium atoms and then shot them at a gold foil target. He observed that some of those helium atoms bounced backwards. Rutherford deduced that this was because the helium atoms were colliding with the gold nuclei in the foil.

8. Refer to Section 2.2, Example 2.1, and the Periodic Table (inside front cover).

Rn: Atomic number (Z) = 86 = 86 protons
   Mass number (A) = 220 = neutrons + 86 protons

   neutrons = 220 − 86 = 134 neutrons

10. Refer to Section 2.2, Example 2.1, and the Periodic Table (inside front cover).

Fe Atomic number (Z) = 26 = 26 protons
   Mass number (A) = 28 neutrons + 26 protons = 54
   Mass number (A) = 30 neutrons + 26 protons = 56

a. $^{54}_{26}\text{Fe}$ $^{56}_{26}\text{Fe}$

b. Fe-54 contains 28 neutrons whereas Fe-56 contains 30 neutrons.

12. Refer to Section 2.2.

a. Na-21 has 11 protons and $21 - 11 = 10$ neutrons. Adding one neutron makes the number 22, thus the symbol is $_{11}^{22}\text{Na}$.

b. An isobar of Na-21 with 10 protons would have an atomic number of 10, thus the element symbol would be Ne. The nuclear symbol is $_{10}^{21}\text{Ne}$.

c. A nucleus with 11 protons is sodium. The mass number is $11 + 12 = 23$. The nuclear symbol is $_{11}^{23}\text{Na}$. This is indeed an isotope of Na-21.

14. Refer to Section 2.2, Example 2.1, and the Periodic Table (inside front cover).

a. Number of protons = the atomic number (written above the chemical symbol)
Number of protons = 34

b. Number of neutrons = mass number – atomic number.
Number of neutrons = $75 - 34 = 41$

c. Number of electrons = number of protons (in a neutral atom) = 34

d. Se^{2-} must have two more electrons than protons [(# of protons) – (# of electrons) = charge]. The number of protons equals the atomic number.
# of neutrons = 41
# of protons = 34
# of electrons = 36

16. Refer to Section 2.2 and Example 2.3.

Remember: Number of protons = atomic number
Number of neutrons = mass number – atomic number
Number of electrons = (# of protons) – charge

Nuclear Symbol	Atomic Charge	# of Protons	# of Neutrons	# of Electrons
$_{35}^{79}\text{Br}$	0	35	44	35
$_{7}^{14}\text{N}^{3-}$	-3	7	7	10
$_{33}^{75}\text{As}^{5+}$	+5	33	42	28
$_{40}^{90}\text{Zr}^{4+}$	+4	40	50	36

18. Refer to Section 2.2 and Example 2.1.

Number of protons = sum of the atomic numbers
Number of electrons = (# of protons) – charge

a. C_{60}: 60 carbons, each with an atomic number of 6, therefore
the number of protons is: 60 x 6 = 360.
This molecule has no charge, thus the number of electrons = 360 – 0 = 360.

b. CN^-: One carbon with 6 protons and one nitrogen with 7 protons, therefore
the number of protons is: 6 + 7 = 13.
This ion has a -1 charge, thus the number of electrons = 13 – (-1) = 14.

c. CO_2: One carbon with 6 protons and two oxygens, each with 8 protons, therefore
the number of protons is: 6 + (2 x 8) = 22.
This molecule has no charge, thus the number of electrons = 22 – 0 = 22.

d. N^{3-}: One nitrogen with an atomic number of 7, therefore
the number of protons is: 7.
This ion has a -3 charge, thus the number of electrons = 7 – (-3) = 10.

20. Refer to Section 2.3, Figure 2.7, and the Table of Atomic Masses (inside front cover).

a. Mn   manganese

b. Na   sodium

c. As   arsenic

d. S   sulfur

e. Pb   lead

22. Refer to Section 2.3, Figure 2.7, and the Periodic Table (inside front cover).

a. Mn   metal, transition metal

b. Na   metal, main group metal

c. As   metalloid

d. S   nonmetal

e. Pb   metal, post-transition metal

A period is a horizontal row, numbered from top to bottom. The first period consists of hydrogen and helium.

a. The second period is the row beginning with Li and ending with Ne. B is a metalloid. The two elements to the left of B are metals, the **five** to the right of B (C, N, O, F, and Ne) are nonmetals.

b. The fourth period is the row beginning with K and ending with Kr. Ge and As are metalloids. The 13 elements to the left of the metalloids are metals, the **three** to the right (Se, Br, and Kr) are nonmetals.

c. The sixth period is the row beginning with Cs and ending with Rn (including the lanthanide series). The 30 elements to the left of the thick black line are metals, the **two** to the right (At and Rn) are the nonmetals.

A period is a horizontal row, numbered from top to bottom. The first period consists of hydrogen and helium.

a. The **first period** has no metals; it consists of hydrogen and helium.

b. The **seventh period**, the one beginning with Fr, has no nonmetals, so long as you don't count element 118.

c. The **fourth period** has one post-transition metal, Ga, and two metalloids, Ge and As.

For the condensed structural formula, write the formula of the highlighted portion separate from the formula of the rest of the molecule. Count up the total number of each element. Write that number as a subscript after the symbol for the element.

a. CH_3COOH, $C_2H_4O_2$

b. CH_3Cl

a. H_2O

b. NH_3

c. N_2H_4

d. SF_6 (the "hexa" prefix indicate that 6 fluorides are attached to the sulfur).

e. PCl_5 (the "penta" prefix indicate that 5 chlorides are attached to the phosphorus).

---

32. Refer to Section 2.6 and Example 2.7.

a. CO   carbon monoxide

b. SiC   silicon carbide

c. XeF_6 xenon hexafluoride

d. P_4O_{10} tetraphosphorus decaoxide

e. C_2H_2 acetylene

---

34. Refer to Section 2.5.

The sum of the charges must equal zero; if not, add extras of the needed ion to "balance" the charges.

a. Ba^{2+} and I^- give BaI_2 $(+2 + 2(-1) = 0)$

   Ba^{2+} and N^{3-} give Ba_3N_2 $(3(+2) + 2(-3) = 0)$

b. O^{2-} and Fe^{2+} give FeO $(-2 + (+2) = 0)$

   O^{2-} and Fe^{3+} give Fe_2O_3 $(3(-2) + 2(+3) = 0)$

---

36. Refer to Section 2.5, Tables 2.2 and 2.3, and Example 2.6.

The sum of the charges of the ions must equal zero. Increase the number of the deficient ion until the charges "balance."

a. Cobalt(II) acetate. The Roman numeral tells us that cobalt is a +2 ion (Co^{2+}).
   Acetate is a -1 ion ($C_2H_3O_2^-$). For the charges to add up to zero, we must have two acetates.
   $Co(C_2H_3O_2)_2$

b. Barium oxide. Barium is in group 2 and thus has a +2 charge. Oxygen is in group 16 and thus has –2 charge. These charges add up to zero.
BaO

b. Aluminum sulfide. Aluminum is in group 1**3** and thus has a +3 charge (Al^{+3}), while sulfide has a –2 charge (S^{-2}). For these charges to add up to zero, we must have 2 aluminum and 3 sulfide ions.
Al_2S_3

d. Potassium permanganate. Potassium, in group 1, has a +1 charge (K^+) and permanganate has a –1 charge (MnO_4^-). These charges add up to zero.
$KMnO_4$

e. Sodium hydrogen carbonate. Sodium is in group 1 and thus has a +1 charge (Na^+). Hydrogen carbonate ion is a -1 ion (HCO_3^-). These charges add up to zero.
$NaHCO_3$

---

38. *Refer to Section 2.6 and Example 2.6.*

Name the compounds by naming the cations (positive ions) followed by the anions (negative ions). For transition metals, indicate the charge of the metal with Roman numerals after the chemical symbol.

a. $ScCl_3$ **scandium(III) chloride** Since each chloride has a –1 charge, the Sc must have a +3 charge for the charges to balance (add up to zero).

b. $Sr(OH)_2$ **strontium hydroxide**

c. $KMnO_4$ **potassium permanganate**

d. Rb_2S **rubidium sulfide**

e. Na_2CO_3 **sodium carbonate**

---

40. *Refer to Sections 2.5 and 2.6, and Example 2.8.*

---

a. hydrochloric acid   **HCl**

b. sodium nitrite   **$NaNO_2$**

c. chromium(III) sulfite   **$Cr_2(SO_3)_3$**

d. potassium chlorate   **$KClO_3$**

e. iron(III) perbromate   **$Fe(BrO_4)_3$**

	Name	Formula
a	sodium dichromate	$Na_2Cr_2O_7$
b	**bromine triiodide**	BrI_3
c	copper(II) hypochlorite	$Cu(ClO)_2$
d	**disulfur dichloride**	S_2Cl_2
e	potassium nitride	K_3N

a. Sodium has a +1 charge (Na^+) while dichromate has a –2 charge ($Cr_2O_7^{2-}$), thus there must be two sodiums to balance the charges.

b. This is a molecular compound since both elements are nonmetals, so it is important to add the proper prefix to denote that there are three iodides.

c. Copper has a +2 charge (Cu^{2+}) while hypochlorite has a –1 charge (ClO^-), thus there must be two hypochlorites to balance the charges.

d. This is a molecular compound since both elements are nonmetals, so it is important to add the proper prefixes to denote that there are two sulfurs and two chlorides.

e. Potassium has a +1 charge (K^+) while nitride has a –3 charge (N^{3-}), thus there must be three potassiums to balance the charges.

44. Refer to Sections 2.2 and 2.3 and Figure 2.7.

a. Tin, Sn, has 50 protons, whereas tellurium, Te, has 52 protons.

b. Vanadium, V.

c. Cesium, Cs.

d. Antimony, Sb, is a metalloid in Group 15 and has 51 protons.

46. Refer to Section 2.4 and Example 2.2.

a. C_2H_7N

b. $C_2H_5NH_2$ Write the formula of the reactive group separately from the rest of the molecule.

48. *Refer to Sections 2.4 and 2.5.*

a. "Compounds containing carbon atoms are molecular" is **usually true**. Carbon, being a nonmetal can combine with other non-metals to form molecular compounds such as those mentioned in Section 2.4 (indeed, a whole field of chemistry, called organic chemistry, is dedicated to the study of such compounds). Carbon can, however, also form ionic compounds such as calcium carbide (used in old miners lamps).

b. "A molecule is made up of nonmetal atoms" is **always true**, by definition.

c. "An ionic compound has at least one metal atom" is **usually true**. Most ionic compounds do contain a metal, but there are ionic compounds in which the cation (positive ion) is not a metal, such as in ammonium chloride, NH_4Cl.

50. *Refer to Section 2.2.*

a. True, the unknown element is sodium, Na, since it has 11 protons.

b. False, the unknown element has 11 protons and vanadium has 23 protons.

c. False, the unknown element has 12 neutrons.

d. False, X^{2+} would have 9 electrons.

e. False, the proton/neutron ratio, 11/12, is about 0.9.

52. *Refer to Section 2.1.*

Look for the box with the same number of squares and circles as the reaction mixture: 5 squares and 6 circles. This product mixture is found in box **#4**.

54. *Refer to Section 2.5 and Figure 2.11.*

The potassium ions (K^+) and chloride ions (Cl^-) would surround each other similar to the representation of NaCl in Figure 2.11.

Since the charge of the ion is –2, the number of electrons in this ion is two more than the number of protons.

$$e^- = p + 2$$

It is given in the problem that the number of neutrons is 20 more than the number of electrons. Therefore, there are 22 more neutrons than protons.

$$n = e^- + 20 = (p + 2) + 20 = p + 22$$

It is also given that the ion has a mass number of 126. Therefore, the sum of the protons and neutrons is equal to 126.

$$p + n = 126$$

Since $n = p + 22$, this information can be used to solve for the number of protons.

$$p + (p + 22) = 126$$

$$2p = 104$$

$p = 52$, therefore the element must be tellerium, Te.

$n = p + 22 = 52 + 22 = 74$, therefore there must be 74 neutrons in this ion.

$e^- = p + 2 = 52 + 2 = 54$, therefore there must be 54 electrons.

The symbol for this ion is therefore $^{126}_{52}Te^{2-}$

Compound A: $\dfrac{\text{mass H}}{\text{mass C}} = \dfrac{2.39\,\text{g}}{28.5\,\text{g}} = 0.0839$.

Compound B: $\dfrac{\text{mass H}}{\text{mass C}} = \dfrac{11.6\,\text{g}}{34.7\,\text{g}} = 0.334$.

a. Compound C: $\dfrac{\text{mass H}}{\text{mass C}} = \dfrac{5.84\,g}{16.2\,g} = 0.360$.

$\dfrac{0.360}{0.0839}$ = 4.30 This example does **not** follow the law of multiple proportions.

b. Compound C: $\dfrac{\text{mass H}}{\text{mass C}} = \dfrac{3.47\,g}{16.2\,g} = 0.214$.

$\dfrac{0.214}{0.0839}$ = 2.55 This example does **not** follow the law of multiple proportions.

c. Compound C: $\dfrac{\text{mass H}}{\text{mass C}} = \dfrac{2.72\,g}{16.2\,g} = 0.168$.

$\dfrac{0.168}{0.0839}$ = 2.00 This example **does** follow the law of multiple proportions since the H/C ratio in compound A is a whole number multiple of the H/C ratio in compound C.

---

59. Refer to Sections 2.1 and 2.4.

a. ethane: $\dfrac{4.53\ g\ H}{18.0\ g\ C}$ = 0.252 g H / g C

ethene: $\dfrac{7.25\ g\ H}{43.20\ g\ C}$ = 0.168 g H / g C

While both these compounds have only hydrogen and carbon, they have different ratios of these elements.

b. $\dfrac{\text{ethane}}{\text{ethene}} = \dfrac{0.252\ g\ H\,/\,g\ C}{0.168\ g\ H\,/\,g\ C}$ = 1.5

Thus ethane has 3 hydrogens for every 2 in ethene.

Reasonable formulae: ethane - CH_3 (actual formula is C_2H_6)
ethene - CH_2 (actual formula is C_2H_4)

---

60. Refer to Sections 1.3 and 2.2.

Calculate the mass and volume of one Al-27, then calculate the density.

mass = 13(mass of a proton) + 14(mass of a neutron) + 13(mass of an electron)
= 13(1.6726 x 10^{-24} g) + 14(1.6749 x 10^{-24} g) + 13(9.1094 x 10^{-28} g)
= 2.1744 x 10^{-23} g + 2.3449 x 10^{-23} g + 1.1842 x 10^{-26} g
= 4.5205 x 10^{-23} g

$$\text{volume} = \frac{4}{3}\pi r^3 = \frac{4}{3}\pi(1.43 \times 10^{-8}\,\text{cm})^3 = 1.22 \times 10^{-23}\,\text{cm}^3$$

$$\text{density} = \frac{\text{mass}}{\text{volume}} = \frac{4.5205 \times 10^{-23}\,\text{g}}{1.22 \times 10^{-23}\,\text{cm}^3} = 3.71\,\text{g/cm}^3$$

The actual density of Al is 2.70 g/cm^3. There is quite a bit of free space around each aluminum atom, much as marbles in a jar have free space around them.

---

61. Refer to Section 2.2.

The difference between a Be atom and a Be^{2+} ion is two electrons.

mass of Be^{2+} = 1.4965 x 10^{-23}g - 2(9.1094 x 10^{-28}g) = 1.4963 x 10^{-23}g

---

62. Refer to Section 1.2.

---

a. $200\text{ breaths} \times \dfrac{500\text{ mL}}{\text{breath}} \times \dfrac{2.5 \times 10^{19}\text{ molecules}}{\text{mL}} = 2.5 \times 10^{24}\text{ molecules}$

b. $\dfrac{2.5 \times 10^{24}}{1.1 \times 10^{44}} = 2.3 \times 10^{-20}$

c. $1\text{ breath} \times \dfrac{500\text{ mL}}{\text{breath}} \times \dfrac{2.5 \times 10^{19}\text{ molecules}}{\text{mL}} \times 2.3 \times 10^{-20} = 2.9 \times 10^{2}\text{ molecules}$

Chapter 3: Mass Relations in Chemistry; Stoichiometry

---

2. *Refer to Section 3.1 and the Periodic Table and Table of Atomic Masses (inside front cover).*

Determine the masses of each and then arrange in order.

a. Na^+ ion: 22.99 amu
b. Se atom: 78.96 amu
c. S_8 molecule: 8(32.07) = 560 amu
d. Sc atom: 44.96 amu

Thus, the order of increasing mass is a < d < b < c

---

4. *Refer to Section 3.1 and Example 3.1.*

Convert from percent to decimal by dividing by 100.
$0.04\% / 100\% = 4 \times 10^{-4}$
$0.20\% / 100\% = 2.0 \times 10^{-3}$
$99.76\% / 100\% = 9.976 \times 10^{-1}$

Atomic mass of oxygen
$= 16.00(9.976 \times 10^{-1}) + 17.00(4 \times 10^{-4}) + 18.00(2.0 \times 10^{-3})$
$= 15.69 + 0.007 + 0.036 = 16.00$ amu

---

6. *Refer of Section 3.1, Example 3.1 and the Table of Atomic Masses.*

If Cu-63 is 69.17% abundant, then the other isotope must be is 30.83% (100-69.17) abundant.
The average mass (called atomic mass) of Cu = 63.546 amu (from the table of atomic masses).
average mass = (fraction of Cu-63)(mass of Cu-63) + (fraction of Cu-?)(mass of Cu-?)
set (mass of Cu-?) = x, and then solve for x.
63.546 amu = (0.6917)(62.9296 amu) + (0.3083)x
63.546 amu = 43.5284 amu + 0.3083x
20.0176 = 0.0.3083x
x = 64.93 amu
The mass number (A) is thus 65 and the number of protons (Z) is 29.
The nuclear symbol is $^{65}_{29}Cu$

8. *Refer to Section 3.1 and Example 3.1.*

Let the abundance of the first isotope be x.
The total abundance must equal 100, thus:
.204% + x + abundance of isotope 2 = 100, and
abundance of isotope 2 = 99.796% - x

Isotope	Mass	Abundance
1	15.9949 amu	x
2	16.9993 amu	99.796% - x
3	17.9992 amu	0.204%

$15.9994 = 15.9949(x) + 16.9993(.99796 - x) + 17.9992(0.00204)$
$15.9994 = 15.9949x + 16.9646 - 16.9993x + 0.0367$
$15.9994 = 17.0013 - 1.0044x$
$1.0044x = 1.0019$
$x = 0.99751 = 99.751\%$

Abundance of isotope 1 = 99.751%
Abundance of isotope 2 = 99.796 – 99.751% = 0.045%
Abundance of isotope 3 = 0.204%

10. *Refer to Section 3.1 and Figure 3.2.*

a.   Two, HCl-35 and HCl-37.

b.   HCl-35: 35 + 1 = 36; and HCl-37: 37 + 1 = 38.

c.

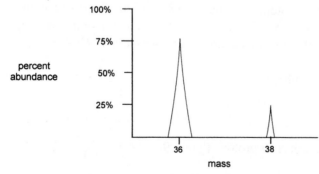

12. *Refer to Section 3.2 and Example 3.3.*

a. $1\times10^{-6}\,\text{mol.}\times\dfrac{6.022\times10^{23}\,\text{hazelnut meats}}{1\,\text{mol.}}\times\dfrac{0.985\text{g}}{1\,\text{hazelnut meat}} = 5.93\times10^{17}\,\text{g}$

b. $1\,lb \times \dfrac{453.6\,g}{1\,lb} \times \dfrac{1\,hazelnut\,meat}{0.985\,g} \times \dfrac{1\,mol.}{6.022 \times 10^{23}\,hazelnut\,meat} = 7.65 \times 10^{-22}\,mol.$

---

14. Refer to Section 3.2.

By definition, one mole of silver (Ag) = 6.022 x 10²³ atoms; one mole of silver = 107.87 g.

a. $\dfrac{1 \times 10^{-8}\,g\,Ag}{L} \times \dfrac{1\,mol.\,Ag}{107.87\,g\,Ag} \times \dfrac{6.022 \times 10^{23}\,atoms\,Ag}{1\,mol.\,Ag} = 6 \times 10^{13}\,atoms\,Ag/L/week$

b. $\dfrac{1 \times 10^{-8}\,g\,Ag}{L} \times \dfrac{1\,mol.\,Ag}{107.87\,g\,Ag} = 9 \times 10^{-11}\,mol.\,Ag/L\,of\,air$

---

16. Refer to Sections 3.1 and 3.2.

a. $10.000\,g\,Si \times \dfrac{1\,mole}{27.977\,g} = 0.35744\,moles$

b. $0.35744\,mol. \times \dfrac{6.022 \times 10^{23}\,atoms}{1\,mole} = 2.152 \times 10^{23}\,atoms$

c. $2.152 \times 10^{23}\,atoms \times \dfrac{14\,protons}{1\,atom} = 3.013 \times 10^{24}\,protons$

$2.152 \times 10^{23}\,atoms \times \dfrac{14\,neutrons}{1\,atom} = 3.013 \times 10^{24}\,neutrons$

$2.152 \times 10^{23}\,atoms \times \dfrac{14\,electrons}{1\,atom} = 3.013 \times 10^{24}\,electrons$

Total = 3(3.013 x 10²⁴) = 9.039 x 10²⁴ particles

---

18. Refer to Sections 3.1 and 3.2 and Example 3.2.

Recall that the number of neutrons = mass number – atomic number, and that only the number of electons (not the number of neutrons) changes going from atom to ion.
Note: Since atoms and neutrons are exact quantities, there are no restrictions in significant figures.

a. $25\,atoms\,^{90}Y \times \dfrac{51\,neutrons}{atom\,of\,^{90}Y} = 1.275 \times 10^{3}\,neutrons$

b. $0.250\,mol.\,^{90}Y \times \dfrac{6.022 \times 10^{23}\,atoms\,^{90}Y}{1\,mol.\,^{90}Y} \times \dfrac{51\,neutrons}{atom\,of\,^{90}Y} = 7.68 \times 10^{24}\,neutrons$

c. 1.0×10^{-9} g ^{90}Y $\times \dfrac{1\,\text{mol.}\ ^{90}\text{Y}}{90\,\text{g}\ ^{90}\text{Y}} \times \dfrac{6.022 \times 10^{23}\ \text{atoms}\ ^{90}\text{Y}}{1\,\text{mol.}\ ^{90}\text{Y}} \times \dfrac{51\,\text{neutrons}}{\text{atom of}\ ^{90}\text{Y}} = 3.4 \times 10^{14}$ neutrons

---

20. Refer to Section 3.2 and Example 3.3.

Multiply the number of each atom in the molecule by the atomic mass of that atom, and then add up the masses to get the molar mass of the molecule.

a. Ga: $1(69.723) = 69.723$ g/mol.

b. $CaSO_4 \cdot \tfrac{1}{2}\, H_2O$: $1(40.078) + 1(32.066) + 4(15.9994) + \tfrac{1}{2}\,[2(1.00794) + 1(15.9994)]$
$= 136.1416 + \tfrac{1}{2}\,(18.01528) = 145.149$ g/mol.

c. $C_{14}H_{10}O_4$: $14(12.011) + 10(1.0079) + 4(15.9994)$
$= 168.154 + 10.0794 + 63.9976 = 242.231$ g/mol.

---

22. Refer to Section 3.2 and Example 3.3.

a. CF_2Cl_2: $12.01 + 2(19.00) + 2(35.45) = 120.91$ g/mol.
$$35.00\,\text{g}\ CF_2Cl_2 \times \dfrac{1\,\text{mol.}\ CF_2Cl_2}{120.91\,\text{g}\ CF_2Cl_2} = 0.2895\,\text{mol.}\ CF_2Cl_2$$

b. iron(II) sulfate ($FeSO_4$): $55.85 + 32.07 + 4(16.00) = 151.92$ g/mol.
$$100.0\,\text{mg}\ FeSO_4 \times \dfrac{1\,\text{g}}{1000\,\text{mg}} \times \dfrac{1\,\text{mol.}\ FeSO_4}{151.92\,\text{g}\ FeSO_4} = 6.582 \times 10^{-4}\,\text{mol.}\ FeSO_4$$

c. $C_{15}H_{13}ClN_2O$: $15(12.01) + 13(1.008) + 35.45 + 2(14.01) + 16.00 = 272.7$ g/mol.
$$2.00\,\text{g}\ C_{15}H_{13}ClN_2O \times \dfrac{1\,\text{mol.}\ C_{15}H_{13}ClN_2O}{272.7\,\text{g}\ C_{15}H_{13}ClN_2O} = 7.33 \times 10^{-3}\,\text{mol.}\ C_{15}H_{13}ClN_2O$$

---

24. Refer to Section 3.2 and Example 3.3.

Determine or calculate the atomic or molar mass of the species of interest, then use that atomic or molar mass as a conversion factor to convert the given moles to grams.

a. 1 mole $C_2H_3Cl = 2(12.01) + 3(1.008) + 35.45 = 62.49$ g
$$13.5\,\text{mol.} \times \dfrac{62.49\,\text{g}}{1\,\text{mol.}} = 844\,\text{g}$$

b. 1 mole $C_{18}H_{27}NO_3 = 18(12.01) + 27(1.008) + 14.01 + 3(16.00) = 305.41$ g
$$13.5\,\text{mol.} \times \dfrac{305.41\,\text{g}}{1\,\text{mol.}} = 4.12 \times 10^3\,\text{g}$$

c. 1 mole $C_{18}H_{36}O_2 = 18(12.01) + 36(1.008) + 2(16.00) = 284.47$ g
$$13.5\,\text{mol.} \times \dfrac{284.47\,\text{g}}{1\,\text{mol.}} = 3.84 \times 10^3\,\text{g}$$

Number of Grams	Number of Moles	Number of Molecules	Number of O Atoms
0.1364	7.100×10^{-4}	4.276×10^{20}	2.993×10^{21}
239.8	1.248	7.516×10^{23}	5.261×10^{24}
13.8	7.17×10^{-2}	4.32×10^{22}	3.02×10^{23}
0.00253	1.32×10^{-5}	7.93×10^{18}	5.55×10^{19}

First, calculate the molar mass of citric acid.

$C_6H_8O_7$: 6(12.01) + 8(1.008) + 7(16.00) = 192.12 g/mol. citric acid.

a. $0.1364 \text{ g} \times \dfrac{1 \text{ mol.}}{192.12 \text{ g}} = 7.100 \times 10^{-4} \text{ mol.}$

$7.100 \times 10^{-4} \text{ mol.} \times \dfrac{6.022 \times 10^{23} \text{ molecules}}{1 \text{ mol.}} = 4.276 \times 10^{20} \text{ molecules}$

$4.276 \times 10^{20} \text{ molecules} \times \dfrac{7 \text{ oxygen atoms}}{1 \text{ molecule}} = 2.993 \times 10^{21} \text{ oxygen atoms}$

b. $1.248 \text{ mol.} \times \dfrac{192.12 \text{ g}}{1 \text{ mol.}} = 239.8 \text{ g}$

$1.248 \text{ mol.} \times \dfrac{6.022 \times 10^{23} \text{ molecules}}{1 \text{ mol.}} = 7.516 \times 10^{23} \text{ molecules}$

$7.516 \times 10^{23} \text{ molecules} \times \dfrac{7 \text{ oxygen atoms}}{1 \text{ molecule}} = 5.261 \times 10^{24} \text{ nitrogen atoms}$

c. $4.32 \times 10^{22} \text{ molecules} \times \dfrac{1 \text{ mole}}{6.022 \times 10^{23} \text{ molecules}} = 7.17 \times 10^{-2} \text{ mol.}$

$7.17 \times 10^{-2} \text{ mol.} \times \dfrac{192.12 \text{ g}}{1 \text{ mol.}} = 13.8 \text{ g}$

$4.32 \times 10^{22} \text{ molecules} \times \dfrac{7 \text{ oxygen atoms}}{1 \text{ molecule}} = 3.02 \times 10^{23} \text{ nitrogen atoms}$

d. $5.55 \times 10^{19} \text{ oxygen atoms} \times \dfrac{1 \text{ molecule}}{7 \text{ oxygen atoms}} = 7.93 \times 10^{18} \text{ molecules}$

$7.93 \times 10^{18} \text{ molecules} \times \dfrac{1 \text{ mol.}}{6.022 \times 10^{23} \text{ molecules}} = 1.32 \times 10^{-5} \text{ mol.}$

$1.32 \times 10^{-5} \text{ mol.} \times \dfrac{192.12 \text{ g}}{1 \text{ mol.}} = 2.54 \times 10^{-3} \text{ g}$

28. *Refer to Section 3.3 and Example 3.4.*

Find the mass of each element in $Al_2(OH)_5Cl$ and the total mass, then calculate the mass percents of each element.

Al:	$2(26.98) =$	53.96 g
O:	$5(16.00) =$	80.00 g
H:	$5(1.008) =$	5.040 g
Cl:	$1(35.45) =$	35.45 g
	Total mass	174.45 g

$$\text{mass \% Al} = \frac{53.96}{174.45} \times 100\% = 30.93\%$$

$$\text{mass \% O} = \frac{80.00}{174.45} \times 100\% = 45.86\%$$

$$\text{mass \% H} = \frac{5.040}{174.45} \times 100\% = 2.889\%$$

$$\text{mass \% Cl} = \frac{35.45}{174.45} \times 100\% = 20.32\%$$

Note that the mass percentages should equal 100%.

$$30.93 + 45.86 + 2.886 + 20.32 = 100.00\%$$

30. *Refer to Section 3.3 and Example 3.4.*

Formula mass of allicin: $C_6H_{10}O_2S = 6(12.01) + 10(1.008) + 2(16.00) + (32.07)$
$$= 146.2 \text{ g/mol.}$$

$$25.0 \text{ g allicin} \times \frac{1 \text{ mol. allicin}}{146.2 \text{ g allicin}} \times \frac{1 \text{ mol. S}}{1 \text{ mol. allicin}} \times \frac{32.07 \text{ g S}}{1 \text{ mol. S}} = 5.48 \text{ g S}$$

32. *Refer to Section 3.3.*

Mass of 1 mol. $C_7H_5BiO_4 = 7(12.01) + 5(1.008) + 209.0 + 4(16.00) = 362.1$ g

$$346 \text{ mg Bi} \times \frac{1 \text{ g}}{1000 \text{ mg}} \times \frac{1 \text{ mol. } C_7H_5BiO_4}{209.0 \text{ g Bi}} \times \frac{362.1 \text{ g}}{1 \text{ mol. } C_7H_5BiO_4} = 0.599 \text{ g } C_7H_5BiO_4$$

$$\frac{0.599 \text{ g } C_7H_5BiO_4}{1.500 \text{ g Pepto} - \text{Bismal}} \times 100\% = 39.9\%$$

34. *Refer to Section 3.3 and Example 3.6.*

Remember that we use grams of CO_2 to determine the mass percent of C, grams H_2O to determine the mass percent of hydrogen and grams of Cl_2 to determine the mass percent of Cl, while oxygen is determined by difference.

$$1.407 \text{ g } CO_2 \times \frac{1 \text{ mol. } CO_2}{44.01 \text{ g}} \times \frac{1 \text{ mol. C}}{1 \text{ mol. } CO_2} \times \frac{12.01 \text{ g C}}{1 \text{ mol. C}} = 0.3840 \text{ g C}$$

$$\text{mass \% C} = \frac{0.3840 \text{ g C}}{1.000 \text{ g sample}} \times 100\% = 38.40 \text{ \% C}$$

$$0.134 \text{ g } H_2O \times \frac{1 \text{ mol. } H_2O}{18.02 \text{ g}} \times \frac{2 \text{ mol. H}}{1 \text{ mol. } H_2O} \times \frac{1.008 \text{ g H}}{1 \text{ mol. H}} = 0.0150 \text{ g H}$$

$$\text{mass \% H} = \frac{0.0150 \text{ g H}}{1.000 \text{ g sample}} \times 100\% = 1.50 \text{ \% H}$$

$$0.523 \text{ g } Cl_2 \times \frac{1 \text{ mol. } Cl_2}{70.90 \text{ g}} \times \frac{2 \text{ mol. Cl}}{1 \text{ mol. } Cl_2} \times \frac{35.45 \text{ g Cl}}{1 \text{ mol. Cl}} = 0.523 \text{ g Cl}$$

$$\text{mass \% Cl} = \frac{0.523 \text{ g Cl}}{1.000 \text{ g sample}} \times 100\% = 52.3 \text{ \% Cl}$$

$$\text{mass O} = 1.000 - 0.3840 - 0.0150 - 0.523 = 0.0780 \text{ g}$$

$$\text{mass \% O} = \frac{0.0780 \text{ g O}}{1.000 \text{ g sample}} \times 100\% = 7.80 \text{ \% O}$$

36. *Refer to Section 3.3 and Example 3.5.*

Calculate the mass of S, convert the masses of Ni and S to moles, calculate the mole ratio by dividing each of the moles by the smaller of the two values, and then name the resulting compound.

$$5.433 \text{ g} - 2.986 \text{ g} = 2.447 \text{ g S}$$

$$2.986 \text{ g Ni} \times \frac{1 \text{ mol. Ni}}{58.69 \text{ g Ni}} = 0.0.05088 \text{ moles Ni}$$

$$2.447 \text{ g S} \times \frac{1 \text{ mol. S}}{32.07 \text{ g S}} = 0.07630 \text{ moles S}$$

0.05088 mol. / 0.05088 mol. = 1 mole Ni
0.07630 mol. / 0.05088 mol. = 1.500 moles S
Mole ratio is 1 mol. Ni: 1.5 mol. S
To find the simplest whole-number atom ratio, multiply throughout by 2
Ni_2S_3 nickel(III) sulfide

Use the percentages as masses to calculate the moles of each element and then calculate the mole ratios of the elements.

a. $45.90 \text{ g C} \times \dfrac{1 \text{ mol. C}}{12.01 \text{ g C}} = 3.822 \text{ mol. C}$

   $2.75 \text{ g H} \times \dfrac{1 \text{ mol. H}}{1.008 \text{ g H}} = 2.73 \text{ mol. H}$

   $26.20 \text{ g O} \times \dfrac{1 \text{ mol. O}}{16.00 \text{ g O}} = 1.638 \text{ mol. O}$

   $17.50 \text{ g S} \times \dfrac{1 \text{ mol. S}}{32.07 \text{ g S}} = 0.5457 \text{ mol. S}$

   $7.65 \text{ g N} \times \dfrac{1 \text{ mol. N}}{14.01 \text{ g N}} = 0.546 \text{ mol. N}$

   3.822 mol. / 0.5457 = 7 mol. C
   2.73 mol. / 0.5457 = 5 mol. H
   1.638 mol. / 0.5457 = 3 mol. O
   0.5457 mol. / 0.5457 = 1 mol. S
   0.546 mol. / 0.5457 = 1 mol. N

   Simplest formula: $C_7H_5O_3SN$

b. $6.21 \text{ g H} \times \dfrac{1 \text{ mol. H}}{1.008 \text{ g H}} = 6.16 \text{ mol. H}$

   $44.4 \text{ g C} \times \dfrac{1 \text{ mol. C}}{12.01 \text{ g C}} = 3.70 \text{ mol. C}$

   $9.86 \text{ g O} \times \dfrac{1 \text{ mol. O}}{16.00 \text{ g O}} = 0.616 \text{ mol. O}$

   $39.51 \text{ g S} \times \dfrac{1 \text{ mol. S}}{32.07 \text{ g S}} = 1.232 \text{ mol. S}$

   6.16 mol. / 0.616 = 10 mol. H
   3.70 mol. / 0.616 = 6 mol. C
   0.616 mol. / 0.616 = 1 mol. O
   1.232 mol. / 0.616 = 2 mol. S

   Simplest formula: $C_6H_{10}OS_2$

c. $30.36 \text{ g O} \times \dfrac{1 \text{ mol. O}}{16.00 \text{ g O}} = 1.898 \text{ mol. O}$

$29.08 \text{ g Na} \times \dfrac{1 \text{ mol. Na}}{22.99 \text{ g Na}} = 1.265 \text{ mol. Na}$

$40.56 \text{ g S} \times \dfrac{1 \text{ mol. S}}{32.07 \text{ g S}} = 1.265 \text{ mol. S}$

1.898 mol. / 1.265 mol. = 1.5 mol. O
1.265 mol. / 1.265 mol. = 1 mol. Na
1.265 mol. / 1.265 mol. = 1 mol. S

Simplest formula has to be whole numbers; thus, multiply by a factor of 2:

1.5 mol. O x 2 = 3 mol. O
1 mol. Na x 2 = 2 mol. Na
1 mol. S x 2 = 2 mol. S

Simplest formula: $O_3Na_2S_2$

---

40. Refer to Section 3.3 and Example 3.6.

Use grams of CO_2 to determine the mass of carbon and grams H_2O to determine the mass of hydrogen, while oxygen is determined by difference. Once the mass of each element is determined, calculate the moles of each and then the mole ratios.

$12.24 \text{ g CO}_2 \times \dfrac{1 \text{ mol. CO}_2}{44.01 \text{ g CO}_2} \times \dfrac{1 \text{ mol. C}}{1 \text{ mol. CO}_2} \times \dfrac{12.01 \text{ g C}}{1 \text{ mol. C}} = 3.340 \text{ g C}$

$2.505 \text{ g H}_2\text{O} \times \dfrac{1 \text{ mol. H}_2\text{O}}{18.02 \text{ g}} \times \dfrac{2 \text{ mol. H}}{1 \text{ mol. H}_2\text{O}} \times \dfrac{1.008 \text{ g H}}{1 \text{ mol. H}} = 0.2802 \text{ g H}$

mass O = 5.287 g − 3.340 g − 0.2802 g = 1.667 g O

$3.340 \text{ g C} \times \dfrac{1 \text{ mol. C}}{12.01 \text{ g C}} = 0.2781 \text{ mol. C}$

$0.2802 \text{ g H} \times \dfrac{1 \text{ mol. H}}{1.008 \text{ g H}} = 0.2780 \text{ mol. H}$

$1.667 \text{ g O} \times \dfrac{1 \text{ mol. O}}{16.00 \text{ g O}} = 0.1042 \text{ mol. O}$

0.2781 mol. / 0.1042 mol. = 2.669 mol. C
0.2780 mol. / 0.1042 mol. = 2.668 mol. H
0.1042 mol. / 0.1042 mol. = 1.00 mol. O

0.67 is the decimal equivalent of 2/3; thus, to get whole numbers, multiply by a factor of 3.

2.669 mol. C x 3 = 8 mol. C
2.668 mol. H x 3 = 8 mol. H
1.00 mol. O x 3 = 3 mol. O

Simplest formula: $C_8H_8O_3$

Use grams of CO_2 to determine the mass of carbon, grams H_2O to determine the mass of hydrogen, and mass percent of nitrogen to determine mass of nitrogen, while oxygen is determined by difference. Once the mass of each element is determined, calculate the moles of each and then the mole ratios.

$$19.88 \text{ g } CO_2 \times \frac{1 \text{ mol. } CO_2}{44.01 \text{ g } CO_2} \times \frac{1 \text{ mol. C}}{1 \text{ mol. } CO_2} \times \frac{12.01 \text{ g C}}{1 \text{ mol. C}} = 5.425 \text{ g C}$$

$$4.79 \text{ g } H_2O \times \frac{1 \text{ mol. } H_2O}{18.02 \text{ g}} \times \frac{2 \text{ mol. H}}{1 \text{ mol. } H_2O} \times \frac{1.008 \text{ g H}}{1 \text{ mol. H}} = 0.536 \text{ g H}$$

$$10.00 \text{ g sample} \times \frac{14.89 \text{ g N}}{100 \text{ g sample}} = 1.489 \text{ g N}$$

mass O = 10.00 g sample − 5.425 g C − 0.536 g H − 1.489 g N = 2.550 g O

$$5.425 \text{ g C} \times \frac{1 \text{ mol. C}}{12.01 \text{ g C}} = 0.4517 \text{ mol. C}$$

$$0.536 \text{ g H} \times \frac{1 \text{ mol. H}}{1.008 \text{ g H}} = 0.532 \text{ mol. H}$$

$$1.489 \text{ g N} \times \frac{1 \text{ mol. N}}{14.01 \text{ g N}} = 0.1063 \text{ mol. N}$$

$$2.550 \text{ g O} \times \frac{1 \text{ mol. O}}{16.00 \text{ g O}} = 0.1594 \text{ mol. O}$$

0.4517 mol. / 0.1063 mol. = 4.25 mol. C
0.532 mol. / 0.1063 mol. = 5.00 mol. H
0.1063 mol. / 0.1063 mol. = 1.00 mol. N
0.1594 mol. / 0.1063 mol. = 1.50 mol. O

To get whole numbers, multiply by a factor of 4.

4.25 mol. C x 4 = 17 mol. C
5.00 mol. H x 4 = 20 mol. H
1.00 mol. N x 4 = 4 mol. N
1.50 mol. O x 4 = 6 mol. O

The simplest formula of vitamin B_6 is therefore $C_{17}H_{20}N_4O_6$.

Use grams of CO_2 to determine the mass of carbon and grams H_2O to determine the mass of hydrogen, while nitrogen is determined by difference. Once the mass of each element is determined, calculate the moles of each and then the mole ratios to determine the simplest formula. Then calculate the molar mass of that formula and divide the actual molar mass by the simplest molar mass to determine the multiplier for the simplest formula.

$$4.190 \text{ g CO}_2 \text{ x } \frac{1 \text{ mol. CO}_2}{44.01 \text{ g CO}_2} \text{ x } \frac{1 \text{ mol. C}}{1 \text{ mol. CO}_2} \text{ x } \frac{12.01 \text{ g C}}{1 \text{ mol. C}} = 1.143 \text{ g C}$$

$$3.428 \text{ g H}_2\text{O x } \frac{1 \text{ mol. H}_2\text{O}}{18.02 \text{ g}} \text{ x } \frac{2 \text{ mol. H}}{1 \text{ mol. H}_2\text{O}} \text{ x } \frac{1.008 \text{ g H}}{1 \text{ mol. H}} = 0.3835 \text{ g H}$$

mass of N = 2.859 g – 1.143 g - 0.3835 g = 1.332 g N

$$1.143 \text{ g C x } \frac{1 \text{ mol. C}}{12.01 \text{ g C}} = 0.09517 \text{ mol. C}$$

$$0.3835 \text{ g H x } \frac{1 \text{ mol. H}}{1.008 \text{ g H}} = 0.3805 \text{ mol. H}$$

$$1.332 \text{ g N x } \frac{1 \text{ mol. N}}{14.01 \text{ g N}} = 0.09507 \text{ mol. N}$$

0.09517 mol. / 0.09507 mol. = 1 mol. C
0.3805 mol. / 0.09507 mol. = 4 mol. H
0.09507 mol. / 0.09507 mol. = 1 mol. N

Simplest formula: CH_4N

Formula mass (of the simplest formula) = 12.01 + 4(1.008) + 14.01 = 30.05 g/mol.

The actual molar mass 60.10 g/mol.

$\frac{60.10}{30.05} = 2$. Thus the molecular formula = $C_{1x2}H_{4x2}N_{1x2} = C_2H_8N_2$

---

46. *Refer to Section 3.3 and Example 3.4.*

Calculate the mass % of anhydrous salt in the hydrated sample using molar masses. Then use that value to determine the mass of anhydrous salt in the sample.

molar mass of $MgSO_4 \bullet 7H_2O$:   246.51 g
molar mass of $MgSO_4$:   120.37 g
mass of $7H_2O$:   126.14 g

$$\text{mass \% of water} = \frac{126.14 \text{ g H}_2\text{O}}{246.51 \text{ g MgSO}_4 \bullet 7\text{H}_2\text{O}} \text{ x } 100\% = 51.17\%$$

$$\text{mass \% of anhydrous salt in hydrate} = \frac{120.37 \text{ g MgSO}_4}{246.51 \text{ g MgSO}_4 \bullet 7\text{H}_2\text{O}} \text{ x } 100\% = 48.83\%$$

$$\text{mass of anhydrous salt } 7.834 \text{ g MgSO}_4 \bullet 7\text{H}_2\text{O} = \frac{48.83 \text{ g MgSO}_4}{100 \text{ g MgSO}_4 \bullet 7\text{H}_2\text{O}} = 3.825 \text{ g MgSO}_4$$

Compare the reactant and product sides of the equation, starting with elements that appear only once on each side. Balance the elements by adjusting the coefficients (*never the subscripts!*). When you have a choice, it is usually advantageous to start with the element that is present in greatest number.

a. $H_2S(g) + SO_2(g) \rightarrow S(s) + H_2O(g)$

  $H_2S(g) + SO_2(g) \rightarrow S(s) + 2H_2O(g)$ balance O's

  $2H_2S(g) + SO_2(g) \rightarrow S(s) + 2H_2O(g)$ balance H's

  $2H_2S(g) + SO_2(g) \rightarrow 3S(s) + 2H_2O(g)$ balance S's

b. $CH_4(g) + NH_3(g) + O_2(g) \rightarrow HCN(g) + H_2O(g)$

  $CH_4(g) + NH_3(g) + O_2(g) \rightarrow HCN(g) + 3H_2O(g)$ balance H's

  $CH_4(g) + NH_3(g) + 3/2O_2(g) \rightarrow HCN(g) + 3H_2O(g)$ balance O's

  $2CH_4(g) + 2NH_3(g) + 3O_2(g) \rightarrow 2HCN(g) + 6H_2O(g)$ multiply x 2

c. $Fe_2O_3(s) + H_2(g) \rightarrow Fe(l) + H_2O(g)$

  $Fe_2O_3(s) + H_2(g) \rightarrow 2Fe(l) + H_2O(g)$ balance Fe's

  $Fe_2O_3(s) + H_2(g) \rightarrow 2Fe(l) + 3H_2O(g)$ balance O's

  $Fe_2O_3(s) + 3H_2(g) \rightarrow 2Fe(l) + 3H_2O(g)$ balance H's

Compare the reactant and product sides of the equation, starting with elements that appear only once on each side. Balance the elements by adjusting the coefficients (*never the subscripts!*). When you have a choice, it is usually advantageous to start with the element that is most abundant.

a. $Al(s) + S(s) \rightarrow Al_2S_3(s)$

  $Al(s) + 3S(s) \rightarrow Al_2S_3(s)$ balance S's

  $2Al(s) + 3S(s) \rightarrow Al_2S_3(s)$ balance Al's

b. $Al(s) + Br_2(l) \rightarrow AlBr_3(s)$

  $Al(s) + 3/2Br_2(l) \rightarrow AlBr_3(s)$ balance Br's

  $2Al(s) + 3Br_2(l) \rightarrow 2AlBr_3(s)$ multiply x 2

c. $Al(s) + N_2(g) \rightarrow AlN(s)$

  $Al(s) + 1/2N_2(g) \rightarrow AlN(s)$ balance N's

  $2Al(s) + N_2(g) \rightarrow 2AlN(s)$ multiply x 2

d. $Al(s) + O_2(g) \rightarrow Al_2O_3(s)$

  $Al(s) + 3/2O_2(g) \rightarrow Al_2O_3(s)$ balance O's

  $2Al(s) + 3/2O_2(g) \rightarrow Al_2O_3(s)$ balance Al's

  $4Al(s) + 3O_2(g) \rightarrow 2Al_2O_3(s)$ multiply x 2

e. $Al(s) + O_2(g) \rightarrow Al_2(O_2)_3(s)$
      $Al(s) + 3O_2(g) \rightarrow Al_2(O_2)_3(s)$ balance O's
      $2Al(s) + 3O_2(g) \rightarrow Al_2(O_2)_3(s)$ balance Al's

---

52. Refer to Sections 2.6 and 3.4, Example 3.8 and Figure 2.10.

Compare the reactant and products side of the equation, starting with elements that appear only once on each side. Balance the elements by adjusting the coefficients (*never the subscripts!*). When you have a choice, it is usually advantageous to start with the element that is present in greatest number.

a. $F_2(g) + H_2O(l) \rightarrow OF_2(g) + HF(g)$
      $F_2(g) + H_2O(l) \rightarrow OF_2(g) + 2HF(g)$ balance H's
      $2F_2(g) + H_2O(l) \rightarrow OF_2(g) + 2HF(g)$ balance F's

b. $O_2(g) + NH_3(g) \rightarrow NO_2(g) + H_2O(l)$
      $O_2(g) + 2NH_3(g) \rightarrow NO_2(g) + 3H_2O(l)$ balance H's
      $O_2(g) + 2NH_3(g) \rightarrow 2NO_2(g) + 3H_2O(l)$ balance N's
      $(7/2)O_2(g) + 2NH_3(g) \rightarrow 2NO_2(g) + 3H_2O(l)$ balance O's
      Since fractions are generally not allowed, multiply all coefficients by 2.
      $7O_2(g) + 4NH_3(g) \rightarrow 4NO_2(g) + 6H_2O(l)$

c. $Au_2S_3(s) + H_2(g) \rightarrow Au(s) + H_2S(g)$
      $Au_2S_3(s) + H_2(g) \rightarrow Au(s) + 3H_2S(g)$ balance the S's
      $Au_2S_3(s) + H_2(g) \rightarrow 2Au(s) + 3H_2S(g)$ balance the Au's
      $Au_2S_3(s) + 3H_2(g) \rightarrow 2Au(s) + 3H_2S(g)$ balance the H's

d. $NaHCO_3(s) \rightarrow Na_2CO_3(s) + H_2O(l) + CO_2(s)$
      $2NaHCO_3(s) \rightarrow Na_2CO_3(s) + H_2O(l) + CO_2(s)$ balance the Na's

e. $SO_2(g) + HF(l) \rightarrow SF_4(g) + H_2O(l)$
      $SO_2(g) + 4HF(l) \rightarrow SF_4(g) + H_2O(l)$ balance F's
      $SO_2(g) + 4HF(l) \rightarrow SF_4(g) + 2H_2O(l)$ balance H's

---

54. Refer to Section 3.4 and Example 3.9.

Use the coefficients from the balanced equation to establish the mole ratios of the species in question.

a. $1.299 \, \text{mol. BF}_3 \times \dfrac{3 \, \text{mol. NaBH}_4}{4 \, \text{mol. BF}_3} = 0.9742 \, \text{mol. NaBH}_4$

b. $0.893 \, \text{mol. NaBH}_4 \times \dfrac{2 \, \text{mol. B}_2\text{H}_6}{3 \, \text{mol. NaBH}_4} = 0.595 \, \text{mol. B}_2\text{H}_6$

c. $1.987 \, \text{mol. B}_2\text{H}_6 \times \dfrac{3 \, \text{mol. NaBF}_4}{2 \, \text{mol. B}_2\text{H}_6} = 2.980 \, \text{mol. NaBF}_4$

d. $4.992\text{ mol. NaBF}_4 \times \dfrac{4\text{ mol. BF}_3}{3\text{ mol. NaBF}_4} = 6.656\text{ mol. BF}_3$

---

56. Refer to Section 3.4 and Example 3.9.

Use the coefficients from the balanced equation to establish the mole ratios of the species in question. Remember your mass to mole and/or mole to gram conversions.

a. $12.43\text{ mol. PH}_3 \times \dfrac{1\text{ mol. P}_4\text{O}_{10}}{4\text{ mol.PH}_3} \times \dfrac{283.88\text{ g P}_4\text{O}_{10}}{1\text{ mol. P}_4\text{O}_{10}} = 882.2\text{ g P}_4\text{O}_{10}$

b. $0.739\text{ mol. H}_2\text{O} \times \dfrac{4\text{ mol.PH}_3}{6\text{ mol.H}_2\text{O}} \times \dfrac{33.994\text{ g PH}_3}{1\text{ mol.PH}_3} = 16.7\text{ g PH}_3$

c. $1.000\text{ g H}_2\text{O} \times \dfrac{1\text{ mol. H}_2\text{O}}{18.02\text{ g H}_2\text{O}} \times \dfrac{8\text{ mol. O}_2}{6\text{ mol.H}_2\text{O}} \times \dfrac{32.00\text{ g O}_2}{1\text{ mol.O}_2} = 2.368\text{ g O}_2$

d. $20.50\text{ g PH}_3 \times \dfrac{1\text{ mol. PH}_3}{33.99\text{ g PH}_3} \times \dfrac{8\text{ mol. O}_2}{4\text{ mol. PH}_3} \times \dfrac{32.00\text{ g O}_2}{1\text{ mol.O}_2} = 38.60\text{ g O}_2$

---

58. Refer to Sections 2.5, 2.6, 3.3 and 3.4 and Example 3.9.

Balance the equation to determine the mole ratios, then calculate the requested masses. Use the coefficients from the balanced equation to establish the mole ratios of the species in question. Remember your mass to mole and/or mole to gram conversions.

a. $SiO_2(s) + C(s) \rightarrow Si(s) + CO(g)$
$SiO_2(s) + C(s) \rightarrow Si(s) + 2CO(g)$ balance O's
$SiO_2(s) + 2C(s) \rightarrow Si(s) + 2CO(g)$ balance C's

b. $12.72\text{ g Si} \times \dfrac{1\text{ mol. Si}}{28.09\text{ g Si}} \times \dfrac{1\text{ mol. SiO}_2}{1\text{ mol.Si}} = 0.4528\text{ mol. SiO}_2$

c. $44.99\text{ g Si} \times \dfrac{1\text{ mol. Si}}{28.09\text{ g Si}} \times \dfrac{2\text{ mol. CO}}{1\text{ mol.Si}} \times \dfrac{28.01\text{ g CO}}{1\text{ mol. CO}} = 89.72\text{ g CO}$

---

60. Refer to Section 3.4 and Example 3.9.

a. First, the mass of the sample of tin, Sn, must be determined. Note: the thickness is given in units of millimeters and must first be converted into cm.

$0.600\text{ mm} \times \dfrac{\text{cm}}{10\text{ mm}} = 0.0600\text{ cm}$

$$(8.25\,\text{cm} \times 21.5\,\text{cm} \times 0.0600\,\text{cm}) \times \frac{7.28\,\text{g Sn}}{\text{cm}^3} = 77.5\,\text{g Sn}$$

$$77.5\,\text{g Sn} \times \frac{1\,\text{mol. Sn}}{118.71\,\text{g Sn}} \times \frac{1\,\text{mol. SnO}_2}{1\,\text{mol. Sn}} \times \frac{150.71\,\text{g SnO}_2}{1\,\text{mol. SnO}_2} = 98.4\,\text{g SnO}_2$$

b. $98.4\,\text{g SnO}_2 \times \dfrac{31.999\,\text{g O}_2}{150.71\,\text{g SnO}_2} \times \dfrac{\text{L O}_2}{1.309\,\text{g O}_2} \times \dfrac{100\,\text{L air}}{21\,\text{L O}_2} = 76\,\text{L air}$

---

62. *Refer to Section 3.4.*

Calculate the grams of CO_2 produced. Then convert from grams CO_2 to moles CO_2. Use the balanced chemical equation to convert from moles of CO_2 to moles of KO_2. Finally, use the molar mass to convert form moles of KO_2 to grams of KO_2.

$$25\,\text{min} \times \frac{0.702\,\text{g CO}_2}{1\,\text{min}} = 17\,\text{g CO}_2$$

$$17\,\text{g CO}_2 \times \frac{1\,\text{mol. CO}_2}{44.01\,\text{g CO}_2} \times \frac{4\,\text{mol KO}_2}{4\,\text{mol. CO}_2} \times \frac{71.10\,\text{g KO}_2}{1\,\text{mol. KO}_2} = 28\,\text{g KO}_2$$

---

64. *Refer to Section 3.4 and Example 3.10.*

Use the balanced equation to calculate the moles of Al_2S_3 based on the amount of Al, and also based on the amount of S. The one giving the lower value will be the limiting reactant, and the yield calculated is the theoretical yield.

a. $2Al(s) + 3S(s) \rightarrow Al_2S_3(s)$

b. $1.18\,\text{mol. Al} \times \dfrac{1\,\text{mol. Al}_2\text{S}_3}{2\,\text{mol. Al}} = 0.590\,\text{mol. Al}_2\text{S}_3$

   $2.25\,\text{mol. S} \times \dfrac{1\,\text{mol. Al}_2\text{S}_3}{3\,\text{mol. S}} = 0.750\,\text{mol. Al}_2\text{S}_3$

   Al produces the least amount of Al_2S_3, and is therefore the limiting reagent.

c. Theoretical yield, based on the limiting reagent, is calculated above as 0.590 mol. Al_2S_3.

d. Calculate the amount of S that would be consumed making the theoretical amount of Al_2S_3 and subtract that from the initial amount.

   $0.590\,\text{mol. Al}_2\text{S}_3 \times \dfrac{3\,\text{mol. S}}{1\,\text{mol. Al}_2\text{S}_3} = 1.77\,\text{mol. S (used in the reaction)}$

   2.25 mol. (initial) − 1.77 mol. (reacted) = 0.48 mol. S (unreacted).

66. *Refer to Section 3.4 and Example 3.11.*

First, the theoretical yield of the reaction must be determined.

$$\frac{\text{Actual Yield}}{\text{Theoretical Yield}} \times 100\% = \% \text{ Yield}$$

$$\frac{897 \text{ g Fe}_3\text{O}_4}{\text{Theoretical Yield}} \times 100\% = 69\%$$

Theoretical Yield $= 1.3 \times 10^3$ g Fe$_3$O$_4$

Now, the mass of iron needed for this reaction can be calculated.

$$1.3 \times 10^3 \text{ g Fe}_3\text{O}_4 \times \frac{1 \text{ mol. Fe}_3\text{O}_4}{231.54 \text{ g Fe}_3\text{O}_4} \times \frac{3 \text{ mol. Fe}}{1 \text{ mol. Fe}_3\text{O}_4} \times \frac{55.85 \text{ g Fe}}{1 \text{ mol. Fe}} = 9.4 \times 10^2 \text{ g Fe}$$

68. *Refer to Section 3.4 and Example 3.10.*

Use the balanced equation to calculate the moles of Al$_2$O$_3$ based on the amount of Al, and also based on the amount of NH$_4$ClO$_4$. The one giving the lower value will be the limiting reactant, and the yield calculated will also be the theoretical yield.

a. $Al(s) + NH_4ClO_4(s) \rightarrow Al_2O_3(s) + AlCl_3(s) + NO(g) + H_2O(g)$
    $3Al(s) + NH_4ClO_4(s) \rightarrow Al_2O_3(s) + AlCl_3(s) + NO(g) + H_2O(g)$ balance Al's
    $3Al(s) + 3NH_4ClO_4(s) \rightarrow Al_2O_3(s) + AlCl_3(s) + NO(g) + H_2O(g)$ balance Cl's
    $3Al(s) + 3NH_4ClO_4(s) \rightarrow Al_2O_3(s) + AlCl_3(s) + 3NO(g) + H_2O(g)$ balance N's
    $3Al(s) + 3NH_4ClO_4(s) \rightarrow Al_2O_3(s) + AlCl_3(s) + 3NO(g) + 6H_2O(g)$ balance O's

b. $$7.00 \text{ g Al} \times \frac{1 \text{ mol. Al}}{26.98 \text{ g Al}} \times \frac{1 \text{ mol. Al}_2\text{O}_3}{3 \text{ mol. Al}} \times \frac{101.96 \text{ g Al}_2\text{O}_3}{1 \text{ mol. Al}_2\text{O}_3} = 8.82 \text{ g Al}_2\text{O}_3$$

$$9.32 \text{ g NH}_4\text{ClO}_4 \times \frac{1 \text{ mol. NH}_4\text{ClO}_4}{117.49 \text{ g NH}_4\text{ClO}_4} \times \frac{1 \text{ mol. Al}_2\text{O}_3}{3 \text{ mol. NH}_4\text{ClO}_4} \times \frac{101.96 \text{ g Al}_2\text{O}_3}{1 \text{ mol. Al}_2\text{O}_3} = 2.70 \text{ g Al}_2\text{O}_3$$

Thus, NH$_4$ClO$_4$ is the limiting reagent and 2.70 g Al$_2$O$_3$ is the theoretical yield.

c. $$\% \text{ yield} = \frac{\text{experimental yield}}{\text{theoretical yield}} \times 100\% = \frac{1.56 \text{ g}}{2.70 \text{ g}} \times 100\% = 57.8\%$$

d. $$2.70 \text{ g Al}_2\text{O}_3 \times \frac{1 \text{ mol. Al}_2\text{O}_3}{101.96 \text{ g Al}_2\text{O}_3} \times \frac{3 \text{ mol. Al}}{1 \text{ mol. Al}_2\text{O}_3} \times \frac{26.98 \text{ g Al}}{1 \text{ mol. Al}} = 2.14 \text{ g Al}$$
    7.00 g (initial) $-$ 2.14 g (used) $=$ 4.86 g (unused)

$$\% \text{ yield} = \frac{\text{experimental yield}}{\text{theoretical yield}} \times 100\% \Rightarrow 75.0\% = \frac{0.250\,\text{L}}{\text{theoretical yield}} \times 100\%$$

theor. yield = (0.250)(100%) / 75.0% = 0.333 L H$_3$PO$_4$.

$$0.333\,\text{L H}_3\text{PO}_4 \times \frac{1000\,\text{mL}}{1\,\text{L}} \times \frac{1\,\text{cm}^3}{1\,\text{mL}} \times \frac{1.651\,\text{g H}_3\text{PO}_4}{1\,\text{cm}^3\,\text{H}_3\text{PO}_4} = 550\,\text{g H}_3\text{PO}_4$$

$$550\,\text{g H}_3\text{PO}_4 \times \frac{1\,\text{mol. H}_3\text{PO}_4}{81.99\,\text{g H}_3\text{PO}_4} \times \frac{1\,\text{mol. PI}_3}{1\,\text{mol. H}_3\text{PO}_4} \times \frac{411.67\,\text{g PI}_3}{1\,\text{mol. PI}_3} = 2.76 \times 10^3\,\text{g PI}_3$$

Weigh out 2.76 × 10^3 g PI$_3$.

$$550\,\text{g H}_3\text{PO}_4 \times \frac{1\,\text{mol. H}_3\text{PO}_4}{81.99\,\text{g H}_3\text{PO}_4} \times \frac{3\,\text{mol. H}_2\text{O}}{1\,\text{mol. H}_3\text{PO}_4} \times \frac{18.02\,\text{g H}_2\text{O}}{1\,\text{mol. H}_2\text{O}} = 363\,\text{g H}_2\text{O}$$

$$363\,\text{g H}_2\text{O} \times \frac{1\,\text{cm}^3\,\text{H}_2\text{O}}{1.00\,\text{g H}_2\text{O}} \times \frac{1\,\text{mL}}{1\,\text{cm}^3} = 363\,\text{mL H}_2\text{O}$$

need 45% excess, thus: 363 mL + (0.45)(363 mL) = 526 mL H$_2$O

72. Refer to Section 3.4.

a. Approximately **0.75 g** as read from the graph.

b. Since the amount of Fe determines the amount of product, **Fe** is the limiting reagent when 2.00 g Fe is used.

c. Since the amount of product stays the same after the mass of Fe passes 3.0 g, the **O$_2$** must be the limiting reagent when 5.00 g Fe is used.

d. According to the graph, **3.0 g Fe** react exactly with the amount of O$_2$ supplied. This is the point on the graph where the mass of the product no longer increases upon addition of more Fe.

e. Determine the masses of product and of Fe from the graph. The difference in mass is the mass of O. From the masses of Fe and O, calculate moles of each, then determine the simplest formula.
3.0 g Fe and 4.3 g product. Thus, there are 1.3 g O.

$$1.3\,\text{g O} \times \frac{1\,\text{mol. O}}{16.00\,\text{g O}} = 0.081\,\text{mol. O}$$

$$3.0\,\text{g Fe} \times \frac{1\,\text{mol. Fe}}{55.85\,\text{g Fe}} = 0.054\,\text{mol. Fe}$$

0.081 mol. /0.054 mol. = 1.5 mol. O x 2 = 3 mol. O

0.054 mol. /0.054 mol. = 1 mol. Fe x 2 = 2 mol. Fe

simplest formula: Fe_2O_3

---

74. *Refer to Sections 1.2, 3.2 and 3.3 and Table 1.3.*

$$6.00 \text{ oz. salami} \times \frac{1 \text{ g}}{0.03527 \text{ oz.}} \times \frac{0.090 \text{ g NaC}_7\text{H}_5\text{O}_2}{100 \text{ g salami}} = 15 \text{ g NaC}_7\text{H}_5\text{O}_2$$

$$15 \text{ g NaC}_7\text{H}_5\text{O}_2 \times \frac{1 \text{ mol. NaC}_7\text{H}_5\text{O}_2}{144.10 \text{ g NaC}_7\text{H}_5\text{O}_2} \times \frac{6.022 \times 10^{23} \text{ molecules}}{1 \text{ mol.}} = 6.4 \times 10^{20} \text{ molecules}$$

$$6.4 \times 10^{20} \text{ molecules} \times \frac{1 \text{ Na atom}}{1 \text{ NaC}_7\text{H}_5\text{O}_2 \text{ molecule}} = 6.4 \times 10^{20} \text{ Na atoms}$$

---

76. *Refer to Section 3.4.*

The question is asking you to calculate which fertilizer has the lowest mass per mole of nitrogen. Since the bags all weigh the same, the answer can be calculated by dividing the molar mass of the fertilizer by the moles of nitrogen in that fertilizer. The one with the smallest number will cost the least.

$$\frac{60.06 \text{ g (NH}_2)_2\text{CO}}{1 \text{ mol. (NH}_2)_2\text{CO}} \times \frac{1 \text{ mol. (NH}_2)_2\text{CO}}{2 \text{ mol. N}} = 30.03$$

$$\frac{17.03 \text{ g NH}_3}{1 \text{ mol. NH}_3} \times \frac{1 \text{ mol. NH}_3}{1 \text{ mol. N}} = 17.03$$

$$\frac{80.05 \text{ g NH}_4\text{NO}_3}{1 \text{ mol. NH}_4\text{NO}_3} \times \frac{1 \text{ mol. NH}_4\text{NO}_3}{2 \text{ mol. N}} = 40.02$$

$$\frac{59.08 \text{ g HNC(NH}_2)_2}{1 \text{ mol. HNC(NH}_2)_2} \times \frac{1 \text{ mol. HNC(NH}_2)_2}{3 \text{ mol. N}} = 19.69$$

Thus, the **NH₃** fertilizer would cost the least since it has the lowest ratio of mass of fertilizer to mol. of nitrogen.

---

78. *Refer to Sections 3.2 and 3.4.*

$$2AB_3 + 3C \rightarrow 3CB_2 + 2A$$

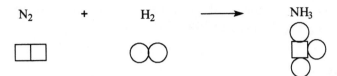

Balance the chemical equation to see what is going to react and what will be left over.

$N_2 + 3H_2 \rightarrow 2NH_3$

Keep this ratio in mind as you write pictorially what is depicted in the balanced equation

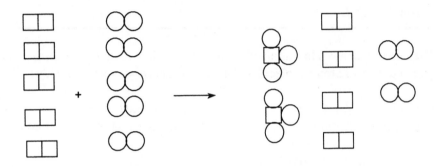

a. **False**. theoretical yield = $4.0 \text{ mol. CCl}_4 \times \dfrac{1 \text{ mol. CCl}_2\text{F}_2}{1 \text{ mol. CCl}_4} = 4.0 \text{ mol. CCl}_2\text{F}_2$

b. **False**. theoretical yield = $4.0 \text{ mol. CCl}_4 \times \dfrac{2 \text{ mol. HCl}}{1 \text{ mol. CCl}_4} \times \dfrac{36.46 \text{ g HCl}}{1 \text{ mol. HCl}} = 2.9 \times 10^2 \text{ g HCl}$

c. **True**. $\% \text{ yield} = \dfrac{\text{experimental yield}}{\text{theoretical yield}} \times 100\% = \dfrac{3.0 \text{ mol. CCl}_2\text{F}_2}{4.0 \text{ mol. CCl}_2\text{F}_2} = 75\%$

d. **False**. The theoretical yield is based on the limiting reagent, which must be CCl_4 since the text states that HF is used in excess.

e. **True**. We do not know how much HF was used (only that an excess was used), so we cannot calculate how much was left unreacted.

f. **False**. Law of conservation of mass states that the mass of products will equal the mass of reactants, but this does not hold true for moles.

g. **False**. Two moles of HF are consumed for every one mole of CCl_4 used.

h. **True**. As the limiting reactant, CCl_4, is completely consumed.

84. Refer to Section 3.1.

Calculate the mass ratio of Si to Li_3N using standard masses. Then use that ratio to calculate the mass of Li_3N assuming a molar mass of 10 for Si.

$$ratio = \frac{34.83 \text{ g } Li_3N}{27.96924 \text{ g Si - 28}} = 1.245$$

$$1.245 = \frac{\text{mass of } Li_3N}{10.00 \text{ g Si - 28}} \Rightarrow \text{mass of } Li_3N = 12.45 \text{ g}$$

86. Refer to Sections 3.1 – 3.4.

a. **False.** The mass of an atom has units of atomic mass units. A mole is determined by dividing the number of atoms/molecules/ions by Avogadro's number, 6.022×10^{23}

b. **False.** In 1 mole of N_2O_4, the mass of nitrogen would be 28.02 grams and the mass of oxygen would be 64.00 grams. The ratio of oxygen/nitrogen 64.00/28.02 is equal to 2.28.

c. **True.** Using the atomic mass from the periodic table, 1 mole of chlorine atoms is equal to 35.45 g.

d. **False.** Since the value 10.81 is closer to 11.01 than to 10.01, this would indicate a higher amount of the heavier isotope. Actually, boron occurs in nature as 20% B-10 and 80% B-11.

e. **False.** The simplest formula MUST contain only whole numbers.

f. **False.** 558.5 grams of Fe is 10 moles of Fe; 0.5200 grams of chromium is only 0.010 moles of Cr. There are 1000 times more atoms of Fe than Cr in these samples.

g. **True.** Only 0.80 moles of nitrogen would be consumed when 1.00 moles of oxygen have been consumed in this reaction.

h. **False. The law of conservation of mass** states that the mass of the products must equal the mass of the reactants, however, the number of moles of products does not always equal the moles of reactants.

87. Refer to Section 3.3.

Calculate the mass of chlorophyll in one mole: Use the chlorophyll:Mg mole ratio as a conversion factor to calculate the moles of Mg in one mole of chlorophyll. Convert moles of Mg to grams and then calculate mass of chlorophyll using the mass % of Mg in chlorophyll.

$$1 \text{ mol. chlorophyll} \times \frac{1 \text{ mol. Mg}}{1 \text{ mol. chlorophyll}} \times \frac{24.30 \text{ g Mg}}{1 \text{ mol. Mg}} \times \frac{100 \text{ g chlorophyll}}{2.72 \text{ g Mg}} = 893 \text{ g chlorophyll}$$

Thus the molar mass of chlorophyll = 893 g/mol.

---

88. *Refer to Sections 1.2, 1.3 and 3.2.*

volume of a cube $= s^3$

$$s = 0.409 \text{ nm} \times \frac{1 \text{ m}}{1 \times 10^9 \text{ nm}} \times \frac{100 \text{ cm}}{1 \text{ m}} = 4.09 \times 10^{-8} \text{ cm}$$

$$V = (4.09 \times 10^{-8} \text{ cm})^3 = 6.84 \times 10^{-23} \text{ cm}^3$$

$$\frac{4 \text{ atoms}}{6.84 \times 10^{-23} \text{ cm}^3} \times \frac{1 \text{ cm3}}{10.5 \text{ g}} \times \frac{107.87 \text{ g Ag}}{1 \text{ mole Ag}} = 6.01 \times 10^{23} \text{ atoms/mole}$$

(Avogadro was right.)

---

89. *Refer to Sections 3.3 and 3.4.*

First write balanced chemical equations. Then calculate the amount of CaO formed and the amount of Ca needed to form the CaO. The remaining Ca will have formed the Ca_3N_2.

Reactions: $2Ca + O_2 \rightarrow 2CaO$
$3Ca + N_2 \rightarrow Ca_3N_2$
$CaO + H_2O \rightarrow Ca(OH)_2$

$$4.832 \text{ g Ca(OH)}_2 \times \frac{1 \text{ mol. Ca(OH)}_2}{74.10 \text{ g Ca(OH)}_2} \times \frac{1 \text{ mol. CaO}}{1 \text{ mol. Ca(OH)}_2} \times \frac{56.08 \text{ g CaO}}{1 \text{ mol. CaO}}$$

$$= 3.657 \text{ g CaO}$$

$$3.657 \text{ g CaO} \times \frac{1 \text{ mol. CaO}}{56.08 \text{ g CaO}} \times \frac{2 \text{ mol. Ca}}{2 \text{ mol. CaO}} \times \frac{40.08 \text{ g Ca}}{1 \text{ mol. Ca}} = 2.614 \text{ g Ca}$$

5.025 g Ca (original) - 2.614 g Ca (to form CaO) = 2.411 g Ca (to form Ca_3N_2)

$$2.411 \text{ g Ca} \times \frac{1 \text{ mol. Ca}}{40.08 \text{ g Ca}} \times \frac{1 \text{ mole Ca}_3\text{N}_2}{3 \text{ mol. Ca}} \times \frac{148.26 \text{ g Ca}_3\text{N}_2}{1 \text{ mol. Ca}_3\text{N}_2} = 2.973 \text{ g Ca}_3\text{N}_2$$

---

90. *Refer to Sections 3.3 and 3.4.*

Calculate the mass lost on converting KBr to KCl. This corresponds to the difference in masses between Br and Cl. Then calculate the mass the of KBr and the percentage.

79.90 g/mol. Br - 35.45 g/mol. Cl = 44.45 g/mol (of Br replaced with Cl)

3.595 g (original mixture) - 3.129 g (final KCl) = 0.466 g

$$0.466 \text{ g} \times \frac{1 \text{ mol.}}{44.45 \text{ g}} = 0.0105 \text{ mol. KBr converted to KCl.}$$

$$0.0105 \text{ mol. KBr} \times \frac{119.0 \text{ g KBr}}{1 \text{ mol. KBr}} = 1.25 \text{ g KBr}$$

$$\text{mass} \% = \frac{1.25 \text{ g}}{3.595 \text{ g}} \times 100\% = 34.8\%$$

91. Refer to Sections 3.3 and 3.4.

a. Both oxides contain 2.573 g V. The first also contains (4.589 g VO_X - 2.573 g V) 2.016 g O, while the second contains (3.782 g VO_Y - 2.573 g V) 1.209 g O.

$$2.573 \text{ g V} \times \frac{1 \text{ mol. V}}{50.94 \text{ g V}} = 0.05051 \text{ mol. V}$$

$$2.016 \text{ g O} \times \frac{1 \text{ mol. O}}{16.00 \text{ g O}} = 0.1260 \text{ mol. O} \quad \text{(1st oxide)}$$

$$1.209 \text{ g O} \times \frac{1 \text{ mol. O}}{16.00 \text{ g O}} = 0.07556 \text{ mol. O} \quad \text{(2nd oxide)}$$

0.05051 mol. / 0.05051 mol. = 1.000 mol. V
0.07556 mol. / 0.05051 mol. = 1.496 mol. O
0.1260 mol. / 0.05051 mol. = 2.495 mol. O

1st oxide: V_2O_3
2nd oxide: V_2O_5

b. Calculate the mass of H_2O formed using the mass of O lost. The second heating converted all the oxygen to water.

$$2.016 \text{ g O} \times \frac{1 \text{ mol. O}}{16.00 \text{ g O}} \times \frac{1 \text{ mol. H}_2\text{O}}{1 \text{ mol. O}} \times \frac{18.02 \text{ g H}_2\text{O}}{1 \text{ mol. H}_2\text{O}} = 2.271 \text{ g H}_2\text{O}$$

92. Refer to Sections 3.3 and 3.4.

Calculate the amount of CO_2 expected from burning cocaine and sucrose using variables for the masses of those compounds. Remember that cocaine has 17 C atoms and will thus produce 17 CO_2 atoms; similarly, sucrose will produce 12 CO_2's Then set up a algebraic equation and solve for the mass of cocaine in the sample. Then calculate the mass percent.

Molar mass of $C_{17}H_{21}O_4N$ = 303.39 g/mol.

Molar mass of $C_{12}H_{22}O_{11}$ = 342.34 g/mol.

$$\text{mass of CO}_2 = 1.00 \text{ mL} \times \frac{1 \text{ L}}{1000 \text{ mL}} \times \frac{1.80 \text{ g}}{1 \text{ L}} = 1.80 \times 10^{-3} \text{ g}$$

mass of sugar + cocaine = 1.00 mg x $\dfrac{1\,g}{1000\,mg}$ = 1.00 x 10^{-3} g

Let X = mass of cocaine, then

$(1.0 \times 10^{-3} - X)$ = mass of sucrose

X g cocaine x $\dfrac{1\,mol.\,cocaine}{303.39\,g\,cocaine}$ x $\dfrac{17\,mol.\,CO_2}{1\,mol.\,cocaine}$ x $\dfrac{44.01\,g\,CO_2}{1\,mol.\,CO_2}$ = 2.466 X g CO_2

$(1.00 \times 10^{-3} - X)$ g sucrose x $\dfrac{1\,mol.\,sucrose}{342.34\,g\,sucrose}$ x $\dfrac{12\,mol.\,CO_2}{1\,mol.\,sucrose}$ x $\dfrac{44.01\,g\,CO_2}{1\,mol.\,CO_2}$

$= (1.54 \times 10^{-3} - 1.54X)$ g CO_2

total CO_2 = CO_2 from cocaine + CO_2 from sucrose

$1.80 \times 10^{-3} = 2.466X + 1.54 \times 10^{-3} - 1.54X$

$2.60 \times 10^{-4} = 0.93X$

$X = 2.8 \times 10^{-4}$ g cocaine

mass % cocaine = $\dfrac{2.8 \times 10^{-4}\,g}{1.00 \times 10^{-3}\,g}$ x 100% = 28%

Chapter 4: Reactions in Aqueous Solution

2. *Refer to Section 4.1, Example 4.1, and Figure 4.1.*

Use molarity to convert from liters of solution to moles of solute and then convert to mass.

a. $2.00 \, L \times \dfrac{0.685 \, mol. \, Ni(NO_3)_2}{1 \, L} \times \dfrac{182.71 \, g \, Ni(NO_3)_2}{1 \, mol. \, Ni(NO_3)_2} = 250 \, g \, Ni(NO_3)_2$

Dissolve 250 g $Ni(NO_3)_2$ in sufficient water to make 2.00 L of solution.

b. $2.00 \, L \times \dfrac{0.685 \, mol. \, CuCl_2}{1 \, L} \times \dfrac{134.45 \, g \, CuCl_2}{1 \, mol. \, CuCl_2} = 184 \, g \, CuCl_2$

Dissolve 184 g $CuCl_2$ in sufficient water to make 2.00 L of solution.

c. $2.00 \, L \times \dfrac{0.685 \, mol. \, C_6H_8O_6}{1 \, L} \times \dfrac{176.12 \, g \, C_6H_8O_6}{1 \, mol. \, C_6H_8O_6} = 241 \, g \, C_6H_8O_6$

Dissolve 241 g $C_6H_8O_6$ in sufficient water to make 2.00 L of solution.

4. *Refer to Section 4.1 and Example 4.1.*

a.   Convert from mL to L, then use molarity to convert from L to moles.

$45.6 \, mL \times \dfrac{1 \, L}{1000 \, mL} \times \dfrac{0.450 \, mol. \, K_2CO_3}{1 \, L} = 0.0205 \, mol. \, K_2CO_3$

b.   Use molarity to convert from moles to L, then convert L to mL.

$0.800 \, mol. \, K_2CO_3 \times \dfrac{1 \, L}{0.450 \, mol. \, K_2CO_3} \times \dfrac{1000 \, mL}{1 \, L} = 1.78 \times 10^3 \, mL$

c.   Calculate moles K_2CO_3 present in the 0.450 *M* solution and the moles needed to make 1 *M* solution. The difference is the moles needed to bring the solution up to 1 *M*.

$2.00 \, L \times \dfrac{0.450 \, mol. \, K_2CO_3}{1 \, L} = 0.900 \, mol. \, K_2CO_3$

$2.00 \, L \times \dfrac{1.000 \, mol. \, K_2CO_3}{1 \, L} = 2.00 \, mol. \, K_2CO_3$

moles needed = 2.00 mol. − 0.900 mol. = 1.10 mol.

$$1.10 \text{ mol. K}_2\text{CO}_3 \times \frac{138.21 \text{ g K}_2\text{CO}_3}{1 \text{ mol. K}_2\text{CO}_3} = 152 \text{ g K}_2\text{CO}_3$$

d. Calculate moles in 50.0 mL of solution, then recalculate molarity with the new volume.

$$50.0 \text{ mL} \times \frac{1 \text{ L}}{1000 \text{ mL}} \times \frac{0.450 \text{ mol. K}_2\text{CO}_3}{1 \text{ L}} = 0.0225 \text{ mol. K}_2\text{CO}_3$$

$$M = \frac{0.0225 \text{ mol. K}_2\text{CO}_3}{0.125 \text{ L}} = 0.180 \text{ } M \text{ K}_2\text{CO}_3$$

Alternatively, you could use the equation: $(M_{init})(V_{init}) = (M_{final})(V_{final})$.

$(0.450 \text{ } M)(50.0 \text{ mL}) = (M_{final})(125 \text{ mL})$

$(M_{final}) = 0.180 \text{ } M$

---

6. *Refer to Sections 2.6 and 4.1 and Example 4.2.*

a. $10.00 \text{ g ScBr}_3 \times \dfrac{1 \text{ mol. ScBr}_3}{284.668 \text{ g ScBr}_3} \times \dfrac{4 \text{ mol. ions}}{1 \text{ mol. ScBr}_3} = 0.1405 \text{ mol. ions}$

b. $10.00 \text{ g Ni}_2(\text{SO}_4)_3 \times \dfrac{1 \text{ mol. Ni}_2(\text{SO}_4)_3}{405.578 \text{ g Ni}_2(\text{SO}_4)_3} \times \dfrac{5 \text{ mol. ions}}{1 \text{ mol. Ni}_2(\text{SO}_4)_3} = 0.1233 \text{ mol. ions}$

c. $10.00 \text{ g BaS} \times \dfrac{1 \text{ mol. BaS}}{169.393 \text{ g BaS}} \times \dfrac{2 \text{ mol. ions}}{1 \text{ mol. BaS}} = 0.1181 \text{ mol. ions}$

d. $10.00 \text{ g (NH}_4)_3\text{PO}_4 \times \dfrac{1 \text{ mol. (NH}_4)_3\text{PO}_4}{149.087 \text{ g (NH}_4)_3\text{PO}_4} \times \dfrac{4 \text{ mol. ions}}{1 \text{ mol. (NH}_4)_3\text{PO}_4}$
$= 0.2683 \text{ mol. ions}$

---

8. *Refer to Sections 2.6 and 4.2 and Figure 4.3.*

a. $BaCl_2$: **Soluble**

b. $Mg(OH)_2$: **Insoluble**

c. $Cr_2(CO_3)_3$: **Insoluble**

d. K_3PO_4: **Soluble**

$$CoCl_3(s) \rightarrow Co^{3+}(aq) + 3Cl^-(aq)$$

a. To precipitate $CoPO_4$, add a soluble phosphate such as sodium phosphate (Na_3PO_4).

b. To precipitate $Co_2(CO_3)_3$, add a soluble carbonate such as sodium carbonate (Na_2CO_3).

c. To precipitate $Co(OH)_3$, add a soluble hydroxide such as sodium hydroxide (NaOH).

Recall that soluble salts ionize when dissolved. Write the reactions for the ionizations. Look at the resulting ions. If there are pairs that would result in insoluble salts, these salts would form and precipitate from solution.

a. Ions present: Ca^{2+}, SO_4^{2-}, Na^+, CO_3^{2-}
Possible precipitates: $CaCO_3$, Na_2SO_4
$CaCO_3$ would precipitate from solution (Na_2SO_4 is soluble and remains in solution).
$Ca^{2+}(aq) + CO_3^{2-}(aq) \rightarrow CaCO_3(s)$ (*net ionic equation*)

b. Ions present: Fe^{3+}, SO_4^{2-}, Ba^{2+}, OH^-
Possible precipitates: $BaSO_4$, $Fe(OH)_3$
$BaSO_4$ and $Fe(OH)_3$ would both precipitate from solution.
$Fe^{3+}(aq) + SO_4^{2-}(aq) + Ba^{2+}(aq) + 3OH^-(aq) \rightarrow BaSO_4(s) + Fe(OH)_3(s)$

Recall that soluble salts ionize when dissolved. Write the reactions for the ionizations. Look at the resulting ions. If there are pairs that would result in insoluble salts, these salts would form and precipitate from solution.

a. Ions present: Cu^{2+}, SO_4^{2-}, Na^+, Cl^-
Possible precipitates: $CuCl_2$, Na_2SO_4
Both compounds are soluble, thus no precipitate forms.

b. Ions present: Mn^{2+}, NO_3^-, Na^+, OH^-
Possible precipitates: $Mn(OH)_2$, $NaNO_3$
$NaNO_3$ is soluble, but $Mn(OH)_2$ is not.
$Mn^{2+}(aq) + 2OH^-(aq) \rightarrow Mn(OH)_2(s)$

c. Ions present: Ag^+, NO_3^-, H^+, Cl^-
Possible precipitates: $AgCl$, HNO_3
HNO_3 is soluble, but $AgCl$ is not.
$Ag^+(aq) + Cl^-(aq) \rightarrow AgCl(s)$

d. Ions present: Co^{2+}, SO_4^{2-}, Ba^{2+}, OH^-
   Possible precipitates: $Co(OH)_2$, $BaSO_4$
   Both compounds are insoluble.
   $Co^{2+}(aq) + 2OH^-(aq) \rightarrow Co(OH)_2(s)$
   $Ba^{2+}(aq) + SO_4^{2-}(aq) \rightarrow BaSO_4(s)$

e. Ions present: NH_4^+, K^+, OH^-, CO_3^{2-}
   Possible precipitates: NH_4OH, K_2CO_3
   Both compounds are soluble, thus no precipitate forms.

---

16. Refer to Sections 2.6, 4.1, and 4.2, Figures 4.4 and 4.5, and Example 4.3.

Recall that soluble salts ionize when dissolved. Write the reactions for the ionizations. Look at the resulting ions. If there are pairs that would result in insoluble salts, these salts would form and precipitate from solution.

a. Ions present: Na^+, PO_4^{3-}, Ba^{2+}, Cl^-
   Possible precipitates: $Ba_3(PO_4)_2$, NaCl
   NaCl is soluble, but $Ba_3(PO_4)_2$ is not.
   $3Ba^{2+}(aq) + 2PO_4^{3-}(aq) \rightarrow Ba_3(PO_4)_2(s)$ *(net ionic equation)*

b. Ions present: Zn^{2+}, SO_4^{2-}, K^+, OH^-
   Possible precipitates: $Zn(OH)_2$, K_2SO_4
   K_2SO_4 is soluble, but $Zn(OH)_2$ is not.
   $Zn^{2+}(aq) + 2OH^-(aq) \rightarrow Zn(OH)_2(s)$ *(net ionic equation)*

c. Ions present: NH_4^+, SO_4^{2-}, Na^+, Cl^-
   Possible precipitates: NH_4Cl, Na_2SO_4
   Both soluble and remain in solution, thus no precipitate forms.

d. Ions present: Na^+, PO_4^{3-}, Co^{3+}, NO_3^-
   Possible precipitates: $CoPO_4$, $NaNO_3$
   $NaNO_3$ is soluble, but $CoPO_4$ is not.
   $Co^{3+}(aq) + PO_4^{3-}(aq) \rightarrow CoPO_4(s)$ *(net ionic equation)*

---

18. Refer to Section 4.2, Figure 4.3, and Example 4.4.

Write the equations for the reactions. Use the volume and molarity of the given solution to calculate moles, then use the mole ratios from the equation to calculate the moles of $BaCl_2$. Finally, use the moles of $BaCl_2$ and the molarity to calculate the volume of $BaCl_2$ solution.

a. $BaCl_2(aq) + H_2SO_4(aq) \rightarrow BaSO_4(s) + 2HCl(aq)$

$$12.45 \, \text{mL H}_2\text{SO}_4(aq) \times \frac{1 \, \text{L H}_2\text{SO}_4(aq)}{1000 \, \text{mL H}_2\text{SO}_4(aq)} \times \frac{1.732 \, \text{mol. H}_2\text{SO}_4(aq)}{1 \, \text{L H}_2\text{SO}_4(aq)}$$

$$\times \frac{1 \, \text{mol. BaCl}_2(aq)}{1 \, \text{mol. H}_2\text{SO}_4(aq)} \times \frac{1 \, \text{L BaCl}_2(aq)}{0.4163 \, \text{mol. BaCl}_2(aq)} \times \frac{1000 \, \text{mL BaCl}_2(aq)}{1 \, \text{L BaCl}_2(aq)} = 51.80 \, \text{mL BaCl}_2(s)$$

b. $3\text{BaCl}_2(aq) + 2(\text{NH}_4)_3\text{PO}_4(aq) \rightarrow \text{Ba}_3(\text{PO}_4)_2(s) + 6\text{NH}_4\text{Cl}(aq)$

$$15.00 \, \text{g (NH}_4)_3\text{PO}_4(aq) \times \frac{1 \, \text{mol. (NH}_4)_3\text{PO}_4(aq)}{149.087 \, \text{g (NH}_4)_3\text{PO}_4(aq)} \times \frac{3 \, \text{mol. BaCl}_2(aq)}{2 \, \text{mol. (NH}_4)_3\text{PO}_4(aq)}$$

$$\times \frac{1 \, \text{L BaCl}_2(aq)}{0.4163 \, \text{mol. BaCl}_2(aq)} \times \frac{1000 \, \text{mL BaCl}_2(aq)}{1 \, \text{L BaCl}_2(aq)} = 362.5 \, \text{mL BaCl}_2(aq)$$

c. $\text{BaCl}_2(aq) + \text{K}_2\text{CO}_3(aq) \rightarrow \text{BaCO}_3(s) + 2 \, \text{KCl}(aq)$

$$35.15 \, \text{mL K}_2\text{CO}_3(aq) \times \frac{1 \, \text{L K}_2\text{CO}_3(aq)}{1000 \, \text{mL K}_2\text{CO}_3(aq)} \times \frac{1.28 \, \text{mol. K}_2\text{CO}_3(aq)}{1 \, \text{L K}_2\text{CO}_3(aq)} \times \frac{1 \, \text{mol. BaCl}_2(aq)}{1 \, \text{mol. K}_2\text{CO}_3(aq)}$$

$$\times \frac{1 \, \text{L BaCl}_2(aq)}{0.4163 \, \text{mol. BaCl}_2(aq)} \times \frac{1000 \, \text{mL BaCl}_2(aq)}{1 \, \text{L BaCl}_2(aq)} = 108 \, \text{mL BaCl}_2(aq)$$

---

20. Refer to Section 4.2, Figure 4.3, and Example 4.4.

---

a. $2\text{Al}^{3+}(aq) + 3\text{CO}_3^{2-}(aq) \rightarrow \text{Al}_2(\text{CO}_3)_3(s)$ (*net ionic equation*)

b. $0.0355 \, \text{L} \times \dfrac{0.137 \, \text{mol. Na}_2\text{CO}_3}{1 \, \text{L}} \times \dfrac{1 \, \text{mol. CO}_3^{2-}}{1 \, \text{mol. Na}_2\text{CO}_3} = 4.86 \times 10^{-3} \, \text{mol. CO}_3^{2-}$

$$4.86 \times 10^{-3} \, \text{mol. CO}_3^{2-} \times \frac{2 \, \text{mol. Al}^{3+}}{3 \, \text{mol. CO}_3^{2-}} = 0.00324 \, \text{mol. Al}^{3+}$$

$$0.00324 \, \text{mol. Al}^{3+} \times \frac{1 \, \text{mol. Al}^{3+}}{1 \, \text{mol. AlCl}_3} = 0.00324 \, \text{mol. AlCl}_3$$

$$M = \frac{\text{moles}}{\text{volume (L)}} \implies \frac{0.00324 \, \text{mol. AlCl}_3}{0.0300 \, \text{L}} = 0.108 M$$

c. 4.86×10^{-3} mol. $CO_3^{2-} \times \dfrac{1\,\text{mol. Al}_2(CO_3)_3}{3\,\text{mol. }CO_3^{2-}} \times \dfrac{234.0\,\text{g Al}_2(CO_3)_3}{1\,\text{mol. Al}_2(CO_3)_3} = 0.379\,\text{g Al}_2(CO_3)_3$

22. *Refer to Sections 2.6 and 4.3 and Table 4.1.*

Acids **not** in Table 4.1 are weak acids. Also, acids containing only C, H and O are organic acids and should also be considered weak.

a. Sulfurous acid (H_2SO_3) is a weak acid, the reacting species is H_2SO_3.

b. Chlorous acid ($HClO_2$) is a weak acid, the reacting species is $HClO_2$.

c. Perchloric acid ($HClO_4$) is a strong acid, the reacting species is H^+.

d. Sulfuric acid (H_2SO_4) is a strong acid, the reacting species is H^+.

e. Formic acid ($HCHO_2$) is a weak acid, the reacting species is $HCHO_2$.

24. *Refer to Sections 2.6 and 4.3 and Table 4.1.*

Bases **not** in Table 4.1 are weak bases. Also, bases containing only C, H and N are organic bases and should also be considered weak.

a. Indol (C_8H_6NH) is a weak base, the reacting species is C_8H_6NH.

b. Potassium hydroxide (KOH) is a strong base, the reacting species is OH^-.

c. Ammonia (NH_3) is a weak base, the reacting species is NH_3.

d. Calcium hydroxide ($Ca(OH)_2$) is a strong base, the reacting species is OH^-.

26. *Refer to Sections 2.6 and 4.3 and Table 4.1.*

Acids and bases **not** in Table 4.1 are weak. Acids containing only C, H and O are organic acids and should also be considered weak. Also, bases containing only C, H and N are organic bases and should also be considered weak.

a. H_2SO_3 is a weak acid

b. NH_3 is a weak organic base

c. $Ba(OH)_2$ is a strong base

d. HI is a strong acid

Identify the acid and the base as weak or strong, then identify the reacting species. Then write the reaction for the two reacting species.

a. Acetic acid ($HC_2H_3O_2$) is a weak acid, the reacting species is $HC_2H_3O_2$.
Strontium hydroxide ($Sr(OH)_2$) is a strong base, the reacting species is OH^-.
$HC_2H_3O_2(aq) + OH^-(aq) \rightarrow H_2O(l) + C_2H_3O_2^-(aq)$

b. Sulfuric acid (H_2SO_4) is a strong acid, the reacting species is H^+.
Diethylamine ($(C_2H_5)_2NH$) is a weak base, the reacting species is $(C_2H_5)_2NH$.
$H^+(aq) + (C_2H_5)_2NH(aq) \rightarrow (C_2H_5)_2NH_2^+(aq)$

c. Hydrogen cyanide (HCN) is a weak acid, the reacting species is HCN.
Sodium hydroxide (NaOH) is a strong base, the reacting species is OH^-.
$HCN(aq) + OH^-(aq) \rightarrow H_2O(l) + CN^-(aq)$

For the net ionic equation $OH^-(aq) + HB(aq) \rightarrow B^-(aq) + H_2O(l)$ to be correct, the reactants must be a weak acid and a strong base.

a. The equation is **not** correct.
Hydrochloric acid (HCl) is a strong acid, the reacting species is H^+.
Pyridine (C_5H_5N) is a weak base, the reacting species is C_5H_5N.
$H^+(aq) + C_5H_5N(aq) \rightarrow C_5H_5NH^+(aq)$

b. The equation is **not** correct.
Sulfuric acid (H_2SO_4) is a strong acid, the reacting species is H^+.
Rubidium hydroxide (RbOH) is a strong base, the reacting species is OH^-.
$H^+(aq) + OH^-(aq) \rightarrow H_2O(l)$

c. The equation **is** correct.
Hydrofluoric acid (HF) is a weak acid, the reacting species is HF.
Potassium hydroxide (KOH) is a strong base, the reacting species is OH^-.
$OH^-(aq) + HF(aq) \rightarrow F^-(aq) + H_2O(l)$

d. The equation is **not** correct.
Hydroiodic acid (HI) is a strong acid, the reacting species is H^+.
Ammonia (NH_3) is a weak base, the reacting species is NH_3.
$H^+(aq) + NH_3(aq) \rightarrow NH_4^+(aq)$

e. The equation is correct.
   Hydrocyanic acid (HCN) is a weak acid, the reacting species is HCN.
   Strontium hydroxide ($Sr(OH)_2$) is a strong base, the reacting species is OH^-.
   $OH^-(aq) + HCN(aq) \rightarrow CN^-(aq) + H_2O(l)$

---

32. Refer to Section 4.3 and Example 4.7.

$Sr(OH)_2(aq) + 2HF(aq) \rightarrow SrF_2(aq) + H_2O(l)$

$$25.00 \text{ mL} \times \frac{1 \text{ L}}{1000 \text{ mL}} \times \frac{0.275 \text{ mol. HF}}{1 \text{ L}} \times \frac{1 \text{ mol. Sr(OH)}_2}{2 \text{ mol. HF}}$$

$$\times \frac{1 \text{ L}}{0.285 \text{ mol. Sr(OH)}_2} = 0.0121 \text{ L Sr(OH)}_2 \text{ solution}$$

---

34. Refer to Section 4.3 and Example 4.7.

Calculate the moles of the given base, use the mole ratio of HCl to base to calculate the moles of HCl and then use molarity to calculate the volume of HCl.

a. $HCl + NH_3 \rightarrow NH_4^+ + Cl^-$

$$0.02500 \text{ L} \times \frac{0.288 \text{ mol. NH}_3}{1 \text{ L}} \times \frac{1 \text{ mol. HCl}}{1 \text{ mol. NH}_3} \times \frac{1 \text{ L}}{0.885 \text{ mol. HCl}} = 8.14 \times 10^{-3} \text{ L HCl}$$

$$= 8.14 \text{ mL HCl}$$

b. $NaOH + HCl \rightarrow NaCl + H_2O$

$$10.00 \text{ g NaOH} \times \frac{1 \text{ mol. NaOH}}{40.00 \text{ g NaOH}} \times \frac{1 \text{ mol. HCl}}{1 \text{ mol. NaOH}} = 0.2500 \text{ mol. HCl}$$

$$M = \frac{\text{moles}}{\text{volume (L)}} \Rightarrow 0.885 \, M = \frac{0.2500 \text{ mol. HCl}}{\text{volume}}$$

volume $= 0.282 \text{ L} = 282 \text{ mL}$

c. Use density to convert volume of solution to mass and then use the mass percent to calculate the mass of methylamine in solution. Then proceed as above.

$CH_3NH_2 + HCl \rightarrow CH_3NH_3^+ + Cl^-$

$$25.0 \text{ mL} \times \frac{1 \text{ cm}^3}{1 \text{ mL}} \times \frac{0.928 \text{ g solution}}{1 \text{ cm}^3} \times \frac{10 \text{ g CH}_3NH_2}{100 \text{ g solution}} = 2.32 \text{ g CH}_3NH_2$$

$$2.32 \text{ g CH}_3\text{NH}_2 \text{ x } \frac{1 \text{ mol. CH}_3\text{NH}_2}{31.06 \text{ g CH}_3\text{NH}_2} \text{ x } \frac{1 \text{ mol. HCl}}{1 \text{ mol. CH}_3\text{NH}_2} = 0.0747 \text{ mol. HCl}$$

$$M = \frac{\text{moles}}{\text{volume (L)}} \Rightarrow 0.885 \, M = \frac{0.0747 \text{ mol. HCl}}{\text{volume}}$$

volume = 0.0844 L = 84.4 mL

---

36. *Refer to Section 4.3 and Example 4.7.*

Write the balanced equation, use this to calculate the mass of $HC_2H_3O_2$, then calculate the mass percent.

$$Ba(OH)_2 + 2HC_2H_3O_2 \rightarrow Ba(C_2H_3O_2)_2 + 2H_2O$$

$$37.50 \text{ mL Ba(OH)}_2 \times \frac{1 \text{ L Ba(OH)}_2}{1000 \text{ mL Ba(OH)}_2} \times \frac{0.1250 \text{ mol. Ba(OH)}_2}{1 \text{ L Ba(OH)}_2} \times \frac{2 \text{ mol. HC}_2\text{H}_3\text{O}_2}{1 \text{ mol. Ba(OH)}_2}$$

$$\times \frac{60.053 \text{ g HC}_2\text{H}_3\text{O}_2}{1 \text{ mol. HC}_2\text{H}_3\text{O}_2} = 0.5630 \text{ g HC}_2\text{H}_3\text{O}_2$$

$$\% \, HC_2H_3O_2 = \frac{\text{mass of HC}_2\text{H}_3\text{O}_2}{\text{mass of sample}} \times 100\% = \frac{0.5630 \text{ g HC}_2\text{H}_3\text{O}_2}{10.00 \text{ g sample}} \times 100\% = 5.630\%$$

Yes, the sample can be considered vinegar.

---

38. *Refer to Section 4.3 and Example 4.7.*

Calculate mass of $C_6H_8O_6$ as in problem 36 above, then calculate the mass percent.

$$C_6H_8O_6 + NaOH \rightarrow H_2O + NaC_6H_8O_6$$

$$0.00620 \text{ L} \times \frac{0.425 \text{ mol. NaOH}}{1 \text{ L}} \times \frac{1 \text{ mol. C}_6\text{H}_8\text{O}_6}{1 \text{ mol. NaOH}} \times \frac{176.12 \text{ g}}{1 \text{ mol. C}_6\text{H}_8\text{O}_6} = 0.464 \text{ g C}_6\text{H}_8\text{O}_6$$

$$\% \, C_6H_8O_6 = \frac{\text{mass of C}_6\text{H}_8\text{O}_6}{\text{mass of sample}} \times 100\% = \frac{0.464 \text{ g C}_6\text{H}_8\text{O}_6}{0.628 \text{ g sample}} \times 100\% = 73.9\%$$

40. *Refer to Section 4.3.*

Determine the mole ratio of lactic acid to NaOH by calculating the moles of each.

$$0.100\,g\,C_3H_6O_3 \times \frac{1\,mol.\,C_3H_6O_3}{90.08\,g\,C_3H_6O_3} = 1.11 \times 10^{-3}\ mol.\,C_3H_6O_3$$

$$0.01295\,L \times \frac{0.0857\,mol.\,NaOH}{1\,L} = 1.11 \times 10^{-3}\ mol.\,NaOH$$

mole ratio = 1:1 (one mole of NaOH is required to neutralize one mole of $C_3H_6O_3$)

42. *Refer to Sections 2.6 and 4.4 and Example 4.8.*

To calculate the oxidation number, recall that oxygen has an oxidation number of –2 (O^{2-}), that hydrogen has an oxidation number of +1 (H^+), and that the sum of the oxidation numbers must equal the overall charge.

a. CH_4

[(C) + 4(H) = charge]
[(C) + 4(+1) = 0]
C = -4

b. H_2SO_3

[(S) + 2(H) + 3(O) = charge]
[(S) + 2(+1) + 3(-2) = 0]
S = +4

c. Na_2O

[2(Na) + (O) = charge]
[2(Na) + (-2) = 0]
Na = +1

d. $H_2PO_4^-$

[(P) + 2(H) + 4(O) = charge]
[(P) + 2(+1) + 4(-2) = -1]
P = +5

44. *Refer to Section 4.4 and Example 4.8.*

To calculate the oxidation number, recall that oxygen has an oxidation number of –2 (O^{2-}), that hydrogen has an oxidation number of +1 (H^+), and that the sum of the oxidation numbers must equal the overall charge. Note that part (e) is calculated differently

a. HIO_3

[(H) + (I) + 3(O) = charge]
[(+1) + (I) + 3(-2) = 0]
I = +5

b. $NaMnO_4$ $[(Na) + (Mn) + 4(O) = charge]$
                         $[(+1) + (Mn) + 4(-2) = 0]$
                         $Mn = +7$

c. SnO_2 $[(Sn) + 2(O) = charge]$
                         $[Sn + 2(-2) = 0]$
                         $Sn = +4$

d. NOF $[(N) + (O) + (F) = charge]$
                         $[(N) + (-2) + (-1) = 0]$
                         $N = +3$

e. NaO_2 Since it is known that the sodium ion has an oxidation number of +1, use this information to calculate the oxidation number of oxygen.
                         $[(Na) + 2(O) = charge]$
                         $[(+1) + 2(O) = 0]$
                         $O = -½$

---

46. *Refer to Section 4.4.*

Determine the oxidation number of the species of interest in each of the half reactions. If the oxidation number decreases, there is a reduction, if it increases, there is an oxidation.

a. TiO_2: Ti^{4+}
      Ti^{3+}
      oxidation number *de*creased from +4 to +3, so this is a **reduction**.

b. Zn^{2+}
      Zn^0
      oxidation number *de*creased from +2 to 0, so this is a **reduction**.

c. NH_4^+: N^{3-}
      N_2: N^0
      oxidation number *in*creased from -3 to 0, so this is an **oxidation**.

d. CH_3OH: $[1(C) + 4(H) + 1(O) = charge]$
                         $[1(C) + 4(+1) + (-2) = 0]$
                         C^{2-}
      CH_2O: $[1(C) + 2(H) + 1(O) = charge]$
                         $[1(C) + 2(+1) + (-2) = 0]$
                         C^0
      oxidation number *in*creased from -2 to 0, so this is an **oxidation**.

Begin by determining the oxidation states of the element being reduced or oxidized. Balance the element being oxidized or reduced and then balance the oxidation number by adding electrons. Balance the charge by adding H^+ (if acidic) or OH^- (if basic), and H_2O to balance H and O.

a. $TiO_2(s) \rightarrow Ti^{3+}(aq)$ [acidic]

     $Ti^{4+} \rightarrow Ti^{3+}$ **Reduction**

     $TiO_2(s) \rightarrow Ti^{3+}(aq)$ Ti balanced

     $TiO_2(s) + e^- \rightarrow Ti^{3+}(aq)$ electrons balanced

     $TiO_2(s) + e^- + 4H^+(aq) \rightarrow Ti^{3+}(aq)$ charges balanced

     $TiO_2(s) + e^- + 4H^+(aq) \rightarrow Ti^{3+}(aq) + 4H_2O(l)$ O and H balanced

b. $Zn^{2+}(aq) \rightarrow Zn(s)$ [acidic]

     $Zn^{2+} \rightarrow Zn^0$ **Reduction**

     $Zn^{2+}(aq) + 2e^- \rightarrow Zn(s)$ completely balanced

c. $NH_4^+(aq) \rightarrow N_2(g)$ [basic]

     $N^{3-} \rightarrow N^0$ **Oxidation**

     $2NH_4^+(aq) \rightarrow N_2(g)$ N balanced

     $2NH_4^+(aq) \rightarrow N_2(g) + 6e^-$ electrons balanced

     $2NH_4^+(aq) + 8OH^-(aq) \rightarrow N_2(g) + 6e^-$ charges balanced

     $2NH_4^+(aq) + 8OH^-(aq) \rightarrow N_2(g) + 6e^- + 8H_2O(l)$ O and H balanced

d. $CH_3OH(aq) \rightarrow CH_2O(aq)$ [basic]

     $C^{2-} \rightarrow C^0 + 2e^-$ **Oxidation**

     $CH_3OH(aq) \rightarrow CH_2O(aq)$ C balanced

     $CH_3OH(aq) \rightarrow CH_2O(aq) + 2e^-$ electrons balanced

     $CH_3OH(aq) + 2OH^-(aq) \rightarrow CH_2O(aq) + 2e^-$ charges balanced

     $CH_3OH(aq) + 2OH^-(aq) \rightarrow CH_2O(aq) + 2e^- + 2H_2O(l)$ O and H balanced

Begin by determining the oxidation numbers of the element being reduced or oxidized. Balance the element being oxidized or reduced and then balance the oxidation number by adding electrons. Balance the charge by adding H^+ (if acidic) or OH^- (if basic), and H_2O to balance H and O.

a. $ClO^-(aq) \rightarrow Cl^-(aq)$ [basic]

     $Cl^+ \rightarrow Cl^-$ **Reduction**

     $ClO^-(aq) + 2e^- \rightarrow Cl^-(aq)$ Cl and electrons balanced

     $ClO^-(aq) + 2e^- \rightarrow Cl^-(aq) + 2OH^-(aq)$ charges balanced

     $ClO^-(aq) + 2e^- + H_2O(l) \rightarrow Cl^-(aq) + 2OH^-(aq)$ O and H balanced

b. $NO_3^-(aq) \rightarrow NO(g)$ [acidic]

    $N^{5+} \rightarrow N^{2+}$ **Reduction**

    $NO_3^-(aq) + 3e^- \rightarrow NO(g)$ N and electrons balanced

    $NO_3^-(aq) + 3e^- + 4H^+(aq) \rightarrow NO(g)$ charges balanced

    $NO_3^-(aq) + 3e^- + 4H^+(aq) \rightarrow NO(g) + 2H_2O(l)$ O and H balanced

c. $Ni^{2+}(aq) \rightarrow Ni_2O_3(s)$ [basic]

    $Ni^{2+} \rightarrow Ni^{3+}$ **Oxidation**

    $2Ni^{2+}(aq) \rightarrow Ni_2O_3(s)$ Ni balanced

    $2Ni^{2+}(aq) \rightarrow Ni_2O_3(s) + 2e^-$ electrons balanced

    $2Ni^{2+}(aq) + 6OH^-(aq) \rightarrow Ni_2O_3(s) + 2e^-$ charges balanced

    $2Ni^{2+}(aq) + 6OH^-(aq) \rightarrow Ni_2O_3(s) + 2e^- + 3H_2O(l)$ O and H balanced

d. $Mn^{2+}(aq) \rightarrow MnO_2(s)$ [acidic]

    $Mn^{2+} \rightarrow Mn^{4+}$ **Oxidation**

    $Mn^{2+}(aq) \rightarrow MnO_2(s) + 2e^-$ Mn and electrons balanced

    $Mn^{2+}(aq) \rightarrow MnO_2(s) + 2e^- + 4H^+(aq)$ charges balanced

    $Mn^{2+}(aq) + 2H_2O(l) \rightarrow MnO_2(s) + 2e^- + 4H^+(aq)$ O and H balanced

---

52. Refer to Section 4.4 and Example 4.10.

---

a. $H_2O_2(aq) \rightarrow H_2O(l)$

    $O^- \rightarrow O^{2-}$

        Oxidation number has decreased, so O is reduced and H_2O_2 is the oxidizing agent.

    $Ni^{2+}(aq) \rightarrow Ni^{3+}(aq)$

        Oxidation number has increased, so Ni is oxidized and Ni^{2+} is the reducing agent.

b. $Cr_2O_7^{2-}(aq) \rightarrow Cr^{3+}(l)$

    $Cr^{6+} \rightarrow Cr^{3+}$

        Oxidation number has decreased, so Cr is reduced and $Cr_2O_7^{2-}$ is the oxidizing agent.

    $Sn^{2+}(aq) \rightarrow Sn^{4+}(aq)$

        Oxidation number has increased, so Sn is oxidized and Sn^{2+} is the reducing agent.

---

54. Refer to Section 4.4 and Example 4.10.

Balance the two half reactions. If the number of electrons in the balanced half reactions are not equal, multiply the equations through with the appropriate coefficient. Combine the two half reactions and cancel those species which appear on both sides of the reaction.

a. $H_2O_2(aq) \rightarrow H_2O(l)$ [basic]

    $O^- \rightarrow O^{2-}$

    $H_2O_2(aq) \rightarrow 2H_2O(l)$ O balanced

$H_2O_2(aq) + 2e^- \rightarrow 2H_2O(l)$ electrons balanced
                                                                                          Note: 2 O thus need $2e^-$

$H_2O_2(aq) + 2e^- \rightarrow 2H_2O(l) + 2OH^-(aq)$ charges balanced
$H_2O_2(aq) + 2e^- + 2H_2O(l) \rightarrow 2H_2O(l) + 2OH^-(aq)$ O and H balanced

$H_2O_2(aq) + 2e^- \rightarrow 2OH^-(aq)$ remove water

$Ni^{2+}(aq) \rightarrow Ni^{3+}(aq)$
$Ni^{2+}(aq) \rightarrow Ni^{3+}(aq) + e^-$ everything is balanced

$2Ni^{2+}(aq) \rightarrow 2Ni^{3+}(aq) + 2e^-$ balance electrons in half-rxns

$2Ni^{2+}(aq) + H_2O_2(aq) + 2e^- \rightarrow 2Ni^{3+}(aq) + 2OH^-(aq) + 2e^-$
$2Ni^{2+}(aq) + H_2O_2(aq) \rightarrow 2Ni^{3+}(aq) + 2OH^-(aq)$

b. $Cr_2O_7^{2-}(aq) \rightarrow Cr^{3+}(aq)$ [basic]
    $Cr^{6+} \rightarrow Cr^{3+}$
    $Cr_2O_7^{2-}(aq) + 6e^- \rightarrow 2Cr^{3+}(aq)$ Cr and electrons balanced
    $Cr_2O_7^{2-}(aq) + 6e^- \rightarrow 2Cr^{3+}(aq) + 14OH^-(aq)$ charges balanced
    $Cr_2O_7^{2-}(aq) + 6e^- + 7H_2O(l) \rightarrow 2Cr^{3+}(aq) + 14OH^-(aq)$ H and O balanced

    $Sn^{2+}(aq) \rightarrow Sn^{4+}(aq)$
    $Sn^{2+} \rightarrow Sn^{4+} + 2e^-$ everything is balanced

    $3Sn^{2+}(aq) \rightarrow 3Sn^{4+}(aq) + 6e^-$ balance electrons in half-rxns

    $3Sn^{2+}(aq) + Cr_2O_7^{2-}(aq) + 6e^- + 7H_2O(l) \rightarrow 2Cr^{3+}(aq) + 14OH^-(aq) + 3Sn^{4+}(aq) + 6e^-$
    $3Sn^{2+}(aq) + Cr_2O_7^{2-}(aq) + 7H_2O(l) \rightarrow 2Cr^{3+}(aq) + 14OH^-(aq) + 3Sn^{4+}(aq)$

---

56. Refer to Section 4.4 and Example 4.10.

---

a. $P_4(s) \rightarrow PH_3(g)$
    $P^0 \rightarrow P^{3-}$
    $P_4(s) \rightarrow 4PH_3(g)$ P balanced
    $P_4(s) + 12e^- \rightarrow 4PH_3(g)$ electrons balanced
                                                                                          Note: 4 P, thus need $12e^-$
    $P_4(s) + 12e^- + 12H^+(aq) \rightarrow 4PH_3(g)$ charges balanced

    $Cl^-(aq) \rightarrow Cl_2(g)$
    $Cl^- \rightarrow Cl^0$
    $2Cl^-(aq) \rightarrow Cl_2(g)$ Cl balanced
    $2Cl^-(aq) \rightarrow Cl_2(g) + 2e^-$ electrons balanced

    $12Cl^-(aq) \rightarrow 6Cl_2(g) + 12e^-$ balance electrons in half-reactions

    $P_4(s) + 12e^- + 12H^+(aq) + 12Cl^-(aq) \rightarrow 4PH_3(g) + 6Cl_2(g) + 12e^-$
    $P_4(s) + 12H^+(aq) + 12Cl^-(aq) \rightarrow 4PH_3(g) + 6Cl_2(g)$

b. $MnO_4^-(aq) \rightarrow Mn^{2+}(aq)$
$Mn^{7+} \rightarrow Mn^{2+}$
$MnO_4^-(aq) \rightarrow Mn^{2+}(aq)$ Mn balanced
$MnO_4^-(aq) + 5e^- \rightarrow Mn^{2+}(aq)$ electrons balanced
$MnO_4^-(aq) + 5e^- + 8H^+(aq) \rightarrow Mn^{2+}(aq)$ charges balanced
$MnO_4^-(aq) + 5e^- + 8H^+(aq) \rightarrow Mn^{2+}(aq) + 4H_2O(l)$ H and O balanced

$NO_2^-(aq) \rightarrow NO_3^-(aq)$
$N^{3+} \rightarrow N^{5+}$
$NO_2^-(aq) \rightarrow NO_3^-(aq)$ N balanced
$NO_2^-(aq) \rightarrow NO_3^-(aq) + 2e^-$ electrons balanced
$NO_2^-(aq) \rightarrow NO_3^-(aq) + 2e^- + 2H^+(aq)$ charges balanced
$NO_2^-(aq) + H_2O(l) \rightarrow NO_3^-(aq) + 2e^- + 2H^+(aq)$ H and O balanced

$2MnO_4^-(aq) + 10e^- + 16H^+(aq) \rightarrow 2Mn^{2+}(aq) + 8H_2O(l)$ balance electrons in half-rxns
$5NO_2^-(aq) + 5H_2O(l) \rightarrow 5NO_3^-(aq) + 10e^- + 10H^+(aq)$

$2MnO_4^-(aq) + 10e^- + 16H^+(aq) + 5NO_2^-(aq) + 5H_2O(l)$
$$\rightarrow 2Mn^{2+}(aq) + 8H_2O(l) + 5NO_3^-(aq) + 10e^- + 10H^+(aq)$$

$2MnO_4^-(aq) + 6H^+(aq) + 5NO_2^-(aq) \rightarrow 2Mn^{2+}(aq) + 3H_2O(l) + 5NO_3^-(aq)$

c. $HBrO_3(aq) \rightarrow HBrO_2(aq)$
$Br^{5+} \rightarrow Br^{3+}$
$HBrO_3(aq) \rightarrow HBrO_2(aq)$ Br balanced
$HBrO_3(aq) + 2e^- \rightarrow HBrO_2(aq)$ electrons balanced
$HBrO_3(aq) + 2e^- + 2H^+(aq) \rightarrow HBrO_2(aq)$ charges balanced
$HBrO_3(aq) + 2e^- + 2H^+(aq) \rightarrow HBrO_2(aq) + H_2O(l)$ H and O balanced

$Bi(s) \rightarrow Bi_2O_3(s)$
$Bi^0 \rightarrow Bi^{3+}$
$2Bi(s) \rightarrow Bi_2O_3(s)$ Bi balanced
$2Bi(s) \rightarrow Bi_2O_3(s) + 6e^-$ electrons balanced
$2Bi(s) \rightarrow Bi_2O_3(s) + 6e^- + 6H^+(aq)$ charges balanced
$2Bi(s) + 3H_2O(l) \rightarrow Bi_2O_3(s) + 6e^- + 6H^+(aq)$ H and O balanced

$3HBrO_3(aq) + 6e^- + 6H^+(aq) \rightarrow 3HBrO_2(aq) + 3H_2O(l)$ balance electrons in half-rxns

$2Bi(s) + 3H_2O(l) + 3HBrO_3(aq) + 6e^- + 6H^+(aq)$
$$\rightarrow Bi_2O_3(s) + 6e^- + 6H^+(aq) + 3HBrO_2(aq) + 3H_2O(l)$$

$2Bi(s) + 3HBrO_3(aq) \rightarrow Bi_2O_3(s) + 3HBrO_2(aq)$

d. $CrO_4^{2-}(aq) \rightarrow Cr^{3+}(aq)$
$Cr^{6+} \rightarrow Cr^{3+}$

$CrO_4^{2-}(aq) \rightarrow Cr^{3+}(aq)$	Cr balanced
$CrO_4^{2-}(aq) + 3e^- \rightarrow Cr^{3+}(aq)$	electrons balanced
$CrO_4^{2-}(aq) + 3e^- + 8H^+(aq) \rightarrow Cr^{3+}(aq)$	charges balanced
$CrO_4^{2-}(aq) + 3e^- + 8H^+(aq) \rightarrow Cr^{3+}(aq) + 4H_2O(l)$	H and O balanced

$SO_3^{2-}(aq) \rightarrow SO_4^{2-}(aq)$
$S^{4+} \rightarrow S^{6+}$

$SO_3^{2-}(aq) \rightarrow SO_4^{2-}(aq)$	S balanced
$SO_3^{2-}(aq) \rightarrow SO_4^{2-}(aq) + 2e^-$	electrons balanced
$SO_3^{2-}(aq) \rightarrow SO_4^{2-}(aq) + 2e^- + 2H^+(aq)$	charges balanced
$SO_3^{2-}(aq) + H_2O(l) \rightarrow SO_4^{2-}(aq) + 2e^- + 2H^+(aq)$	H and O balanced

$2CrO_4^{2-}(aq) + 6e^- + 16H^+(aq) \rightarrow 2Cr^{3+}(aq) + 8H_2O(l)$	balance electrons in half-rxns
$3SO_3^{2-}(aq) + 3H_2O(l) \rightarrow 3SO_4^{2-}(aq) + 6e^- + 6H^+(aq)$	

$2CrO_4^{2-}(aq) + 6e^- + 16H^+(aq) + 3SO_3^{2-}(aq) + 3H_2O(l)$
$\qquad \rightarrow 2Cr^{3+}(aq) + 8H_2O(l) + 3SO_4^{2-}(aq) + 6e^- + 6H^+(aq)$

$2CrO_4^{2-}(aq) + 10H^+(aq) + 3SO_3^{2-}(aq) \rightarrow 2Cr^{3+}(aq) + 5H_2O(l) + 3SO_4^{2-}(aq)$

---

58. Refer to Section 4.4 and Example 4.10.

---

a. $Ni(OH)_2(s) \rightarrow Ni(s)$
$Ni^{2+} \rightarrow Ni^0$

$Ni(OH)_2(s) \rightarrow Ni(s)$	Ni balanced
$Ni(OH)_2(s) + 2e^- \rightarrow Ni(s)$	electrons balanced
$Ni(OH)_2(s) + 2e^- \rightarrow Ni(s) + 2OH^-(aq)$	charges, H and O balanced

$N_2H_4(aq) \rightarrow N_2(g)$
$N^{2-} \rightarrow N^0$

$N_2H_4(aq) \rightarrow N_2(g)$	N balanced
$N_2H_4(aq) \rightarrow N_2(g) + 4e^-$	electrons balanced
	note: 2 N, thus $4e^-$
$N_2H_4(aq) + 4OH^-(aq) \rightarrow N_2(g) + 4e^- + 4H_2O(l)$	charges and H balanced
$2Ni(OH)_2(s) + 4e^- \rightarrow 2Ni(s) + 4OH^-(aq)$	balance electrons in half-rxns

$N_2H_4(aq) + 4OH^-(aq) + 2Ni(OH)_2(s) + 4e^- \rightarrow 2Ni(s) + 4OH^-(aq) + N_2(g) + 4e^- + 4H_2O(l)$
$N_2H_4(aq) + 2Ni(OH)_2(s) \rightarrow 2Ni(s) + N_2(g) + 4H_2O(l)$

b. $Fe(OH)_3(s) \rightarrow Fe(OH)_2(s)$
$Fe^{3+} \rightarrow Fe^{2+}$

$Fe(OH)_3(s) \rightarrow Fe(OH)_2(s)$	Fe balanced

$Fe(OH)_3(s) + e^- \rightarrow Fe(OH)_2(s)$ electrons balanced
$Fe(OH)_3(s) + e^- \rightarrow Fe(OH)_2(s) + OH^-(aq)$ charges, H and O balanced

$Cr^{3+}(aq) \rightarrow CrO_4^{2-}(aq)$
$Cr^{3+}(aq) \rightarrow Cr^{6+}$
$Cr^{3+}(aq) \rightarrow CrO_4^{2-}(aq)$ Cr balanced
$Cr^{3+}(aq) \rightarrow CrO_4^{2-}(aq) + 3\ e^-$ electrons balanced
$Cr^{3+}(aq) + 8OH^-(aq) \rightarrow CrO_4^{2-}(aq) + 3e^-$ charges balanced
$Cr^{3+}(aq) + 8OH^-(aq) \rightarrow CrO_4^{2-}(aq) + 3e^- + 4H_2O(l)$ H and O balanced

$3Fe(OH)_3(s) + 3e^- \rightarrow 3Fe(OH)_2(s) + 3OH^-(aq)$ balance electrons in half-rxns

$3Fe(OH)_3(s) + 3e^- + Cr^{3+}(aq) + 8OH^-(aq)$
$\quad \rightarrow CrO_4^{2-}(aq) + 3e^- + 4H_2O(l) + 3Fe(OH)_2(s) + 3OH^-(aq)$
$3Fe(OH)_3(s) + Cr^{3+}(aq) + 5OH^-(aq) \rightarrow CrO_4^{2-}(aq) + 4H_2O(l) + 3Fe(OH)_2(s)$

c. $MnO_4^-(aq) \rightarrow MnO_2(s)$
$Mn^{7+} \rightarrow Mn^{4+}$
$MnO_4^-(aq) + 3e^- \rightarrow MnO_2(s)$ Mn and electrons balanced
$MnO_4^-(aq) + 3e^- + 2H_2O(l) \rightarrow MnO_2(s) + 4OH^-(aq)$ charges, H and O balanced

$BrO_3^-(aq) \rightarrow BrO_4^-(aq)$
$Br^{5+} \rightarrow Br^{+7}$
$BrO_3^-(aq) \rightarrow BrO_4^-(aq) + 2e^-$ Br and electrons balanced
$BrO_3^-(aq) + 2OH^-(aq) \rightarrow BrO_4^-(aq) + 2e^-$ charges balanced
$BrO_3^-(aq) + 2OH^-(aq) \rightarrow BrO_4^-(aq) + 2e^- + H_2O(l)$ H and O balanced

                                                               balance electrons in half-rxns

$2MnO_4^-(aq) + 6e^- + 4H_2O(l) \rightarrow 2MnO_2(s) + 8OH^-(aq)$
$3BrO_3^-(aq) + 6OH^-(aq) \rightarrow 3BrO_4^-(aq) + 6e^- + 3H_2O(l)$

$2MnO_4^-(aq) + 6e^- + 4H_2O(l) + 3BrO_3^-(aq) + 6OH^-(aq)$
$\quad \rightarrow 3BrO_4^-(aq) + 6e^- + 3H_2O(l) + 2MnO_2(s) + 8OH^-(aq)$
$2MnO_4^-(aq) + 3BrO_3^-(aq) + H_2O(l) \rightarrow 3BrO_4^-(aq) + 2MnO_2(s) + 2OH^-(aq)$

d. $H_2O_2(aq) \rightarrow O_2(g)$
$O^- \rightarrow O^0$
$H_2O_2(aq) \rightarrow O_2(g) + 2e^-$ O and electrons balanced
$H_2O_2(aq) + 2OH^-(aq) \rightarrow O_2(g) + 2e^-$ charges balanced
$H_2O_2(aq) + 2OH^-(aq) \rightarrow O_2(g) + 2e^- + 2H_2O(l)$ H and O balanced

$IO_4^-(aq) \rightarrow IO_2^-(aq)$
$I^{7+} \rightarrow I^{3+}$
$IO_4^-(aq) + 4e^- \rightarrow IO_2^-(aq)$ I and electrons balanced
$IO_4^-(aq) + 4e^- \rightarrow IO_2^-(aq) + 4OH^-(aq)$ charges balanced
$IO_4^-(aq) + 4e^- + 2H_2O(l) \rightarrow IO_2^-(aq) + 4OH^-(aq)$ H and O balanced

$2H_2O_2(aq) + 4OH^-(aq) \rightarrow 2O_2(g) + 4e^- + 4H_2O(l)$ balance electrons in half-rxns

$$2H_2O_2(aq) + 4OH^-(aq) + IO_4^-(aq) + 4e^- + 2H_2O(l)$$
$$\rightarrow IO_2^-(aq) + 4OH^-(aq) + 2O_2(g) + 4e^- + 4H_2O(l)$$
$$2H_2O_2(aq) + IO_4^-(aq) \rightarrow IO_2^-(aq) + 2O_2(g) + 2H_2O(l)$$

---

60. Refer to Sections 2.6 and 4.4 and Example 4.10.

Rewrite the reaction, deleting those species that do not change oxidation number. Balance the two half reactions. Balance the element being oxidized or reduced and then balance the oxidation number by adding electrons. Balance the charge by adding H^+, and H_2O to balance H and O. If the number of electrons in the balanced half reactions are not equal, multiply the equations through with the appropriate coefficient. Combine the two half reactions and cancel those species which appear on both sides of the reaction.

a. $NO(g) + H_2(g) \rightarrow NH_3(g) + H_2O(g)$ (unbalanced equation)

    $H_2(g) \rightarrow H_2O(g)$ H balanced
    oxid. no. H: $0 \rightarrow +1$
    $H_2(g) \rightarrow H_2O(g) + 2e^-$ electrons balanced
                                            Note: 2H, thus $2e^-$

    $H_2(g) \rightarrow H_2O(g) + 2e^- + 2H^+(aq)$ charges balanced
    $H_2O(g) + H_2(g) \rightarrow H_2O(g) + 2e^- + 2H^+(aq)$ cancel water on each side
    $H_2(g) \rightarrow 2e^- + 2H^+(aq)$

    $NO(g) \rightarrow NH_3(g)$ N balanced
    oxid. no. N: $+2 \rightarrow -3$
    $NO(g) + 5e^- \rightarrow NH_3(g)$ electrons balanced
    $NO(g) + 5e^- + 5H^+(aq) \rightarrow NH_3(g)$ charges balanced
    $NO(g) + 5e^- + 5H^+(aq) \rightarrow NH_3(g) + H_2O(g)$ H and O balanced

    $5H_2(g) \rightarrow 10e^- + 10H^+(aq)$ balance electrons in half-reactions
    $2NO(g) + 10e^- + 10H^+(aq) \rightarrow 2NH_3(g) + 2H_2O(g)$

    $2NO(g) + 5H_2(g) + 10e^- + 10H^+(aq) \rightarrow 10e^- + 10H^+(aq) + 2NH_3(g) + 2H_2O(g)$
    $2NO(g) + 5H_2(g) \rightarrow 2NH_3(g) + 2H_2O(g)$

b. $H_2O_2(aq) + NaClO(aq) \rightarrow O_2(g) + Cl_2(g)$
    $H_2O_2(aq) + ClO^-(aq) \rightarrow O_2(g) + Cl_2(g)$ (unbalanced ionic equation)

    $H_2O_2(aq) \rightarrow O_2(g)$ O balanced
    oxid. no. O: $-1 \rightarrow 0$
    $H_2O_2(aq) \rightarrow O_2(g) + 2e^-$ electrons balanced
                                          Note: 2 O, thus $2e^-$

    $H_2O_2(aq) \rightarrow O_2(g) + 2e^- + 2H^+(aq)$ H and charges balanced

$ClO^-(aq) \rightarrow Cl_2(g)$

$2ClO^-(aq) \rightarrow Cl_2(g)$ Cl balanced

oxid. no. Cl: $+1 \rightarrow 0$

$2ClO^-(aq) + 2e^- \rightarrow Cl_2(g)$ electrons balanced
                                                    Note: 2Cl, thus, $2e^-$

$2ClO^-(aq) + 2e^- + 4H^+(aq) \rightarrow Cl_2(g)$ charges balanced

$2ClO^-(aq) + 2e^- + 4H^+(aq) \rightarrow Cl_2(g) + 2H_2O(l)$ H and O balanced

$H_2O_2(aq) + 2ClO^-(aq) + 2e^- + 4H^+(aq) \rightarrow O_2(g) + 2e^- + 2H^+(aq) + Cl_2(g) + 2H_2O(l)$

$H_2O_2(aq) + 2ClO^-(aq) + 2H^+(aq) \rightarrow O_2(g) + Cl_2(g) + 2H_2O(l)$

c. $Zn(s) + VO^{2+}(aq) \rightarrow Zn^{2+}(aq) + V^{3+}(aq)$ (unbalanced ionic equation)

$Zn(s) \rightarrow Zn^{2+}(aq)$ Zn balanced

oxid. no. Zn: $0 \rightarrow +2$

$Zn(s) \rightarrow Zn^{2+}(aq) + 2e^-$ electrons balanced

$VO^{2+}(aq) \rightarrow V^{3+}(aq)$ V balanced

oxid. no. V: $+4 \rightarrow +3$

$VO^{2+}(aq) + e^- \rightarrow V^{3+}(aq)$ electrons balanced

$VO^{2+}(aq) + e^- + 2H^+(aq) \rightarrow V^{3+}(aq)$ charges balanced

$VO^{2+}(aq) + e^- + 2H^+(aq) \rightarrow V^{3+}(aq) + H_2O(l)$ H and O balanced

$Zn(s) \rightarrow Zn^{2+}(aq) + 2e^-$

$2VO^{+2}(aq) + 2e^- + 4H^+(aq) \rightarrow 2V^{3+}(aq) + 2H_2O(l)$ balance electrons in half-reactions

$Zn(s) + 2VO^{2+}(aq) + 2e^- + 4H^+(aq) \rightarrow Zn^{2+}(aq) + 2e^- + 2V^{3+}(aq) + 2H_2O(l)$

$Zn(s) + 2VO^{2+}(aq) + 4H^+(aq) \rightarrow Zn^{2+}(aq) + 2V^{3+}(aq) + 2H_2O(l)$

---

62. Refer to Section 4.4 and Example 4.11.

---

a. $I_2(aq) \rightarrow I^-(aq)$

oxid. no. I: $0 \rightarrow -1$

$I_2(aq) \rightarrow 2I^-(aq)$ I balanced

$I_2(aq) + 2e^- \rightarrow 2I^-(aq)$ electrons balanced
                                                Note: 2I, thus $2e^-$

$S_2O_3^{2-}(aq) \rightarrow S_4O_6^{2-}(aq)$

oxid. no. S: $+2 \rightarrow +2.5$

$2S_2O_3^{2-}(aq) \rightarrow S_4O_6^{2-}(aq)$ S balanced

$2S_2O_3^{2-}(aq) \rightarrow S_4O_6^{2-}(aq) + 2e^-$ electrons balanced
                                                Note: 4S, thus $2e^-$

$I_2(aq) + 2e^- + 2S_2O_3^{2-}(aq) \rightarrow 2I^-(aq) + S_4O_6^{2-}(aq) + 2e^-$

$I_2(aq) + 2S_2O_3^{2-}(aq) \rightarrow 2I^-(aq) + S_4O_6^{2-}(aq)$

b. $25.0 \text{ g I}_2 \times \dfrac{1 \text{ mol. I}_2}{253.8 \text{ g}} = 0.0985 \text{ mol. I}_2$

$0.985 \text{ mol. I}_2 \times \dfrac{2 \text{ mol. S}_2\text{O}_3^{2-}}{1 \text{ mol. I}_2} \times \dfrac{1 \text{ L}}{0.244 \text{ mol.}} \times \dfrac{1000 \text{ mL}}{1 \text{ L}} = 807 \text{ mL}$

---

64. *Refer to Section 4.4 Example 4.11.*

$Ag(s) \rightarrow Ag^+(aq)$

oxid. no. $0 \rightarrow +1$

$Ag(s) \rightarrow Ag^+(aq) + e^-$ everything balanced

$NO_3^-(aq) \rightarrow NO_2(g)$
oxid. no. $+5 \rightarrow +4$
$NO_3^-(aq) + e^- \rightarrow NO_2(g)$ electrons balanced
$NO_3^-(aq) + e^- + 2H^+(aq) \rightarrow NO_2(g)$ charges balanced
$NO_3^-(aq) + e^- + 2H^+(aq) \rightarrow NO_2(g) + H_2O(l)$ H and O balanced

$Ag(s) + NO_3^-(aq) + e^- + 2H^+(aq) \rightarrow NO_2(g) + H_2O(l) + Ag^+(aq) + e^-$
$Ag(s) + NO_3^-(aq) + 2H^+(aq) \rightarrow NO_2(g) + H_2O(l) + Ag^+(aq)$

$0.04250 \text{ L} \times \dfrac{12.0 \text{ mol. HNO}_3}{1 \text{ L}} \times \dfrac{1 \text{ mol. H}^+}{1 \text{ mol. HNO}_3} \times \dfrac{1 \text{ mol. Ag}}{2 \text{ mol. H}^+} \times \dfrac{107.9 \text{ g Ag}}{1 \text{ mol. Ag}} = 27.5 \text{ g Ag}$

---

66. *Refer to Section 4.4 Example 4.11.*

Balance the two half reactions. If the number of electrons in the balanced half reactions are not equal, multiply the equations through with the appropriate coefficient. Combine the two half reactions and cancel those species which appear on both sides of the reaction.

$H_3AsO_3(aq) + I_2(aq) \rightarrow I^-(aq) + H_3AsO_4(aq)$

$H_3AsO_3(aq) \rightarrow H_3AsO_4(aq)$ [acidic]
oxid. no. $+3 \rightarrow +5$
$H_3AsO_3(aq) \rightarrow H_3AsO_4(aq) + 2e^-$ electrons balanced
$H_3AsO_3(aq) + H_2O(l) \rightarrow H_3AsO_4(aq) + 2e^-$ O balanced
$H_3AsO_3(aq) + H_2O(l) \rightarrow H_3AsO_4(aq) + 2H^+(aq) + 2e^-$ charges and H balanced

$I_2(aq) \rightarrow I^-(aq)$
oxid. no. $0 \rightarrow -1$
$I_2(aq) \rightarrow 2I^-(aq)$ I balanced
$I_2(aq) + 2e^- \rightarrow 2I^-(aq)$ electrons balanced

$$H_3AsO_3(aq) + H_2O(l) + I_2(aq) + 2e^- \rightarrow 2I^-(aq) + H_3AsO_4(aq) + 2H^+(aq) + 2e^-$$

$$H_3AsO_3(aq) + H_2O(l) + I_2(aq) \rightarrow 2I^-(aq) + H_3AsO_4(aq) + 2H^+(aq)$$

To determine the molarity of the iodine solution

$$\frac{45.00 \text{ mL } H_3AsO_3}{125 \text{ mL } I_2} \times \frac{1000 \text{ mL } I_2}{1 \text{ L } I_2} \times \frac{1 \text{ L } H_3AsO_3}{1000 \text{ mL } H_3AsO_3} \times \frac{0.2317 \text{ mol. } H_3AsO_3}{1 \text{ L } H_3AsO_3} \times \frac{1 \text{ mol. } I_2}{1 \text{ mol. } H_3AsO_3}$$

$$= 0.0834 M \text{ } I_2$$

To determine the mass of I_2 crystals

$$125 \text{ mL } I_2 \times \frac{1 \text{ L } I_2}{1000 \text{ mL } I_2} \times \frac{0.0834 \text{ mol. } I_2}{1 \text{ L } I_2} \times \frac{253.809 \text{ g } I_2}{1 \text{ mol. } I_2} = 2.65 \text{ g } I_2 \text{ crystals}$$

---

68. *Refer to Section 4.4 and Example 4.13.*

$Cr_2O_7^{-2}(aq) \rightarrow Cr^{+3}(aq)$
oxid. no. $+6 \rightarrow +3$
$Cr_2O_7^{2-}(aq) \rightarrow 2Cr^{3+}(aq)$ Cr balanced
$Cr_2O_7^{2-}(aq) + 6e^- \rightarrow 2Cr^{3+}(aq)$ electrons balanced
$Cr_2O_7^{2-}(aq) + 6e^- + 14H^+(aq) \rightarrow 2Cr^{3+}(aq)$ charges balanced
$Cr_2O_7^{2-}(aq) + 6e^- + 14H^+(aq) \rightarrow 2Cr^{3+}(aq) + 7H_2O(l)$ H and O balanced

$C_2H_5OH(aq) \rightarrow CO_2(g)$
oxid. no. $-2 \rightarrow +4$
$C_2H_5OH(aq) \rightarrow 2CO_2(g)$ C balanced
$C_2H_5OH(aq) \rightarrow 2CO_2(g) + 12e^-$ electrons balanced
$C_2H_5OH(aq) \rightarrow 2CO_2(g) + 12e^- + 12H^+(aq)$ charges balanced
$3H_2O(l) + C_2H_5OH(aq) \rightarrow 2CO_2(g) + 12e^- + 12H^+(aq)$ H and O balanced

$2Cr_2O_7^{2-}(aq) + 12e^- + 28H^+(aq) \rightarrow 4Cr^{3+}(aq) + 14H_2O(l)$ balance electrons in half-rxns

$2Cr_2O_7^{2-}(aq) + 12e^- + 28H^+(aq) + 3H_2O(l) + C_2H_5OH(aq)$
$\qquad\qquad \rightarrow 4Cr^{3+}(aq) + 14H_2O(l) + 2CO_2(g) + 12e^- + 12H^+(aq)$

$2Cr_2O_7^{2-}(aq) + 16H^+(aq) + C_2H_5OH(aq) \rightarrow 4Cr^{3+}(aq) + 11H_2O(l) + 2CO_2(g)$

$$0.03894 \text{ L} \times \frac{0.0723 \text{ mol. Cr}_2\text{O}_7^{2-}}{1 \text{ L}} \times \frac{1 \text{ mol. C}_2\text{H}_5\text{OH}}{2 \text{ mol. Cr}_2\text{O}_7^{2-}} \times \frac{46.07 \text{ g C}_2\text{H}_5\text{OH}}{1 \text{ mol. C}_2\text{H}_5\text{OH}} = 0.0649 \text{ g C}_2\text{H}_5\text{OH}$$

$$\text{mass \%} = \frac{0.0649 \text{ g C}_2\text{H}_5\text{OH}}{50.0 \text{ g blood}} \times 100\% = 0.130\%$$

Yes, the person is legally drunk.

---

70. Refer to Section 4.4 and Example 4.11.

$MnO_4^-(aq) \rightarrow Mn^{2+}(aq)$

$Mn^{7+} \rightarrow Mn^{2+}$

$MnO_4^-(aq) \rightarrow Mn^{2+}(aq)$	Mn balanced
$MnO_4^-(aq) + 5e^- \rightarrow Mn^{2+}(aq)$	electrons balanced
$MnO_4^-(aq) + 5e^- + 8H^+(aq) \rightarrow Mn^{2+}(aq)$	charges balanced
$MnO_4^-(aq) + 5e^- + 8H^+(aq) \rightarrow Mn^{2+}(aq) + 4H_2O(l)$	H and O balanced

$Fe^{2+}(aq) \rightarrow Fe^{3+}(aq)$	
$Fe^{2+}(aq) \rightarrow Fe^{3+}(aq) + e^-$	electrons balanced
$5Fe^{2+}(aq) \rightarrow 5Fe^{3+}(aq) + 5e^-$	balance electrons in half-rxns

$MnO_4^-(aq) + 5e^- + 8H^+(aq) + 5Fe^{2+}(aq) \rightarrow 5Fe^{3+}(aq) + 5e^- + Mn^{2+}(aq) + 4H_2O(l)$

$MnO_4^-(aq) + 8H^+(aq) + 5Fe^{2+}(aq) \rightarrow 5Fe^{3+}(aq) + Mn^{2+}(aq) + 4H_2O(l)$

$$32.3 \text{ mL} \times \frac{1 \text{ L}}{1000 \text{ mL}} \times \frac{0.002100 \text{ mol. KMnO}_4^-}{1 \text{ L}} \times \frac{1 \text{ mol. MnO}_4^-}{1 \text{ mol. KMnO}_4} = 6.78 \times 10^{-5} \text{ mol. MnO}_4^-$$

$$6.78 \times 10^{-5} \text{ mol. MnO}_4^- \times \frac{5 \text{ mol. Fe}^{2+}}{1 \text{ mol. MnO}_4^-} \times \frac{55.85 \text{ g Fe}^{2+}}{1 \text{ mol. Fe}^{2+}} = 0.0189 \text{ g Fe}^{2+}$$

$$\text{mass \%} = \frac{0.0189 \text{ g Fe}^{2+}}{5.00 \text{ g hemoglobin}} \times 100\% = 0.378\%$$

---

72. Refer to Section 4.4.

---

a. $Au(s) + HCl(aq) + HNO_3(aq) \rightarrow AuCl_4^-(aq) + NO(g) + H_2O(l)$
   Strong acids ionize completely, thus the equation becomes:

   $Au(s) + H^+(aq) + Cl^-(aq) + NO_3^-(aq) \rightarrow AuCl_4^-(aq) + NO(g) + H_2O(l)$

$Au^0 \rightarrow Au^{3+} + 3e^-$	Oxidation half-reaction
$Au(s) + 4Cl^-(aq) \rightarrow AuCl_4^-(aq) + 3e^-$	eq. 1

$N^{5+} + 3e^- \rightarrow N^{2+}$ Reduction half-reaction

$NO_3^- (aq) + 3e^- \rightarrow NO (g)$

$4H^+ (aq) + NO_3^- (aq) + 3e^- \rightarrow NO (g) + 2H_2O (l)$ eq .2

Since the electrons are balanced, add eq. 1 to eq. 2.

$Au (s) + 4Cl^- (aq) + 4H^+ (aq) + NO_3^- (aq) + 3e^-$
$$\rightarrow NO (g) + 2H_2O (l) + AuCl_4^- (aq) + 3e^-$$

$Au (s) + 4Cl^- (aq) + 4H^+ (aq) + NO_3^- (aq) \rightarrow NO (g) + 2H_2O (l) + AuCl_4^- (aq)$

b. The ratio above is 4 Cl^- to 1 NO_3^-, thus a mixture of 4 moles HCl to 1 mole HNO_3 should be used.

c. $25.0 \text{ g Au} \times \dfrac{1 \text{ mol. Au}}{197 \text{ g Au}} = 0.127 \text{ mol. Au}$

$0.127 \text{ mol. Au} \times \dfrac{4 \text{ mol. Cl}^-}{1 \text{ mol. Au}} \times \dfrac{1 \text{ mol. HCl}}{1 \text{ mol. Cl}^-} \times \dfrac{1 \text{ L}}{12 \text{ mol. HCl}} = 0.042 \text{ L HCl}$

$0.127 \text{ mol. Au} \times \dfrac{1 \text{ mol. NO}_3^-}{1 \text{ mol. Au}} \times \dfrac{1 \text{ mol. HNO}_3}{1 \text{ mol. NO}_3^-} \times \dfrac{1 \text{ L}}{16 \text{ mol. HNO}_3} = 0.0079 \text{ L HNO}_3$

---

74. *Refer to Section 4.3 and Table 4.1.*

---

a. **SA/WB**
Hydrochloric acid (HCl) is a strong acid, the reacting species is H^+.
Ethylamine ($CH_3CH_2NH_2$) is a weak base, the reacting species is $CH_3CH_2NH_2$.
$H^+(aq) + CH_3CH_2NH_2 (aq) \rightarrow CH_3CH_2NH_3^+(aq)$

b. **WA/SB**
Hydrofluoric acid (HF) is a weak acid, the reacting species is HF.
Calcium hydroxide ($Ca(OH)_2$) is a strong base, the reacting species is OH^-.
$OH^-(aq) + HF aq) \rightarrow F^-(aq) + H_2O(l)$

c. **PPT**
$3Ca(OH)_2(aq) + 2Na_3PO_4(aq) \rightarrow Ca_3(PO_4)_2 (s)\downarrow + 6 NaOH(aq)$

d. **PPT**
$Ag_2SO_4(aq) + BaCl_2(aq) \rightarrow BaSO_4(s)\downarrow + 2AgCl(s)\downarrow$

e. **NR**
$Mg(NO_3)_2(aq) + NaCl(aq) \rightarrow Mg^{2+}(aq) + 2NO_3^-(aq) + Na^+(aq) + Cl^-(aq)$
All ions are in solution and do not react with each other.

76. Refer to Section 4.2.

Write the products of the reactions and determine if there is a precipitate. If so, determine the ratio of cations to anions.

1. no reaction (product is soluble)
2. $Mg(OH)_2$
3. $BaCO_3$

a. This picture represents no precipitation, thus equation (**1**) matches.

b. This picture represents a precipitation reaction where the cations and anions are in a 1:1 ratio, thus equation (**3**) matches.

c. This picture represents a precipitation reaction where the cations and anions are in a 1:2 ratio, thus equation (**2**) matches.

78. Refer to Section 4.3.

a. The species partially dissociates, thus it is a **weak** electrolyte.

b. The species does not dissociate, thus it is a **non**electrolyte.

c. The species completely dissociates, thus it is a **strong** electrolyte.

d. The species partially dissociates, thus it is a **weak** electrolyte.

80. Refer to Section 4.4 and Problem 79.

From the given equations, we deduce the following:
Z^+ is stronger oxidizing agent than W^+, because W^+ is not strong enough to react with Z.
W^+ is a stronger oxidizing agent than X^+, because X^+ is not strong enough to react with W.
Y^+ is a stronger oxidizing agent than X^+, because Y^+ reacts with X.
Y^+ is a stronger oxidizing agent than Z^+, because Y^+ reacts with Z.
Thus, the order of increasing strength as oxidizing agents is
$X^+ < W^+ < Z^+ < Y^+$

82. Refer to Section 4.4 and Example 4.11.

First, write a balanced redox equation for the oxidation of oxalate to CO_2. Then use the amount of $KMnO_4^-$ to calculate the amount of CaC_2O_4 present in the sample.

$$CaC_2O_4(s) + 2H^+(aq) \rightarrow H_2C_2O_4(aq) + Ca^{2+}(aq)$$
$$H_2C_2O_4(aq) + MnO_4^-(aq) \rightarrow CO_2(g) + Mn^{2+}(aq)$$

$C^{3+} \rightarrow C^{4+}$ Oxidation half-reaction

$$H_2C_2O_4(aq) \rightarrow 2CO_2(g) + 2e^-$$
$$H_2C_2O_4(aq) \rightarrow 2CO_2(g) + 2e^- + 2H^+(aq) \qquad \text{eq. 1}$$

$Mn^{7+} \rightarrow Mn^{2+}$ Reduction half-reaction

$$MnO_4^-(aq) + 5e^- \rightarrow Mn^{2+}(aq)$$
$$8H^+(aq) + MnO_4^-(aq) + 5e^- \rightarrow Mn^{2+}(aq)$$
$$8H^+(aq) + MnO_4^-(aq) + 5e^- \rightarrow Mn^{2+}(aq) + 4H_2O(l) \qquad \text{eq. 2}$$

To balance the electrons, multiply eq. 1 by 5 and eq. 2 by 2 and add the results.

$$16H^+(aq) + 2MnO_4^-(aq) + 10e^- + 5H_2C_2O_4(aq)$$
$$\rightarrow 10CO_2(g) + 10e^- + 10H^+(aq) + 2Mn^{2+}(aq) + 8H_2O(l)$$

$$6H^+(aq) + 2MnO_4^-(aq) + 5H_2C_2O_4(aq) \rightarrow 10CO_2(g) + 2Mn^{2+}(aq) + 8H_2O(l)$$

$$26.2 \text{ mL} \times \frac{1 \text{ L}}{1000 \text{ mL}} \times \frac{0.0946 \text{ mol. KMnO}_4}{1 \text{ L}} \times \frac{1 \text{ mol. MnO}_4^-}{1 \text{ mol. KMnO}_4} \times \frac{5 \text{ mol. C}_2O_4^{2-}}{2 \text{ mol. MnO}_4^-}$$

$$\times \frac{1 \text{ mol. CaC}_2O_4}{1 \text{ mol. C}_2O_4^{2-}} \times \frac{128.10 \text{ g CaC}_2O_4}{1 \text{ mol. CaC}_2O_4} = 0.794 \text{ g CaC}_2O_4$$

$$0.794 \text{ g} \times \frac{1 \text{ mol. CaC}_2O_4}{128.10 \text{ g}} \times \frac{1 \text{ mol. Ca}^{2+}}{1 \text{ mol. CaC}_2O_4} \times \frac{40.08 \text{ g Ca}^{2+}}{1 \text{ mol. Ca}^{2+}} = 0.248 \text{ g Ca}^{2+}$$

$$= 248 \text{ mg Ca}^{2+} \qquad \textbf{Yes}, \text{ this is within the normal range.}$$

---

83. Refer to Section 4.3.

Calculate the amount of $Mg(OH)_2$ and $NaHCO_3$ present and the moles of acid each could neutralize. Then calculate the volume of acid this corresponds to.

330 mg = 0.330 g
mass $Mg(OH)_2$ = 0.410 x 0.330 g = 0.135 g
mass $NaHCO_3$ = 0.362 x 0.330 g = 0.119 g

$$0.135 \text{ g Mg(OH)}_2 \times \frac{1 \text{ mol. Mg(OH)}_2}{58.3 \text{ g Mg(OH)}_2} \times \frac{2 \text{ mol. H}^+}{1 \text{ mol. Mg(OH)}_2} = 0.00463 \text{ mol. H}^+$$

$$0.119 \text{ g NaHCO}_3 \times \frac{1 \text{ mol. NaHCO}_3}{84.0 \text{ g NaHCO}_3} \times \frac{1 \text{ mol. H}^+}{1 \text{ mol. NaHCO}_3} = 0.00142 \text{ mol. H}^+$$

Total moles H^+ neutralized = 0.00463 + 0.00142 = 0.00605 mol. H^+

$$0.00605 \, \text{mol. H}^+ \times \frac{1 \, \text{L acid}}{0.020 \, \text{mol. H}^+} = 0.30 \, \text{L acid}$$

84. *Refer to Section 4.4.*

The difference in mass between the original copper strip and the silver coated strip corresponds to the mass of copper lost and the mass of silver gained. Bear in mind that two moles of silver are gained for each mole of copper lost. Set up an algebraic equation.

Let x = mass of Cu lost,
then $2.00 - x$ = mass of Cu remaining.

$$\text{mass of Ag} = x \, \text{g Cu} \times \frac{1 \, \text{mol. Cu}}{63.55 \, \text{g Cu}} \times \frac{2 \, \text{mol. Ag}}{1 \, \text{mol. Cu}} \times \frac{107.9 \, \text{g Ag}}{1 \, \text{mol. Ag}} = 3.40 \, x$$

$4.18 \, \text{g} = (2.00 - x) + 3.40x$
$4.18 \, \text{g} = 2.00 + 2.40x$
$x = 0.908 \, \text{g}$

Mass of Cu remaining = 2.00 - 0.908 = 1.09 g.
Mass of Ag remaining = 3.40 x 0.908 = 3.09 g.

85. *Refer to Section 4.4.*

The first permanganate titration tells us the concentration of ferrous ions, while the second titration tells us the total iron concentration. The difference is the ferric ion concentration.

$MnO_4^-(aq) + Fe^{2+}(aq) \rightarrow Fe^{3+}(aq) + Mn^{2+}(aq)$

$MnO_4^-(aq) \rightarrow Mn^{2+}(aq)$
$Mn^{7+} \rightarrow Mn^{2+}$

$MnO_4^-(aq) \rightarrow Mn^{2+}(aq)$	Mn balanced
$MnO_4^-(aq) + 5e^- \rightarrow Mn^{2+}(aq)$	electrons balanced
$MnO_4^-(aq) + 5e^- + 8H^+(aq) \rightarrow Mn^{2+}(aq)$	charges balanced
$MnO_4^-(aq) + 5e^- + 8H^+(aq) \rightarrow Mn^{2+}(aq) + 4H_2O(l)$	H and O balanced

$Fe^{2+}(aq) \rightarrow Fe^{3+}(aq)$	
$Fe^{2+}(aq) \rightarrow Fe^{3+}(aq) + e^-$	electrons balanced
$5Fe^{2+}(aq) \rightarrow 5Fe^{3+}(aq) + 5e^-$	balance electrons in half-rxns

$MnO_4^-(aq) + 5e^- + 8H^+(aq) + 5Fe^{2+}(aq) \rightarrow 5Fe^{3+}(aq) + 5e^- + Mn^{2+}(aq) + 4H_2O(l)$
$MnO_4^-(aq) + 8H^+(aq) + 5Fe^{2+}(aq) \rightarrow 5Fe^{3+}(aq) + Mn^{2+}(aq) + 4H_2O(l)$

$$35.0 \, \text{mL} \times \frac{1 \, \text{L}}{1000 \, \text{mL}} \times \frac{0.0280 \, \text{mol. KMnO}_4^-}{1 \, \text{L}} \times \frac{1 \, \text{mol. MnO}_4^-}{1 \, \text{mol. KMnO}_4} \times \frac{5 \, \text{mol. Fe}^{2+}}{1 \, \text{mol. MnO}_4^-} = 0.00490 \, \text{mol. Fe}^{2+}$$

$$48.0 \, \text{mL} \times \frac{1 \, \text{L}}{1000 \, \text{mL}} \times \frac{0.0280 \, \text{mol. KMnO}_4^-}{1 \, \text{L}} \times \frac{1 \, \text{mol. MnO}_4^-}{1 \, \text{mol. KMnO}_4} \times \frac{5 \, \text{mol. Fe}^{3+}}{1 \, \text{mol. MnO}_4^-} = 0.00672 \, \text{mol. Fe}^{3+}$$

mol. $Fe^{3+} = 0.00672 - 0.00490 = 0.00182 \, \text{mol. Fe}^{3+}$

$$[Fe^{2+}] = \frac{0.00490 \, \text{mol. Fe}^{2+}}{0.05000 \, \text{L}} = 0.0980 \, M$$

$$[Fe^{3+}] = \frac{0.00182 \, \text{mol. Fe}^{3+}}{0.05000 \, \text{L}} = 0.0364 \, M$$

Chapter 5: Gases

2. *Refer to Section 5.1 and Example 5.1.*

$$6.00 \text{ ft} \times \frac{12 \text{ in}}{1 \text{ ft}} \times \frac{2.54 \text{ cm}}{1 \text{ in}} = 183 \text{ cm}$$

$$26 \text{ in} \times \frac{2.54 \text{ cm}}{1 \text{ in}} = 66 \text{ cm}$$

$V = \pi r^2 h$

$V = \pi (66 \text{ cm})^2 (183 \text{ cm})$

$V = 2.5 \times 10^6 \text{ cm}^3$

$$2.5 \times 10^6 \text{ cm}^3 \times \frac{1 \text{ L}}{10^3 \text{ cm}^3} = 2.5 \times 10^3 \text{ L}$$

$$189 \text{ lb He} \times \frac{453.6 \text{ g}}{1 \text{ lb}} \times \frac{1 \text{ mol. He}}{4.003 \text{ g}} = 2.14 \times 10^4 \text{ mol.}$$

$25 \,°C + 273.15 \text{ K} = 298 \text{ K}$

4. *Refer to Section 5.1.*

mm Hg	atmospheres	kilopascals	bar
1215	**1.600**	**161.9**	**1.619**
543	0.714	**72.3**	**0.723**
1.07 x 10³	**1.41**	143	**1.43**
678	**0.892**	90.4	0.904

$$1215 \text{ mm Hg} \times \frac{1 \text{ atm}}{760.0 \text{ mm Hg}} = 1.600 \text{ atm}$$

$$1.600 \text{ atm} \times \frac{1.013 \text{ bar}}{1 \text{ atm}} = 1.619 \text{ bar}$$

$$1.619 \text{ bar} \times \frac{1 \times 10^5 \text{ Pa}}{1 \text{ atm}} \times \frac{1 \text{ kPa}}{1000 \text{ Pa}} = 161.9 \text{ kPa}$$

$$0.714 \text{ atm} \times \frac{760 \text{ mm Hg}}{1 \text{ atm}} = 543 \text{ mm Hg}$$

$$0.714 \text{ atm} \times \frac{1.013 \times 10^5 \text{ Pa}}{1 \text{ atm}} \times \frac{1 \text{ kPa}}{1000 \text{ Pa}} = 72.3 \text{ kPa}$$

$$0.714 \text{ atm} \times \frac{1.013 \text{ bar}}{1 \text{ atm}} = 0.723 \text{ bar}$$

$$143 \text{ kPa} \times \frac{1000 \text{ Pa}}{1 \text{ kPa}} \times \frac{1 \text{ bar}}{1.01 \times 10^5 \text{ Pa}} = 1.43 \text{ bar}$$

$$1.43 \text{ bar} \times \frac{1 \text{ atm}}{1.013 \text{ bar}} = 1.41 \text{ atm}$$

$$1.41 \text{ atm} \times \frac{760 \text{ mm Hg}}{1 \text{ atm}} = 1.07 \times 10^3 \text{ mm Hg}$$

$$0.904 \text{ bar} \times \frac{1 \text{ atm}}{1.013 \text{ bar}} = 0.892 \text{ atm}$$

$$0.904 \text{ bar} \times \frac{1 \times 10^5 \text{ Pa}}{1 \text{ atm}} \times \frac{1 \text{ kPa}}{1000 \text{ Pa}} = 90.4 \text{ kPa}$$

$$0.892 \text{ atm} \times \frac{760 \text{ mm Hg}}{1 \text{ atm}} = 678 \text{ mm Hg}$$

---

6. Refer to Section 5.1.

$$P_1 = 977 \text{ mm Hg} \times \frac{1 \text{ atm}}{760.0 \text{ mm Hg}} = 1.29 \text{ atm}$$

$$T_1 = 25^\circ\text{C} = 25 + 273 = 298 \text{ K}$$

$$P_2 = 1.50 \text{ atm}$$

$$T_2 = ?$$

$$\frac{P_1}{T_1} = \frac{P_2}{T_2}$$

$$\frac{1.29 \text{ atm}}{298 \text{ K}} = \frac{1.50 \text{ atm}}{T_2}$$

$$T_2 = 347 \text{ K} = 74^\circ\text{C}$$

---

8. Refer to Section 5.3 and Example 5.2.

$$T_1 = 32^\circ\text{C} = 22 + 273 = 305 \text{ K}$$

a. $V_2 = 0.75 V_1$

$$\frac{T_2}{T_1} = \frac{V_2}{V_1} \Rightarrow T_2 = T_1 \frac{V_2}{V_1} = 305 \text{ K} \times \frac{0.75 V_1}{V_1} = 229 \text{ K} = -44^\circ\text{C}$$

b. $V_2 = 0.25 V_1$

$$\frac{T_2}{T_1} = \frac{V_2}{V_1} \Rightarrow T_2 = T_1 \frac{V_2}{V_1} = 305\,\text{K} \times \frac{0.25 V_1}{V_1} = 76\,\text{K} = -197^0\text{C}$$

---

10. Refer to Sections 1.2 and 5.3.

First convert the temperatures to Kelvin and calculate the actual pressure in the tire. Then calculate the new pressure using the ratios of pressures to temperatures.

$$T_1 = \frac{71^\circ\text{F} - 32}{1.8} = 22^\circ\text{C} = 22 + 273 = 295\,\text{K}$$

$$T_2 = \frac{115^\circ\text{F} - 32}{1.8} = 46^\circ\text{C} = 46 + 273 = 319\,\text{K}$$

$$P_{(\text{actual})} = P_{(\text{gauge})} + 14.7 = 28.0 + 14.7 = 42.7\,\text{psi}$$

$$\frac{P_1}{P_2} = \frac{T_1}{T_2} \Rightarrow \frac{42.7\,\text{psi}}{P_2} = \frac{295\,\text{K}}{319\,\text{K}} \Rightarrow P_2 = 46.2\,\text{psi} \quad \text{(final actual pressure)}$$

$$P_{(\text{gauge})} = P_{(\text{actual})} - 14.7 = 46.2 - 14.7 = 31.5\,\text{psi} \quad \text{(final gauge pressure)}$$

---

12. Refer to Section 5.1 and Example 5.3.

$V_1 = 25.0\,\text{mL} = 0.0250\,\text{L}$
$T_1 = 23^\circ\text{C} = 23 + 273 = 296\,\text{K}$

$$P_1 = 745\,\text{mm Hg} \times \frac{1\,\text{atm}}{760.0\,\text{mm Hg}} = 0.980\,\text{atm}$$

Since volume increased by 8%, multiply V_1 by 1.08 to get V_2.
$V_2 = (0.0250\,\text{L})(1.08) = 0.0270\,\text{L}$
$T_2 = 82^\circ\text{C} = 82 + 273 = 355\,\text{K}$
$P_2 = ?$

$$\frac{P_1 V_1}{T_1} = \frac{P_2 V_2}{T_2}$$

$$\frac{(0.980\,\text{atm})(0.0250\,\text{L})}{(296\,\text{K})} = \frac{(P_2)(0.0270\,\text{L})}{355\,\text{K}}$$

$P_2 = 1.09\,\text{atm} = 828\,\text{mm Hg}$

Yes, the pressure increased.

14. Refer to Section 5.2 and Example 5.2.

$$\frac{n_2}{n_1} = \frac{P_2}{P_1}$$

$$\frac{n_2}{1.35 \text{ mol.}} = \frac{1.64 \text{ atm}}{1.05 \text{ atm}}$$

$n_2 = 2.11$ mol.

mol. total = mol. N_2 + mol. H_2
2.11 mol. = mol. N_2 + 1.35 mol. H_2
mol. N_2 = 0.76 mol.

16. Refer to Section 5.3 and Example 5.3.

Calculate the moles of O_2 from the mass, then use the ideal gas law to calculate pressure.

$T = 37°C + 273 = 310$ K

$$n = \frac{0.25 \text{ mg } O_2}{1 \text{ mL}} \times \frac{1000 \text{ mL}}{1 \text{ L}} \times \frac{1 \text{ g}}{1000 \text{ mg}} \times \frac{1 \text{ mol. } O_2}{32.00 \text{ g } O_2} = 7.8 \times 10^{-3} \text{ mol. } O_2 \text{ per liter of blood}$$

$$P = \frac{nRT}{V} = \frac{(7.8 \times 10^{-3} \text{ mol.})(0.0821 \text{ L} \cdot \text{atm/mol.} \cdot \text{K})(310 \text{ K})}{(1.00 \text{ L})} = 0.20 \text{ atm}$$

18. Refer to Section 5.2 and Example 5.3.

$V = 2.00$ L
$T = 22°C = 22 + 273 = 295$ K
$P = 1.78$ atm

$PV = nRT$

$(1.78 \text{ atm})(2.00 \text{ L}) = n (0.0821 \text{ L} \cdot \text{atm/mol.} \cdot \text{K})(295 \text{ K})$

$n = 0.147$ moles butane

$$0.147 \text{ mol. butane} \times \frac{58.123 \text{ g butane}}{1 \text{ mol. butane}} = 8.54 \text{ g butane}$$

Mass of tank after filling
Total mass = mass empty tank + mass butane = 725.6 g + 8.54 g = 734.1 g

Start by converting all units to those used in the ideal gas law. Calculated moles using the ideal gas equation if *P*, *V* and *T* are given, otherwise use the molar mass and the given mass. Then apply the ideal gas law to fill in any other missing data.

Pressure	Volume	Temperature	Moles	Grams
22.7 atm	1.75 L	19°C	1.66	**96.5**
0.895 atm	**6.17 L**	6°C	**0.241**	14.0
433 mm Hg	92.4 mL	**1.62 K**	0.395	**23.0**
1.74 bar	8.66 L	98°F	**0.585**	**34.0**

$T = 19°C + 273 = 292\ K$

$$P = \frac{nRT}{V} = \frac{(1.66\ \text{mol.})(0.0821\ L \cdot atm/mol. \cdot K)(292\ K)}{(1.75\ L)} = 22.7\ atm$$

$$1.66\ \text{mol.}\ C_4H_{10} \times \frac{58.12\ g}{1\ \text{mol.}} = 96.5\ g$$

$T = 6°C + 273 = 279\ K$

$$14.0\ g \times \frac{1\ \text{mol.}}{58.12\ g} = 0.241\ \text{mol.}$$

$$V = \frac{nRT}{P} = \frac{(0.241\ \text{mol.})(0.0821\ L \cdot atm/mol. \cdot K)(279\ K)}{(0.895\ atm)} = 6.17\ L$$

$$433\ \text{mm Hg} \times \frac{1\ atm}{760\ \text{mm Hg}} = 0.570\ atm$$

$$T = \frac{PV}{nR} = \frac{(0.570\ atm)(0.0924\ L)}{(0.0821\ L \cdot atm/mol. \cdot K)(0.395\ \text{mol.})} = 1.62\ K$$

$$1.74\ \text{bar} \times \frac{1\ atm}{1.013\ \text{bar}} = 1.72\ atm$$

$$T = \frac{98°F - 32}{1.8} + 273 = 310\ K$$

$$n = \frac{PV}{RT} = \frac{(1.72\ atm)(8.66\ L)}{(0.0821\ L \cdot atm/mol. \cdot K)(310\ K)} = 0.585\ \text{mol.}$$

$$0.585\ \text{mol.} \times \frac{58.12\ g}{1\ \text{mol.}} = 34.0\ g$$

Calculate the moles of gas in one liter at the given conditions. Then calculate the mass and finally, the density. Remember to convert T to Kelvin and P to atms.

$T = 97°C + 273 = 370 \text{ K}$

$$755 \text{ mm Hg} \times \frac{1 \text{ atm}}{760.0 \text{ mm Hg}} = 0.993 \text{ atm}$$

$$n = \frac{PV}{RT} = \frac{(0.993 \text{ atm})(1.00 \text{ L})}{(0.0821 \text{ L} \cdot \text{atm/mol.} \cdot \text{K})(370 \text{ K})} = 0.0327 \text{ mol. gas}$$

a. $0.0327 \text{ mol. HCl} \times \dfrac{36.46 \text{ g HCl}}{1 \text{ mol. HCl}} = 1.19 \text{ g HCl}$

$$d = \frac{\text{g}}{\text{L}} \times \frac{1.19 \text{ g HCl}}{1.00 \text{ L}} = 1.19 \text{ g/L}$$

b. $0.0327 \text{ mol. SO}_2 \times \dfrac{64.07 \text{ g SO}_2}{1 \text{ mol. SO}_2} = 2.10 \text{ g SO}_2$

$$d = \frac{\text{g}}{\text{L}} \times \frac{2.10 \text{ g SO}_2}{1 \text{ L}} = 2.10 \text{ g/L}$$

c. $0.0327 \text{ mol. C}_4\text{H}_{10} \times \dfrac{58.12 \text{ g C}_4\text{H}_{10}}{1 \text{ mol. C}_4\text{H}_{10}} = 1.90 \text{ g C}_4\text{H}_{10}$

$$d = \frac{\text{g}}{\text{L}} \times \frac{1.90 \text{ g C}_4\text{H}_{10}}{1 \text{ L}} = 1.90 \text{ g/L}$$

$P_1 = 755 \text{ mm Hg} \times \dfrac{1 \text{ atm}}{760.0 \text{ mm Hg}} = 0.993 \text{ atm}$

$T_1 = 0°C = 0 + 273 = 273 \text{ K}$

$P_2 = 210 \text{ mm Hg} \times \dfrac{1 \text{ atm}}{760.0 \text{ mm Hg}} = 0.276 \text{ atm}$

$T_2 = 0°C = 0 + 273 = 273 \text{ K}$

$d = \dfrac{(MM)P}{RT}$

$$d_1 = \frac{(29.0 \text{ g/mol.})(0.993 \text{ atm})}{(0.0821 \text{ L} \cdot \text{atm/mol.} \cdot \text{K})(273 \text{ K})} = 1.28 \text{ g/L}$$

$$d_2 = \frac{(29.0 \text{ g/mol.})(0.276 \text{ atm})}{(0.0821 \text{ L} \cdot \text{atm/mol.} \cdot \text{K})(273 \text{ K})} = 0.357 \text{ g/L}$$

$\dfrac{1.28 \text{ g/L}}{0.357 \text{ g/L}} = 3.59$ At sea level, the density of air is 3.59 times more dense than the air at the top of Mount Everest.

---

26. *Refer to Sections 3.3 and 5.3 and Examples 3.7, 3.9, and 5.4.*

Consider your units. Molar mass has units of g/mol., while density has units of g/L. Therefore, you can calculate the moles of phosgene in 1 L, and then divide the mass of 1 L by the number of moles. Finally, use the percent composition and molar mass to determine the simplest and molecular formula.

a. $T = 25°C + 273 = 298 \text{ K}$

$$n = \frac{PV}{RT} = \frac{(1.05 \text{ atm})(1.00 \text{ L})}{(0.0821 \text{ L} \cdot \text{atm/mol.} \cdot \text{K})(298 \text{ K})} = 0.0429 \text{ mol.}$$

$$\text{molar mass} = \frac{4.24 \text{ g}}{0.0429 \text{ mol.}} = 98.8 \text{ g/mol.}$$

b. $12.01 \text{ g C} \times \dfrac{1 \text{ mol. C}}{12.01 \text{ g}} = 1.000 \text{ mol. C}$

$16.2 \text{ g O} \times \dfrac{1 \text{ mol. O}}{16.00 \text{ g}} = 1.012 \text{ mol. O}$

$71.7 \text{ g Cl} \times \dfrac{1 \text{ mol. Cl}}{35.45 \text{ g}} = 2.022 \text{ mol. Cl}$

C: 1.000 mol. / 1.000 mol. = 1
O: 1.012 mol. / 1.000 mol. = 1
Cl: 1.012 mol. / 1.000 mol. = 2

Thus, the empirical formula is: $COCl_2$
Empirical mass = 12.01 + 16.00 + 2(35.45) = 98.9
Molar mass = Empirical mass
Molecular formula = empirical formula = **$COCl_2$**

28. *Refer to Section 5.3.*

The molar mass of exhaled air is simply the sum of the molar masses of the individual components multiplied by their abundances. Calculate the moles of air in 1 L, then the mass of 1 L to get density.

a. $(28.02 \text{ g/mol. } N_2)(0.745) + (32.00 \text{ g/mol. } O_2)(0.157) + (44.01 \text{ g/mol. } CO_2)(0.036)$
$+ (18.02 \text{ g/mol. } H_2O)(0.062) = 28.6 \text{ g/mol.}$

b. $T = 37°C + 273 = 310 \text{ K}$

$$757 \text{ mm Hg} \times \frac{1 \text{ atm}}{760.0 \text{ mm Hg}} = 0.996 \text{ atm}$$

$$n = \frac{PV}{RT} = \frac{(0.996 \text{ atm})(1.00 \text{ L})}{(0.0821 \text{ L} \cdot \text{atm/mol.} \cdot \text{K})(310 \text{ K})} = 0.0391 \text{ mol.}$$

$$0.0391 \text{ mol.} \times \frac{28.6 \text{ g air}_{\text{Exhaled}}}{1 \text{ mol. air}_{\text{Exhaled}}} = 1.12 \text{ g air}_{\text{Exhaled}}$$

$$d_{\text{air}_{\text{Exhaled}}} = \frac{g}{L} \times \frac{1.12 \text{ g air}_{\text{Exhaled}}}{1.00 \text{ L}} = 1.12 \text{ g/L}$$

$$0.0391 \text{ mol.} \times \frac{29.0 \text{ g air}_{\text{ordinary}}}{1 \text{ mol. air}_{\text{ordinary}}} = 1.13 \text{ g air}_{\text{ordinary}}$$

$$d_{\text{air}_{\text{ordinary}}} = \frac{g}{L} \times \frac{1.13 \text{ g air}_{\text{ordinary}}}{1.00 \text{ L}} = 1.13 \text{ g/L}$$

Ordinary air is slightly more dense than exhaled air.

30. *Refer to Section 5.3.*

Use the ideal gas equation to calculate the number of moles in the sample, then the molar mass of the compound. Then calculate the atomic mass of X and you have the identity.

$T = 20°C + 273 = 293 \text{ K}$

$$329.5 \text{ cm}^3 \times \frac{1 \text{ mL}}{1 \text{ cm}^3} \frac{1 \text{ L}}{1000 \text{ mL}} = 0.3295 \text{ L}$$

$$n = \frac{PV}{RT} = \frac{(1.00 \text{ atm})(0.3295 \text{ L})}{(0.0821 \text{ L} \cdot \text{atm/mol.} \cdot \text{K})(293 \text{ K})} = 0.0137 \text{ mol.}$$

$$\text{molar mass} = \frac{2.00\text{ g}}{0.0137\text{ mol.}} = 146\text{ g/mol.}$$

146 g/mol. = 1(S) + 6(X)
146 g/mol. = (32.07 g/mol.) + 6(X)
6X = 114 g/mol.
X = 19 g/mol.
X = F
Thus the compound is SF_6, sulfur hexafluoride.

---

32. Refer to Section 5.4 and Example 5.5.

Since R is a constant, we can solve this problem using the ideal gas law, by solving for R, setting the equations equal to one another and solving for V_{N2}.

$$\frac{PV}{nT} = R, \text{ thus } \frac{P_{NO}V_{NO}}{n_{NO}T_{NO}} = R = \frac{P_{N_2}V_{N_2}}{n_{N_2}T_{N_2}}. \text{ Solving for } V_{N_2}, \text{ we get}:$$

$$V_{N_2} = \frac{P_{NO}\,V_{NO}\,n_{N_2}T_{N_2}}{n_{NO}\,T_{NO}\,P_{N_2}}$$

$$\text{Since } T_{NO} = T_{N_2} \text{ and } P_{NO} = P_{N_2}, V_{N_2} = \frac{V_{NO}\,n_{N_2}}{n_{NO}} = \frac{(297\text{ L})(5\text{ mol. N}_2)}{(6\text{ mol. NO})} = 248\text{ L N}_2$$

---

34. Refer to Section 5.4 and Example 5.5.

---

a. This equation can be balanced by inspection.

$NH_4NO_3(s) \rightarrow N_2O(g) + 2H_2O(l)$

b. $T = 250°C + 273 = 523$ K

$$V = \frac{nRT}{P} = \frac{(0.0625\text{ mol.})(0.0821\text{ L}\cdot\text{atm/mol.}\cdot\text{K})(298\text{ K})}{1.0\text{ atm}} = 2.7\text{ L}$$

---

36. Refer to Sections 3.4 and 5.4 and Example 5.7.

After balancing the equation, calculate the moles of NH_4NO_3 and the total moles of gas (using the mole ratio of NH_4NO_3 to all gases). Calculate the pressure for the given conditions.

a. $NH_4NO_3(s) \rightarrow N_2(g) + O_2(g) + H_2O(g)$ balance H's
$NH_4NO_3(s) \rightarrow N_2(g) + O_2(g) + 2H_2O(g)$

There are too many O's on the product side. Multiply NH_4NO_3 by two and rebalance.

$$2NH_4NO_3(s) \rightarrow 2N_2(g) + O_2(g) + 4H_2O(g)$$

b. $1.00 \text{ kg } NH_4NO_3 \times \dfrac{1000 \text{ g}}{1 \text{ kg}} \times \dfrac{1 \text{ mol. } NH_4NO_3}{80.05 \text{ g } NH_4NO_3} \times \dfrac{7 \text{ mol. gas}}{2 \text{ mol. } NH_4NO_3} = 43.7 \text{ mol. gas}$

$T = 787°C + 273 = 1060 \text{ K}$

$$P = \dfrac{nRT}{V} = \dfrac{(43.7 \text{ mol.})(0.0821 \text{ L} \cdot \text{atm/mol.} \cdot \text{K})(1060 \text{ K})}{50.0 \text{ L}} = 76.1 \text{ atm}$$

---

38. Refer to Section 5.5 and Example 5.9.

$P_{tot} = 1.20 \text{ atm}$
$P_a = P_{tot}X_a$
$P_{CH_4} = (1.20 \text{ atm})(0.886) = 1.06 \text{ atm}$
$P_{C_2H_6} = (1.20 \text{ atm})(0.089) = 0.11 \text{ atm}$
$P_{C_3H_8} = (1.20 \text{ atm})(0.025) = 0.030 \text{ atm}$

---

40. Refer to Sections 5.3 and 5.5 and Example 5.8.

Calculate the partial pressure of O_2 to determine the moles of O_2; then calculate the pressure of the O_2.

$$1.25 \text{ bar} \times \dfrac{1 \text{ atm}}{1.013 \text{ bar}} \times \dfrac{760.0 \text{ mm Hg}}{1 \text{ atm}} = 938 \text{ mm Hg}$$

$$P_{O_2} = (938 - 23.8) \text{ mm Hg} \times \dfrac{1 \text{ atm}}{760.0 \text{ mm Hg}} = 1.20 \text{ atm}$$

$$n = \dfrac{PV}{RT} = \dfrac{(1.20 \text{ atm})(7.28 \text{ L})}{(0.0821 \text{ L} \cdot \text{atm/mol.} \cdot \text{K})(298 \text{ K})} = 0.357 \text{ mol. } O_2$$

$$V_{O_2} = \dfrac{nRT}{P} = \dfrac{(0.357 \text{ mol. } O_2)(0.0821 \text{ L} \cdot \text{atm/mol.} \cdot \text{K})(37 + 273 \text{ K})}{1.07 \text{ atm}} = 8.49 \text{ L}$$

Calculate the moles of each gas (using any value for T other than zero), then the total moles of gas and the total volume of both bulbs. Then calculate the pressure using the combined volumes and moles (and the same temperature).

$$n_{Ar} = \frac{PV}{RT} = \frac{(2.50 \text{ atm})(4.00 \text{ L})}{(0.0821 \text{ L} \cdot \text{atm/mol.} \cdot \text{K})(300 \text{ K})} = 0.406 \text{ mol. wet gas}$$

$$n_{Cl} = \frac{PV}{RT} = \frac{(1.00 \text{ atm})(1.00 \text{ L})}{(0.0821 \text{ L} \cdot \text{atm/mol.} \cdot \text{K})(300 \text{ K})} = 0.0406 \text{ mol. dry gas}$$

$n_{total} = 0.406 + 0.0406 = 0.447$

$V_{total} = 4.00 \text{ L} + 1.00 \text{ L} = 5.00 \text{ L}$

$$P = \frac{nRT}{V} = \frac{(0.447 \text{ mol.})(0.0821 \text{ L} \cdot \text{atm/mol.} \cdot \text{K})(300 \text{ K})}{5.00 \text{ L}} = 2.20 \text{ atm}$$

a. $5N_2(g) + 6H_2O(g) \rightarrow 4NH_3(g) + 6NO(g)$

b. $V = 20.0 \text{ L}$
    $T = 173^\circ\text{C} = 173 + 273 = 446 \text{ K}$

$$P = 772 \text{ mm Hg} \times \frac{1 \text{ atm}}{760.0 \text{ mm Hg}} = 1.02 \text{ atm}$$

$PV = nRT$
$(1.02 \text{ atm})(20.0 \text{ L}) = n \ (0.0821 \text{ L} \cdot \text{atm/mol} \cdot \text{K})(446 \text{ K})$
$n = 0.556 \text{ moles N}_2$

Products: $0.556 \text{ mol. N}_2 \times \dfrac{4 \text{ mol. NH}_3}{5 \text{ mol. N}_2} = 0.444 \text{ mol. NH}_3$

$0.556 \text{ mol. N}_2 \times \dfrac{6 \text{ mol. NO}}{5 \text{ mol. N}_2} = 0.666 \text{ mol. NO}$

Total moles (products) = 0.444 moles NH_3 + 0.666 moles NO = 1.11 moles products
Final conditions are $V = 15.0 \text{ L}$ and $T = 25^\circ\text{C} = 298 \text{ K}$.
$PV = nRT$
$P_{tot} \times 15.0 \text{ L} = 1.11 \text{ mol.} \times 0.0821 \text{ L} \cdot \text{atm/mol.} \cdot \text{K} \times 298 \text{ K}$
$P_{tot} = 1.81 \text{ atm}$

c. $P_{NH_3} = 1.81 \text{ atm} \times \dfrac{0.444 \text{ mol. NH}_3}{1.11 \text{ total mol.}} = 0.724 \text{ atm NH}_3$

$P_{NO} = 1.81 \text{ atm} \times \dfrac{0.666 \text{ mol. NO}}{1.11 \text{ total mol.}} = 1.09 \text{ atm NO}$

46. Refer to Section 5.6 and Example 5.11.

$$\frac{\text{rate of effusion of B}}{\text{rate of effusion of A}} = \left(\frac{MM_A}{MM_B}\right)^{\frac{1}{2}}$$

$$\frac{\text{rate of effusion of N}_2}{\text{rate of effusion of H}_2} = \left(\frac{2.016}{28.02}\right)^{\frac{1}{2}} = 0.2682$$

48. Refer to Section 5.6 and Example 5.11.

Calculate the ratio of the effusion rates of the two gases.

$$\frac{\text{rate of effusion of H}_2}{\text{rate of effusion of N}_2} = \left(\frac{MM_{N_2}}{MM_{H_2}}\right)^{\frac{1}{2}} = \left(\frac{28.02}{2.016}\right)^{\frac{1}{2}} = 3.73$$

Thus the hydrogen balloon will deflate 3.73 times faster than the nitrogen balloon.

50. Refer to Section 5.6 and Example 5.11.

The key to solving this problem is to realize that because P, T, and V are the same in both experiments, the number of moles of CO and CO_2 effusing through the pinhole is the same.

$$n_{CO} = n_{CO2} = n = 1.73 \times 10^{-3} \text{ mol.}$$

Comparing the two rates in moles per second

$$\frac{\text{rate CO}}{\text{rate CO}_2} = \frac{n/12.6\,s}{n/\,x\,s} = \frac{12.6}{x}$$

Applying Graham's law

$$\frac{12.6}{x} = \left(\frac{MM_{CO}}{MM_{CO_2}}\right)^{1/2} = \left(\frac{28.01\,\text{g/mol. CO}}{44.01\,\text{g/mol. CO}_2}\right)^{1/2} = 0.798$$

Solving for x = 12.6 / 0.798 = **15.8 s** to effuse the same amount of CO_2 through the same pinhole.

52. Refer to Section 5.6 and Example 5.10.

a. $\mu_{Br_2} = \left(\dfrac{3RT}{MM}\right)^{\frac{1}{2}} = \left(\dfrac{(3)(8.31 \times 10^3 \text{ g} \cdot \text{m}^2/\text{s}^2 \cdot \text{mol.} \cdot \text{K})(301 \text{ K})}{159.8 \text{ g/mol.}}\right)^{\frac{1}{2}} = 217 \text{ m/s}$

b. $\mu_{Kr} = \left(\dfrac{3RT}{MM}\right)^{\frac{1}{2}} = \left(\dfrac{(3)(8.31 \times 10^3 \text{ g} \cdot \text{m}^2/\text{s}^2 \cdot \text{mol.} \cdot \text{K})(245 \text{ K})}{83.80 \text{ g/mol.}}\right)^{\frac{1}{2}} = 270 \text{ m/s}$

54. Refer to Section 5.7.

Deviations from ideal behavior tend to be largest at high pressures and low temperatures.

a. If pressure is reduced from 20 atm to 1 atm, CH_4 should behave **more** ideally.

b. If temperature is reduced from 50°C to −50°C, CH_4 should behave **less** ideally.

56. Refer to Sections 5.3 and 5.7 and Figure 5.10.

Use the table in Figure 5.10 to estimate the ratio of the real molar volume to the ideal molar volume. Calculate the ideal molar volume. Calculate the density, substituting the ideal molar volume and the ratio for the real molar volume. Note that a different estimate of the V_m/V_m° ratio will result a slightly different density.

a. According to Figure 5.10, at 100 atm, V_m/V_m° is approximately 0.78.

Thus, $V_m = 0.78 V_m^\circ$.

$V_m^\circ = \dfrac{RT}{P} = \dfrac{(0.0821 \text{ L} \cdot \text{atm/mol.} \cdot \text{K})(298 \text{ K})}{100 \text{ atm}} = 0.245 \text{ L/mol.}$

$d = \dfrac{MM}{V_m} = \dfrac{MM}{(0.78)(V_m^\circ)} = \dfrac{16.04 \text{ g/mol.}}{(0.78)(0.245 \text{ L/mol.})} = 84 \text{ g/L}$

b. $d = \dfrac{(MM)P}{RT} = \dfrac{(16.04 \text{ g/mol.})(100 \text{ atm})}{(0.0821 \text{ L} \cdot \text{atm/mol.} \cdot \text{K})(298 \text{ K})} = 65.6 \text{ g/L}$

c. The densities will be equal when $V_m^\circ = V_m$, which occurs when $P = 340$ atm.

d = density
h = height of tube

$d_{Hg} h_{Hg} = d_{oil} h_{oil}$

$(13.6 \text{ g/cm}^3)(760 \text{ mm}) = (2.73 \text{ g/mL})(h_{oil})$

$h_{oil} = 3790 \text{ mm} = 3.79 \text{ meters long}$

60. Refer to Section 5.3.

a. If n, P, and T are constant, then V will also be constant. Thus, the volumes are the same and the ratio is one.

b. 1.00 mol. Ne = 20.18 g
   1.00 mol. HCl = 36.46 g
   d = g/L
   $$\frac{\text{density Ne}}{\text{density HCl}} = \frac{20.18 \text{ g/L}}{36.46 \text{ g/L}} = 0.5535$$

c. The translational energy depends only on the temperature. Since the molecules are at the same temperature, they must also have the **same** translational energy. Thus, the ratio is one.

d. 1 mol of Ne has 6.022×10^{23} atoms
   1 mol of HCl has 6.022×10^{23} atoms
   Thus, the ratio is one.

62. Refer to Sections 5.3 and 5.4.

Use the ideal gas equation to calculate the moles of air, and then the mole percent to calculate moles of O_2. Then use the mole ratio and the molar mass of gasoline to calculate the mass of gasoline needed.

a. $n_{Air} = \dfrac{PV}{RT} = \dfrac{(1.00 \text{ atm})(0.618 \text{ L})}{(0.0821 \text{ L} \cdot \text{atm/mol.} \cdot \text{K})(348 \text{ K})} = 0.0216 \text{ mol. air}$

   $0.0216 \text{ mol. air} \times \dfrac{21.0 \text{ mol. } O_2}{100 \text{ mol. air}} = 4.54 \times 10^{-3} \text{ mol. } O_2$

b. $4.54 \times 10^{-3} \text{ mol. } O_2 \times \dfrac{1 \text{ mol. gasoline}}{12 \text{ mol. } O_2} \times \dfrac{1.0 \times 10^2 \text{ g}}{1 \text{ mol. gasoline}} = 0.038 \text{ g gasoline}$

Use the ideal gas equation to calculate the moles of CO_2. Then calculate the mass of C and H and the mass percent based on the initial amount of glycine. Do the same for N_2 and then calculate the percentage of O by difference. With the mass percents in hand, calculate the empirical formula.

$$n_{CO_2} = \frac{PV}{RT} = \frac{(1.00\,\text{atm})(0.1329\,\text{L})}{(0.0821\,\text{L}\cdot\text{atm/mol.}\cdot\text{K})(298\,\text{K})} = 5.43 \times 10^{-3}\,\text{mol. CO}_2$$

$$5.43 \times 10^{-3}\,\text{mol. CO}_2 \times \frac{1\,\text{mol. C}}{1\,\text{mol. CO}_2} \times \frac{12.01\,\text{g C}}{1\,\text{mol. C}} = 0.0652\,\text{g C}$$

$$\frac{0.0652\,\text{g C}}{0.2036\,\text{g glycine}} \times 100\% = 32.0\%\,\text{C}$$

$$0.122\,\text{g H}_2\text{O} \times \frac{1\,\text{mol. H}_2\text{O}}{18.02\,\text{g H}_2\text{O}} \times \frac{2\,\text{mol. H}}{1\,\text{mol. H}_2\text{O}} \times \frac{1.008\,\text{g H}}{1\,\text{mol. H}} = 0.0136\,\text{g H}$$

$$\frac{0.0136\,\text{g H}}{0.2036\,\text{g glycine}} \times 100\% = 6.68\%\,\text{H}$$

$$n_{N_2} = \frac{PV}{RT} = \frac{(1.00\,\text{atm})(0.0408\,\text{L})}{(0.0821\,\text{L}\cdot\text{atm/mol.}\cdot\text{K})(298\,\text{K})} = 1.67 \times 10^{-3}\,\text{mol. N}_2$$

$$1.67 \times 10^{-3}\,\text{mol. N}_2 \times \frac{2\,\text{mol. N}}{1\,\text{mol. N}_2} \times \frac{14.01\,\text{g N}}{1\,\text{mol. N}} = 0.0468\,\text{g N}$$

$$\frac{0.0468\,\text{g N}}{0.2500\,\text{g glycine}} \times 100\% = 18.7\%\,\text{N}$$

mass % O = 100% − 32.0% − 6.68% − 18.7% = 42.6% O

$$32.0\,\text{g C} \times \frac{1\,\text{mol. C}}{12.01\,\text{g C}} = 2.66\,\text{mol. C}$$

$$6.70\,\text{g H} \times \frac{1\,\text{mol. H}}{1.008\,\text{g H}} = 6.65\,\text{mol. H}$$

$$18.7\,\text{g N} \times \frac{1\,\text{mol. N}}{14.01\,\text{g N}} = 1.33\,\text{mol. N}$$

$$42.6\,\text{g O} \times \frac{1\,\text{mol. O}}{16.00\,\text{g O}} = 2.66\,\text{mol. O}$$

C: 2.66 / 1.33 = 2
H: 6.65 / 1.33 = 5
N: 1.33 / 1.33 = 1
O: 2.66 / 1.33 = 2

Empirical formula = $C_2H_5NO_2$

---

66. *Refer to Sections 5.2, 5.3, and 5.6.*

a. GT, since all of the tanks have the same number of moles of each gas, the mass of SO_2 in the tank will be nearly twice the mass of the oxygen in the tank due to its higher molar mass.

b. EQ, at a given temperature, molecules of different gases must all have the same average kinetic energy of translational motion.

c. LT, since the effusion rate is equal to the square root of the ratio of the molar masses, it will only take SO_2 28 sec to effuse out of the pinhole in the tank.

$$\left(\frac{64.0648 \, g/mol. \, SO_2}{31.9988 \, g/mol. \, O_2} \right)^{1/2} = 1.41$$

$$1.41 \times 20\,s = 28\,s$$

d. GT, since density has units of g/L and the volume of each tank is the same, then the density is directly related to mass of each gas. The density in the O_2 tank will be greater than the density in the CH_4 tank since the molar mass of O_2, 32.00 g/mol is greater than the molar mass of CH_4, 16.00 g/mol.

e. GT, since pressure is directly related to temperature, if the temperature is doubled then the pressure will also double. The pressure in tank A will be double the pressure in tank B.

---

68. *Refer to Sections 5.2, 5.3, and 5.5.*

Before   After

a.

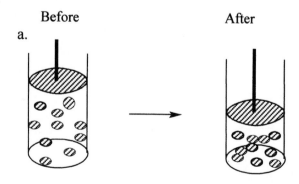

b.

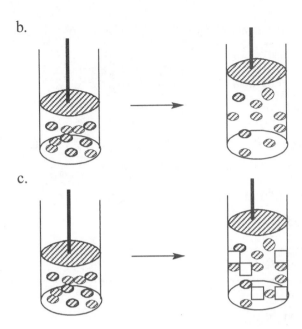

c.

70. Refer to Section 5.2.

Since R is a constant, we can solve these problems using the ideal gas law, by solving for R, and setting the equations equal to one another.

$$\frac{PV}{nT} = R, \text{ thus } \frac{P_A V_A}{n_A T_A} = R = \frac{P_B V_B}{n_B T_B}. \text{ Since } V_A = V_B \text{ and } T_A = T_B, \text{ then: } \frac{P_A}{n_A} = \frac{P_B}{n_B}.$$

a.   If $n_A = n_B$, then: $P_A = P_B$

b. CO_2 has a much larger mass than He, thus if there are equal grams of each, there will be much more He than CO_2, and $n_A < n_B$. Thus $P_A < P_B$.

72. Refer to Sections 5.2 and 5.5.

a.   Pressure is directly proportional to the number of moles, so **bulb C** (the one with the fewest molecules) would have the lowest pressure.

b.   The relative pressures are directly proportional to the relative moles. Bulb C contains half as many molecules (or moles) as bulb A, so the pressure would be half that of A: **1.00 atm**.

c.   As mentioned in part (b) above, the pressure in bulb C is 1.00 atm. From similar reasoning, pressure in bulb B is [(6/8)(2.00)] 1.50 atm.
Total pressure = 2.00 + 1.50 + 1.00 = **4.50 atm**.

d. Bulbs A and B now contain 7 molecules each, so $P_A = P_B = (7/8)(2.00 \text{ atm}) = 1.75 \text{ atm}$.

$P_A + P_B = 1.75 + 1.75 = \textbf{3.50 atm}$

$P_A + P_B + P_C = 1.75 + 1.75 + 1.00 = \textbf{4.50 atm}$

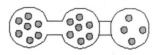

Thus the total pressure is unchanged.

e. Each bulb now contains 6 molecules, so $P_A = P_B = P_C = (6/8)(2.00 \text{ atm}) = 1.50 \text{ atm}$.

$P_A + P_B = 1.50 + 1.50 = \textbf{3.00 atm}$

$P_A + P_B + P_C = 1.50 + 1.50 + 1.50 = \textbf{4.50 atm}$

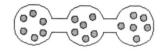

Thus the total pressure is unchanged.

---

74. Refer to Sections 5.3 and 5.7 and Appendix 1.

$$23.76 \text{ mm Hg} \times \frac{1 \text{ atm}}{760.0 \text{ mm Hg}} = 0.03126 \text{ atm}$$

$$\frac{P_1}{T_1} = \frac{P_2}{T_2} \Rightarrow \frac{0.03126}{298} = \frac{P_2}{313} \Rightarrow P_2 = 0.0329 \text{ atm}$$

$$0.0329 \text{ atm} \times \frac{760.0 \text{ mm Hg}}{1 \text{ atm}} = 25.0 \text{ mm Hg}$$

$$\frac{P_1}{T_1} = \frac{P_3}{T_3} \Rightarrow \frac{0.0313}{298} = \frac{P_3}{343} \Rightarrow P_3 = 0.0360 \text{ atm}$$

$$0.0360 \text{ atm} \times \frac{760.0 \text{ mm Hg}}{1 \text{ atm}} = 27.4 \text{ mm Hg}$$

$$\frac{P_1}{T_1} = \frac{P_4}{T_4} \Rightarrow \frac{0.0313}{298} = \frac{P_4}{373} \Rightarrow P_4 = 0.0392 \text{ atm}$$

$$0.0392 \text{ atm} \times \frac{760.0 \text{ mm Hg}}{1 \text{ atm}} = 29.8 \text{ mm Hg}$$

Temperature	P (ideal gas)	P (water vapor)
$T_1 = 25°C$	23.76 mm Hg	23.76 mm Hg
$T_2 = 40°C$	25.0 mm Hg	55.3 mm Hg
$T_3 = 70°C$	27.4 mm Hg	233.7 mm Hg
$T_4 = 100°C$	29.8 mm Hg	760.0 mm Hg

These numbers are so different because the number of moles (n) is not constant for the water vapor. As the temperature increases, the amount of water that is vaporized increases. Since pressure is proportional to the moles of gas, the pressure also increases with increasing temperature.

Use the equation for average speed of a gas to calculate the molar mass. Compare that value to the molar masses of H_2, He, and Ar.

$$u^2 = \frac{3RT}{MM} \Rightarrow MM = \frac{3RT}{u^2}$$

$$MM = \frac{3RT}{u^2} = \frac{(3)(8.31 \times 10^3 \text{ g} \cdot \text{m}^2/\text{s}^2 \cdot \text{mol.} \cdot \text{K})(288 \text{ K})}{(1.12 \times 10^3 \text{ m/s})^2} = 5.72 \text{ g/mol.}$$

Note that molar mass and speed are inversely proportional, thus heavier atoms and molecules will move more slowly. Those atoms and molecules that will have sufficient speed to escape gravity are those with a molar mass of less than 5.72 g/mol., such as H_2 and He. Ar's molar mass is greater than 5.72 g/mol., thus it's speed will be insufficient.

76. Refer to Section 5.6.

Set up an algebraic equation, using x for the distance ammonia travels and $(5 - x)$ for the distance HCl travels, correlating distance traveled with molar mass.

$$\frac{\text{distance}_{NH_3}}{\text{distance}_{HCl}} = \left(\frac{MM_{HCl}}{MM_{NH_3}}\right)^{\frac{1}{2}} \Rightarrow \frac{x}{5 - x} = \left(\frac{35.5}{17}\right)^{\frac{1}{2}}$$

$$\frac{x}{5 - x} = 1.45 \Rightarrow x = 7.25 - 1.45x$$

$$2.45x = 7.25$$

$$x = 3.0 \text{ ft}$$

77. Refer to Sections 1.2 and 5.2.

Kelvin and °C are the same size and °R and °F are also the same size. Thus if 1.8°C = 1°F, then 1.8K = 1°R.

273.15 K = 491.67°R

$$R = \frac{PV}{nT} = \frac{(1.00 \text{ atm})(22.4 \text{ L})}{(1.00 \text{ mol.})(491.67 \text{ °R})} = 0.0456 \text{ L} \cdot \text{atm/mol.} \cdot \text{°R}$$

Calculate the total amount of H_2 produced, then calculate the moles of H_2 produced by each metal using x grams Zn and $(0.2500 - x)$ grams Al. Solve for x to get the grams of Zn, then calculate the mass percent.

$$P = 755 \text{ mm Hg} \times \frac{1 \text{ atm}}{760.0 \text{ mm Hg}} = 0.993 \text{ atm}$$

$$n_{H_2} = \frac{PV}{RT} = \frac{(0.993 \text{ atm})(0.147 \text{ L})}{(0.0821 \text{ L} \cdot \text{atm/mol} \cdot \text{K})(298 \text{ K})} = 0.00597 \text{ mol. } H_2$$

$$x \text{ g Zn} \times \frac{1 \text{ mol. Zn}}{65.39 \text{ g Zn}} \times \frac{1 \text{ mol. } H_2}{1 \text{ mol. Zn}} = 0.0153x \text{ mol. } H_2$$

$$(0.2500 - x) \text{ g Al} \times \frac{1 \text{ mol. Al}}{26.98 \text{ g Al}} \times \frac{\frac{3}{2} \text{ mol. } H_2}{1 \text{ mol. Al}} = (0.0139 - 0.0556x) \text{ mol. } H_2$$

$0.00597 = 0.0153x + 0.0139 - 0.0556x$

$0.0403x = 0.00793$

$x = 0.197 \text{ g Zn}$

$$\text{mass \% Zn} = \frac{0.197 \text{ g Zn}}{0.2500 \text{ g total}} \times 100\% = 78.7\%$$

To get off the ground, the buoyant force of the balloon must exceed the gravitational force holding it down. We can find the volume of the balloon by equating the two forces, converting mass to $(d \times V)$ and solving for V. Although we were not given densities, we can solve the ideal gas law for d and substitute. Now we can calculate a numerical value for V and calculate the radius and diameter of the balloon.

$$\text{mass}_{(air)} = \text{mass}_{(H_2)} + \text{mass}_{(man + balloon)} = \text{mass}_{(H_2)} + 1.68 \times 10^5 \text{ g}$$

$$d_{(air)} \times V = d_{(H_2)} \times V + 1.68 \times 10^5 \text{ g}$$

$$V = \frac{1.68 \times 10^5 \text{ g}}{d_{air} - d_{H_2}} = \frac{1.68 \times 10^5 \text{ g}}{\dfrac{P\,MM_{air}}{RT} - \dfrac{P\,MM_{H_2}}{RT}} = \frac{(1.68 \times 10^5 \text{ g})\,RT}{P(MM_{air} - MM_{H_2})}$$

$$V = \frac{(1.68 \times 10^5 \text{ g})(0.0821 \text{ L} \cdot \text{atm/mol.} \cdot \text{K})(295 \text{ K})}{\left(758 \text{ mm Hg} \times \dfrac{1 \text{ atm}}{760 \text{ mm Hg}}\right)(29.0 \text{ g/mol.} - 2.01 \text{ g/mol.})} = 1.51 \times 10^5 \text{ L}$$

$$V = 1.51 \times 10^5 \text{ L} \times \frac{1 \text{ m}^3}{1000 \text{ L}} = 1.51 \times 10^2 \text{ m}^3$$

$$V = 1.51 \times 10^2 \text{ m}^3 = \frac{4}{3}\pi r^3$$

$$r = 3.30 \text{ m}$$

$$\text{diameter} = 2r = 6.60 \text{ m}$$

---

80. *Refer to Section 5.4.*

Since R is a constant, we can solve this problem using the ideal gas law, by solving for R, and setting the equations equal to one another. (Remember to convert temperature to K.)

$$\frac{PV}{nT} = R, \text{ thus } \frac{P_i V_i}{n_i T_i} = R = \frac{P_f V_f}{n_f T_f}. \text{ Since } V \text{ is constant, we get: } \frac{P_i}{n_i T_i} = \frac{P_f}{n_f T_f}.$$

Solving for P_f, we get: $P_f = \dfrac{P_i \, n_f \, T_f}{n_i \, T_i}$

n_f = unreacted moles of H_2 and O_2 + moles of product.
  (Remember that 3 moles of reactants yields only 2 moles of product.)

Thus, $n_f = (0.120)(n_i) + (0.880)\left(\dfrac{2}{3}n_i\right) = 0.707\, n_i$

$$P_f = \frac{(0.950 \text{ atm})(0.707 n_i)(398 \text{ K})}{(n_i)(298 \text{ K})} = 0.897 \text{ atm}$$

---

81. *Refer to Sections 5.2 and 5.5.*

Use the ideal gas law, solving for V_A and V, and substitute into the equation for volume fraction. Since R, T, and P are constant, they cancel out.

$$\text{volume fraction A} = \frac{V_A}{V} = \frac{\dfrac{n_A \, RT}{P}}{\dfrac{nRT}{P}} = \frac{n_A}{n} = \text{mole fraction A}$$

Volume fraction (and mole fraction) differ from mass fraction because different gases have different molar masses.

Chapter 6: Electronic Structure and the Periodic Table

---

2. Refer to Section 6.1 and Examples 6.1 and 6.2.

a. $c = \lambda v$

$2.998 \times 10^8 \text{m/s} = \lambda(4.00 \times 10^8/\text{s})$

$\lambda = 0.750 \text{ m}$

b. $E = hv = (6.626 \times 10^{-34} \text{ J·s})(4.00 \times 10^8/\text{s}) = 2.65 \times 10^{-25} \text{ J}$

Note that this is for one photon, thus units are J/photon.

c. $\dfrac{2.65 \times 10^{-25} \text{ J}}{1 \text{ photon}} \times \dfrac{6.02 \times 10^{23} \text{ photons}}{1 \text{ mol.}} = 1.60 \times 10^{-1} \text{ J/mol.} = 1.60 \times 10^{-4} \text{ kJ/mol.}$

---

4. Refer to Section 6.1, Example 6.1, and Figure 6.2.

a. $1498 \text{ nm} \times \dfrac{1 \text{ m}}{1 \times 10^9 \text{ nm}} = 1.498 \times 10^{-6} \text{ m}$

1.498×10^{-6} m is just outside the visible region in the **infrared** region.

b. $c = \lambda v$

$v = \dfrac{2.998 \times 10^8 \text{ m/s}}{1498 \times 10^{-6} \text{ m}} = 2.001 \times 10^{14} \text{ s}^{-1} = 2.001 \times 10^{14} \text{ Hz}$

c. $E = hv = (6.626 \times 10^{-34} \text{ J·s})(2.001 \times 10^{14} \text{ s}^{-1}) = 1.326 \times 10^{-19} \text{ J}$

---

6. Refer to Example 6.1.

$\dfrac{941 \text{ kJ}}{1 \text{ mol.}} \times \dfrac{1000 \text{ J}}{1 \text{ kJ}} \times \dfrac{1 \text{ mol.}}{6.022 \times 10^{23} \text{ photons}} = 1.56 \times 10^{-18} \text{ J}$

$\lambda = \dfrac{hc}{E} = \dfrac{(6.626 \times 10^{-34} \text{ J·s})(2.998 \times 10^8 \text{ m/s})}{1.56 \times 10^{-18} \text{ m}} = 1.27 \times 10^{-7} \text{ m} = 127 \text{ nm}$

8. *Refer to Section 6.1 and Example 6.2.*

Calculate the amount of energy in one photon. Use that and the total amount of energy to calculate the number of photons emitted.

$$\lambda = 633 \text{ nm} \times \frac{1 \text{ m}}{1 \times 10^9 \text{ nm}} = 6.33 \times 10^{-7} \text{ m}$$

$$E = \frac{hc}{\lambda} = \frac{(6.626 \times 10^{-34} \text{ J} \cdot \text{s})(2.998 \times 10^8 \text{ m/s})}{6.33 \times 10^{-7} \text{ m}} = 3.14 \times 10^{-19} \text{ J}$$

$$12 \text{ kJ} \times \frac{1000 \text{ J}}{1 \text{ kJ}} \times \frac{1 \text{ photon}}{3.14 \times 10^{-19} \text{ J}} = 3.8 \times 10^{22} \text{ photons}$$

10. *Refer to Section 6.2, Example 6.3, and Figure 6.2.*

Use the Rydberg equation to calculate the frequency. Then calculate the wavelength to determine the region of the spectrum associated with that frequency.

$$v = \frac{R_H}{h} \left[\frac{1}{(n_{lo})^2} - \frac{1}{(n_{hi})^2} \right] = \frac{2.180 \times 10^{-18} \text{ J}}{6.626 \times 10^{-34} \text{ J} \cdot \text{s}} \left[\frac{1}{1^2} - \frac{1}{3^2} \right] = 2.925 \times 10^{15} \text{ s}^{-1}$$

$$\lambda = \frac{c}{v} = \frac{2.998 \times 10^8 \text{ m/s}}{2.925 \times 10^{15} \text{ s}^{-1}} \times \frac{1 \times 10^9 \text{ nm}}{1 \text{ m}} = 102.5 \text{ nm}; \textbf{ ultraviolet}.$$

Yes, in the transition from a low level to a high one, energy is absorbed.

12. *Refer to Section 6.2.*

$$E_n = \frac{-R_H}{n^2}$$

$$E_1 = \frac{-R_H}{n^2} = \frac{2.180 \times 10^{-18} \text{ J}}{1^2} = -2.180 \times 10^{-18} \text{ J}$$

$$E_2 = \frac{-R_H}{n^2} = \frac{2.180 \times 10^{-18} \text{ J}}{2^2} = -5.450 \times 10^{-19} \text{ J}$$

$$E_3 = \frac{-R_H}{n^2} = \frac{2.180 \times 10^{-18} \text{ J}}{3^2} = -2.422 \times 10^{-19} \text{ J}$$

$$E_4 = \frac{-R_H}{\mathbf{n}^2} = \frac{2.180 \times 10^{-18} \text{ J}}{4^2} = -1.363 \times 10^{-19} \text{ J}$$

Lyman   Balmer
Series   Series

Energy (J)

-1.363 x 10⁻¹⁹ ———— n=4
-2.442 x 10⁻¹⁹ ———— n=3
-5.450 x 10⁻¹⁹ ———— n=2

-2.180 x 10⁻¹⁸ ———— n=1

Obviously, this graph assumes that transitions are from E_{hi} to E_{lo}. If one assumes the transitions are in the opposite direction, the only change would be in the directions of the arrows.

---

14. Refer to Section 6.2, Example 6.3, and Figure 6.2.

a. Use the Rydberg equation to calculate the frequency. Then calculate the wavelength to determine the region of the spectrum associated with that frequency.

For the transition n = 5 → n = 4

$$\nu = \frac{R_H}{h}\left[\frac{1}{(n_{lo})^2} - \frac{1}{(n_{hi})^2}\right]$$

$$\nu = \frac{2.180 \times 10^{-18} \text{ J}}{6.626 \times 10^{-34} \text{ J} \cdot \text{s}}\left[\frac{1}{(4)^2} - \frac{1}{(5)^2}\right]$$

$$\nu = 7.403 \times 10^{13} / \text{s}$$

$$\lambda = \frac{c}{\nu} = \frac{2.998 \times 10^8 \text{ m/s}}{7.403 \times 10^{13} / \text{s}} \times \frac{10^9 \text{ nm}}{1 \text{ m}} = 4050 \text{ nm}$$

b. This wavelength of light occurs in the **infrared** region of the spectrum.

---

16. Refer to Section 6.2.

The longest wavelength corresponds to the smallest energy difference

$$\nu = \frac{R_H}{h}\left[\frac{1}{(\mathbf{n}_{lo})^2} - \frac{1}{(\mathbf{n}_{hi})^2}\right] = \frac{2.180 \times 10^{-18} \text{ J}}{6.626 \times 10^{-34} \text{ J} \cdot \text{s}}\left[\frac{1}{3^2} - \frac{1}{4^2}\right] = 1.599 \times 10^{14} \text{ s}^{-1}$$

$$\lambda = \frac{c}{\nu} = \frac{2.998 \times 10^8 \text{ m/s}}{1.599 \times 10^{14} / \text{s}} \times \frac{10^9 \text{ nm}}{1 \text{ m}} = 1875 \text{ nm}$$

18. Refer to Section 6.3.

Values of m_ℓ vary from $-\ell$ to $+\ell$ for any given ℓ value.

a. p-sublevel: $\ell = 1$
$$m_\ell = -1, 0, +1.$$

b. f-sublevel: $\ell = 3$
$$m_\ell = -3, -2, -1, 0, +1, +2, +3.$$

c. For the **n**=3 shell, there are 3 values for $\ell = 0, 1, 2$.
$\ell = 0$: s-sublevel: $m_\ell = 0$.
$\ell = 1$: p-sublevel: $m_\ell = -1, 0, +1$.
$\ell = 2$: d-sublevel: $m_\ell = -2, -1, 0, +1, +2$.

20. Refer to Section 6.3 and Figure 6.8.

Look at Figure 6.8 to determine the order of filling. The orbital that fills last is higher in energy.

a. 3s

b. 4d

c. 4f is higher for atoms with $Z < 58$ and the 6s orbital is higher for atoms with $Z > 57$.

d. 2s

22. Refer to Section 6.3.

The type of orbital is determined by the ℓ value.

a. $\ell = 1$, therefore: p-orbital.

b. $\ell = 0$, therefore: s-orbital.

c. $\ell = 2$, therefore: d-orbital

24. Refer to Section 6.3.

The number of orbitals is equal to $2\ell + 1$. If there is more than one sublevel, add up the orbitals for each sublevel.

a. $n = 3$ has 3 sublevels, $\ell = 0$, $\ell = 1$ and $\ell = 2$
   $\ell = 0$:   1 orbital
   $\ell = 1$:   3 orbitals
   $\ell = 2$:   5 orbitals, thus the $n = 3$ shell has **9 orbitals**.

b. 4p: $\ell = 1$:   **3 orbitals**

c. f: $\ell = 3$:   **7 orbitals**

d. d: $\ell = 2$:   **5 orbitals**

---

26. *Refer to Section 6.3, Example 6.5, and Table 6.3.*

a. These quantum numbers indicate a 1s orbital. The maximum number of electrons in a 1s orbital can be 2 electrons.

b. These quantum numbers indicate one of the 5f orbitals. The maximum number of electrons that can be in a single 5f orbital is 2 electrons.

c. These quantum numbers indicate a 3d orbital. Since there are 5 possible 3d orbitals and each can hold a maximum of 2 electrons, the total number of electrons that can reside in orbitals with these quantum numbers is 10 electrons.

---

28. *Refer to Section 6.3.*

Refer to the rules for quantum numbers. If a rule is violated, the set cannot occur.

a. $n = 1 \Rightarrow \ell = 0$
   $\ell = 0 \Rightarrow m_\ell = 0$
   $m_\ell = 0 \Rightarrow m_s = -\frac{1}{2}, +\frac{1}{2}$
   None of the rules are violated, so this set **can occur.**

b. $n = 1 \Rightarrow \ell = 0, 1$
   $\ell = 2$ This is not possible given the allowed values, this set **cannot occur.**

c. $n = 3 \Rightarrow \ell = 0, 1, 2$
   $\ell = 2 \Rightarrow m_\ell = -2, -1, 0, +1, +2$
   $m_\ell = -2 \Rightarrow m_s = -\frac{1}{2}, +\frac{1}{2}$
   None of the rules are violated, so this set **can occur.**

d. $n = 2 \Rightarrow \ell = 0, 1$
   $\ell = 1 \Rightarrow m_\ell = -1, 0, +1$
   $m_\ell = -2$ This is not possible given the allowed values, this set **cannot occur.**

e. $n = 4 \Rightarrow \ell = 0, 1, 2, 3$

   $\ell = 0 \Rightarrow m_\ell = 0$

   $m_\ell = 2$ This is not possible given the allowed values, this set **cannot occur.**

---

30. *Refer to Section 6.5 and Examples 6.6 and 6.7.*

Determine the number of electrons. Then write down the filling order and start filling the orbitals until you run out of electrons.

a. B ($5e^-$): $1s^2 \, 2s^2 \, 2p^1$

b. Ba ($56e^-$): $1s^2 \, 2s^2 \, 2p^6 \, 3s^2 \, 3p^6 \, 4s^2 \, 3d^{10} \, 4p^6 \, 5s^2 \, 4d^{10} \, 5p^6 \, 6s^2$

c. Be ($4e^-$): $1s^2 \, 2s^2$

d. Bi ($83e^-$): $1s^2 \, 2s^2 \, 2p^6 \, 3s^2 \, 3p^6 \, 4s^2 \, 3d^{10} \, 4p^6 \, 5s^2 \, 4d^{10} \, 5p^6 \, 6s^2 \, 4f^{14} \, 5d^{10} \, 6p^3$

e. Br ($35e^-$): $1s^2 \, 2s^2 \, 2p^6 \, 3s^2 \, 3p^6 \, 4s^2 \, 3d^{10} \, 4p^5$

---

32. *Refer to Section 6.4, Examples 6.6 and 6.7, Table 6.4, and Figure 6.9.*

Locate the element of interest on the periodic table. Move up one period and then move to the extreme right to find the appropriate noble gas. Which period is the element of interest in? That is the s-orbital from which you are to start filling. Then fill until you run out of electrons.

a. Os, $[Xe] \, 6s^2 \, 4f^{14} \, 5d^6$

b. Mg, $[Ne] \, 3s^2$

c. Ge, $[Ar] \, 4s^2 \, 3d^{10} \, 4p^2$

d. V, $[Ar] \, 4s^2 \, 3d^3$

e. At, $[Xe] \, 6s^2 \, 4f^{14} \, 5d^{10} \, 6p^5$

---

34. *Refer to Section 6.5.*

Start writing an electron configuration, filling up the orbitals until you meet the criterion given. Then determine the element that configuration corresponds to.

a. $[Xe]6s^2 \, 4f^{14}$

   This corresponds to **Yb.**

b. $1s^2\,2s^2\,2p^6\,3s^2\,3p^6\,4s^2\,3d^{10}\,4p^6\,5s^2\,4d^{10}\,5p^2$
   This corresponds to **Sn.**

c. $1s^2\,2s^2\,2p^6\,3s^2\,3p^6\,4s^2\,3d^{10}\,4p^6\,5s^2\,4d^2$
   This corresponds to **Zr.**

d. $1s^2\,2s^2\,2p^6\,3s^2\,3p^6\,4s^2\,3d^{10}\,4p^6\,5s^2\,4d^{10}\,5p^5$
   This corresponds to **I.**

---

36. *Refer to Section 6.5.*

Write out the electron configuration, count up the number of electrons in p subshells and divide that number by the total number of electrons.

a. Mg: $1s^2\,2s^2\,2p^6\,3s^2$
   (6 p electrons) / (12 total electrons) = **0.5** (or one half)

b. Mn: $1s^2\,2s^2\,2p^6\,3s^2\,3p^6\,4s^2\,3d^5$
   (12 p electrons) / (25 total electrons) = **0.48** (or almost one half)

c. Mo: $1s^2\,2s^2\,2p^6\,3s^2\,3p^6\,4s^2\,3d^{10}\,4p^6\,5s^2\,4d^4$
   (18 p electrons) / (42 total electrons) = **0.43** (little more than two fifths)

---

38. *Refer to Section 6.5, Figure 6.8, and Example 6.6.*

First determine if the given configuration violates one of the rules for quantum numbers. If it does not, determine the ground state configuration for an atom with the given number of electrons and compare that with the configuration given.

a. 3 electrons, ground state: $1s^2\,2s^1$
   Thus the given configuration is an **excited state**.

b. 8 electrons, ground state: $1s^2\,2s^2\,2p^4$
   Thus the given configuration is the **ground state**.

c. 10 electrons, ground state: $1s^2\,2s^2\,2p^6$
   Thus the given configuration is the **excited state**.

d. The p orbital can have a maximum of 6 electrons.
   Impossible configuration.

e. The d orbital can have a maximum of 10 electrons.
   Impossible configuration.

40. Refer to Section 6.6 and Example 6.8.

a. Na: 1s   2s   2p   3s
       ($\uparrow\downarrow$) ($\uparrow\downarrow$) ($\uparrow\downarrow$)($\uparrow\downarrow$)($\uparrow\downarrow$) ($\uparrow$)

b. O: 1s   2s   2p
      ($\uparrow\downarrow$) ($\uparrow\downarrow$) ($\uparrow\downarrow$)($\uparrow$)($\uparrow$)

c. Co: 1s   2s   2p   3s   3p   4s   3d
      ($\uparrow\downarrow$) ($\uparrow\downarrow$) ($\uparrow\downarrow$)($\uparrow\downarrow$)($\uparrow\downarrow$) ($\uparrow\downarrow$) ($\uparrow\downarrow$)($\uparrow\downarrow$)($\uparrow\downarrow$) ($\uparrow\downarrow$) ($\uparrow\downarrow$)($\uparrow\downarrow$)($\uparrow$)($\uparrow$)($\uparrow$)

d. Cl: 1s   2s   2p   3s   3p
      ($\uparrow\downarrow$) ($\uparrow\downarrow$) ($\uparrow\downarrow$)($\uparrow\downarrow$)($\uparrow\downarrow$) ($\uparrow\downarrow$) ($\uparrow\downarrow$)($\uparrow\downarrow$)($\uparrow$)

42. Refer to Section 6.6.

Count the number of electrons. Since the number of electrons must equal the number of protons (atomic number) in a neutral atom, you can then identify the element from the periodic table.

a. 12 electrons, $Z = 12$, therefore: **Mg**.

b. 15 electrons, $Z = 15$, therefore: **P**.

c. 8 electrons, $Z = 8$, therefore: **O**.

44. Refer to Section 6.6 and Figure 6.9.

a. Sn has 2 half-filled 5p orbitals, Sb has 3 half-filled 5p orbitals, and Te has 2 half-filled 5p orbitals.

b. K, Rb, Cs, Fr; every Group 1 element past Ar which has filled 3p orbitals.

c. Ge, As, Sb, Te; every metalloid past S that has paired 3p electrons.

d. None, the only element that has filled 3d orbitals and 3 half-filled 3p orbitals is As, but As is a metalloid, not a nonmetal.

46. Refer to Section 6.6.

All the inner sublevels will be filled, so only the outermost sublevel will contain unpaired electrons.

a. Hg: $[Xe]6s^2 4f^{14} 5d^{10}$ $5d^{10}$: $(\uparrow\downarrow)(\uparrow\downarrow)(\uparrow\downarrow)(\uparrow\downarrow)(\uparrow\downarrow)$ **0 unpaired electrons.**

b. Mn: $[Ar]4s^2 3d^5$ $3d^5$: $(\uparrow\ )(\uparrow\ )(\uparrow\ )(\uparrow\ )(\uparrow\ )$ **5 unpaired electrons.**

c. Mg: $[Ne]3s^2$ $3s^2$: $(\uparrow\downarrow\ )$ **0 unpaired electrons.**

---

48. *Refer to Section 6.6.*

The only main group metals in the 4th period are K, Ca and Ga, the remainder of the elements are transition metals, nonmetals or metalloids.

$$\begin{array}{cccc} & 4s & 4p & \\ K & (\uparrow\ ) & (\ )(\ )(\ ) & \\ Ca & (\uparrow\downarrow) & (\ )(\ )(\ ) & \\ Ga & (\uparrow\downarrow) & (\uparrow\ )(\ )(\ ) & \end{array}$$

a. Ca

b. K and Ga

c. none

d. none

---

50. *Refer to Section 6.7 and Example 6.9.*

Write the ground state electron configuration for the atom, then remove (for cations) or add (for anions) the number of electrons indicated by the charge.

a. S $1s^2\, 2s^2\, 2p^6\, 3s^2\, 3p^4$
   S^{2-} $1s^2\, 2s^2\, 2p^6\, 3s^2\, 3p^6$

b. Al $1s^2\, 2s^2\, 2p^6\, 3s^2\, 3p^1$
   Al^{3+} $1s^2\, 2s^2\, 2p^6$

c. V $1s^2\, 2s^2\, 2p^6\, 3s^2\, 3p^6\, 4s^2\, 3d^3$
   V^{4+} $1s^2\, 2s^2\, 2p^6\, 3s^2\, 3p^6\, 3d^1$
   For transition metals, electrons are lost first from the outermost s sublevel.

d. Cu^+ $1s^2\, 2s^2\, 2p^6\, 3s^2\, 3p^6\, 3d^{10}$
   Cu^{2+} $1s^2\, 2s^2\, 2p^6\, 3s^2\, 3p^6\, 3d^9$
   Remember that the electron configuration of Cu is $[Ar]\, 4s^1\, 3d^{10}$ and that for transition metals, electrons are lost first from the outermost s sublevel.

Write the ground state electron configuration for the atom, then remove (for cations) or add (for anions) the number of electrons indicated by the charge. Then write the orbital diagram for the ion.

a. Al: [Ne] $2s^2 2p^1$
    Al^{3+}: [Ne] $2s^0 2p^0$ Thus, there are 0 unpaired electrons

b. Cl: [Ne] $3s^2 3p^5$
    Cl^-: [Ne] $3s^2 3p^6$
    Cl^-: [Ne] 3s   3p
             ($\uparrow\downarrow$) ($\uparrow\downarrow$)($\uparrow\downarrow$)($\uparrow\downarrow$)
             Thus, there are 0 unpaired electrons

c. Sr: [Kr] $5s^2$
    Sr^{2+}: [Kr] $5s^0$ Thus, there are 0 unpaired electrons

d. Ag: [Kr] $5s^1 4d^{10}$
    Ag^+: [Kr] $5s^0 4d^{10}$
    Ag^+: [Kr] 5s   4d
             ( ) ($\uparrow\downarrow$)($\uparrow\downarrow$)($\uparrow\downarrow$)($\uparrow\downarrow$)($\uparrow\downarrow$)
             Thus, there are 0 unpaired electrons

---

54. Refer to Section 6.8, Figures 6.13 and 6.15, and Example 6.11.

a. Atomic radius increases from right to left across a period, therefore:
    S < Si < Na

b. Ionization energy increases from left to right across a period, therefore:
    Na < Si < S

c. Electronegativity decreases from right to left across a period, therefore:
    S > Si > Na

---

56. Refer to Section 6.8 and Example 6.10.

a. Atomic radius decreases from left to right across a period and increases down a group, thus the largest atom would be the lower leftmost: **K**.

b. Ionization energy increases from left to right across a period and decreases down a group, thus the atom with the highest ionization energy would be the upper rightmost: **Cl**.

c. Electronegativity increases from left to right across a period and decreases down a group, thus the most electronegative atom would be the upper rightmost: **Cl**.

58. Refer to Section 6.8 and Example 6.10.

Cations are smaller than the corresponding atoms, anions are larger.

a. N

b. Ba^{+2}

c. Se

d. Co^{+3}

60. Refer to Section 6.8, Figure 6.13, and Example 6.10.

Atomic radii decrease from left to right across a period and increases down a group. Thus, the ordering of radii from smallest to largest is:

a. $Kr < K < Rb < Cs$

a. $Ar < Si < Al < Cs$

62. Refer to Section 6.1 and Figure 6.3.

a. Both wavelengths fall into the green region of visible light.

b. $486 \text{ nm} \times \dfrac{1 \text{ m}}{1 \times 10^9 \text{ nm}} = 4.85 \times 10^{-7} \text{ m}$

$E = \dfrac{hc}{\lambda} = \dfrac{(6.626 \times 10^{-34} \text{ J} \cdot \text{s})(2.998 \times 10^8 \text{ m/s})}{4.85 \times 10^{-7} \text{ m}} = 4.10 \times 10^{-19} \text{ J}$

$\dfrac{4.10 \times 10^{-19} \text{ J}}{1 \text{ photon}} \times \dfrac{6.02 \times 10^{23} \text{ photons}}{1 \text{ mol.}} = 2.47 \times 10^5 \text{ J/mol.} = 2.47 \times 10^2 \text{ kJ/mol.}$

$512 \text{ nm} \times \dfrac{1 \text{ m}}{1 \times 10^9 \text{ nm}} = 5.12 \times 10^{-7} \text{ m}$

$E = \dfrac{hc}{\lambda} = \dfrac{(6.626 \times 10^{-34} \text{ J} \cdot \text{s})(2.998 \times 10^8 \text{ m/s})}{5.12 \times 10^{-7} \text{ m}} = 3.88 \times 10^{-19} \text{ J}$

$\dfrac{3.88 \times 10^{-19} \text{ J}}{1 \text{ photon}} \times \dfrac{6.02 \times 10^{23} \text{ photons}}{1 \text{ mol.}} = 2.34 \times 10^5 \text{ J/mol.} = 2.34 \times 10^2 \text{ kJ/mol.}$

c.   [Ne]$3s^2 3p^5$:   3s   3p
                                ($\uparrow\downarrow$)   ($\uparrow\downarrow$)($\uparrow\downarrow$)($\uparrow$)

64. *Refer to Sections 6.5, 6.7, and 6.8.*

a. 17 electrons, therefore 17 protons, thus the element is **Cl**.

b. Since ionization energy decreases going down a group, the element is **Pb**.

c. The +2 ion has 23 electrons, therefore the parent atom has 25 and $Z = 25$, which is **Mn**.

d. Since atomic radii increase going down a group, the smallest would be **Li**.

e. Ionization energy increases from left to right across a period, so the element with the greatest ionization energy is **Kr**.

66. *Refer to Section 6.2.*

a. Energy is absorbed when the electrons are excited from lower levels to a higher ones, thus energy is absorbed for **transitions 2 and 4**.

b. Energy is emitted when the electron relaxes from higher level to a lower one, thus energy is emitted for **transitions 1 and 3**.

c. We cannot answer this question without knowing what the element is since one must know the number of electrons to determine the ground state configuration. If we assume the atom to be hydrogen, then **transition 1** involves the ground state.

d. The transition with largest energy difference (see part (a) and problem 12 above) will absorb the most energy, thus **transition 2**.

e. The transition with largest energy difference (see part (b) and problem 12 above) will emit the most energy, thus **transition 1**.

68. *Refer to Section 6.5.*

Locate the element of interest on the periodic table. Move up one period and then move to the extreme right to find the appropriate noble gas. Which period is the element of interest in? That is the s-orbital from which you are to start filling. Then fill until you run out of electrons.

a. Li   [He] $2s^1$

b. Ra   [Rn] $7s^2$

c. Sc   [Ar] $4s^2 3d^1$

d. Sb   [Kr] $5s^2 4d^{10} 5p^3$

e. Ca   [Ar] $4s^2$

---

70. Refer to Section 6.2 and Figures 6.1 and 6.7.

a. There are two main differences between the Bohr model and the quantum mechanical model.
1.   The kinetic energy of an electron is inversely related to the space it occupies.
2.   You cannot know the position of an electron at any given instant.

b. Frequency and wavelength are inversely related. If frequency is short, wavelength is long and vice versa.

c. The three p orbitals lie along a different axis, x, y, or z. They are all shaped the same, like two small balloons with the knot ends joined. This shape is often referred to as dumbbell shaped.

---

72. Refer to Sections 6.1 – 6.5 and Example 6.7.

a. This statement is **true**. Photons with very short wavelengths are high energy.

b. This statements is **false**. The energy of an electron is inversely proportional to n^2, not ℓ.
$$E_n = -\frac{R_H}{n^2}$$

c. This statements is **false**. Electrons start entering the 5th principal level **before** the fourth is filled. The order of filling is [Kr] 5s 4d 5p …

---

74. Refer to Sections 6.6 and 6.7.

a. True

b. True

c. False   Energy is absorbed when an electron is removed from an atom.

---

76. Refer to Section 6.2.

For a one electron species, the ground state is $n = 1$, so the first excited state is $n = 2$.

$$E = \frac{-BZ^2}{n^2} = \frac{-(2.180 \times 10^{-18} \text{ J})(3)^2}{2^2} = -4.905 \times 10^{-18} \text{ J}$$

$$E = -4.905 \times 10^{-18} \text{ J} \times \frac{1 \text{ kJ}}{1000 \text{ J}} \times \frac{6.022 \times 10^{23}}{1 \text{ mol.}} = -2.954 \times 10^3 \text{ kJ/mol.}$$

E is the energy of the electron. The energy needed to ionize (or remove) an electron is $+2.954 \times 10^3$ kJ/mol.

---

77. Refer to Sections 6.1 and 6.2.

$$\lambda = \frac{hc}{\Delta E} \quad \text{and} \quad \Delta E = -R_H \left[\frac{1}{n_{hi}^2} - \frac{1}{n_{lo}^2} \right]$$

Substituting for ΔE and $n_{lo} = 2$, we get: $\lambda = \dfrac{hc}{-R_H \left[\dfrac{1}{n_{hi}^2} - \dfrac{1}{2^2} \right]} = -\dfrac{hc}{R_H} \dfrac{1}{\left[\dfrac{1}{n_{hi}^2} - \dfrac{1}{4} \right]}$

$$\lambda = -\frac{(6.626 \times 10^{-34} \text{ J} \cdot \text{s})(2.998 \times 10^8 \text{ m/s})}{2.180 \times 10^{-18} \text{ J}} \frac{1}{\left[\dfrac{1}{n_{hi}^2} - \dfrac{1}{4} \right]} = -9.112 \times 10^{-8} \text{ m} \frac{1}{\left[\dfrac{1}{n_{hi}^2} - \dfrac{1}{4} \right]}$$

$$\lambda = 9.112 \times 10^{-8} \text{ m} \frac{1}{\left[\dfrac{1}{4} - \dfrac{1}{n_{hi}^2} \right]} = 9.112 \times 10^{-8} \text{ m} \frac{4n_{hi}^2}{\dfrac{4n_{hi}^2}{4} - \dfrac{4n_{hi}^2}{n_{hi}^2}}$$

$$\lambda = 9.112 \times 10^{-8} \text{ m} \frac{4n_{hi}^2}{n_{hi}^2 - 4} = \frac{(3.645 \times 10^{-7} \text{ m})n_{hi}^2}{n_{hi}^2 - 4}$$

Converting to nanometers: $\lambda = \dfrac{(3.645 \times 10^2 \text{ nm})n_{hi}^2}{n_{hi}^2 - 4}$

---

78. Refer to Section 6.3 and Table 6.3.

---

n	1		2		
ℓ	0	1	0	1	2
sublevel	1s	1p	2s	2p	2d
m_ℓ	0 1	0 1 2	0 1	0 1 2	0 1 2 3

electron configuration with eight electrons: $1s^4 1p^4$

a. If there were 3 values for $\mathbf{m}_s$, then each orbital could hold 3 electrons.
    s-sublevel: one orbital ($\mathbf{m}_\ell = 0$), 3 electrons.
    p-sublevel: 3 orbitals ($\mathbf{m}_\ell = -1, 0, 1$), 9 electrons.
    d-sublevel: 5 orbitals ($\mathbf{m}_\ell = -2, -1, 0, 1, 2$), 15 electrons.

b. The $\mathbf{n} = 3$ level could hold 27 electrons; 3 in the s-sublevel, 9 in the p-sublevel and 15 in the d-sublevel.

c. AN = 8: $1s^3\ 2s^3\ 2p^2$
    AN = 17: $1s^3\ 2s^3\ 2p^9\ 3s^2$

Calculate the energy of the light. The difference between that energy and the kinetic energy is the energy needed to eject the electron (E_{min}).

a. $$E = \frac{hc}{\lambda} = \frac{(6.626 \times 10^{-34}\ \text{J} \cdot \text{s})(2.998 \times 10^8\ \text{m/s})}{5.40 \times 10^{-7}\ \text{m}} = 3.68 \times 10^{-19}\ \text{J}$$

$E_{min} = 3.68 \times 10^{-19}\ \text{J} - 2.60 \times 10^{-20}\ \text{J} = 3.42 \times 10^{-19}\ \text{J}$

While the calculations above used the longer wavelength, we also could have used the shorter wavelength and gotten the same answer (within significance).

$$E = \frac{hc}{\lambda} = \frac{(6.626 \times 10^{-34}\ \text{J} \cdot \text{s})(2.998 \times 10^8\ \text{m/s})}{4.00 \times 10^{-7}\ \text{m}} = 4.97 \times 10^{-19}\ \text{J}$$

$E_{min} = 4.97 \times 10^{-19}\ \text{J} - 1.54 \times 10^{-19}\ \text{J} = 3.43 \times 10^{-19}\ \text{J}$

b. Since longer wavelengths correspond to lower energies, E_{min} would correspond to the longest wavelength.

$$\lambda = \frac{hc}{E_{min}} = \frac{(6.626 \times 10^{-34}\ \text{J} \cdot \text{s})(2.998 \times 10^8\ \text{m/s})}{3.42 \times 10^{-19}\ \text{J}} = 5.81 \times 10^{-7}\ \text{m} = 581\ \text{nm}$$

Chapter 7: Covalent Bonding

> 2. *Refer to Section 7.1 and Examples 7.1 and 7.2.*

Add up the total number of valence electrons. Draw the skeletal structure, then add the electrons (remember that each bond represents two electrons).

a. N: 5 valence electrons
3H: 3 x 1 valence electrons
total: 8 electrons

$$H-\underset{\underset{H}{|}}{N}-H \longrightarrow H-\underset{\underset{H}{|}}{\overset{\cdot\cdot}{N}}-H$$

b. Kr: 8 valence electrons
2F: 2 x 7 valence electrons
total: 22 electrons

$$F-Kr-F \longrightarrow :\overset{\cdot\cdot}{\underset{\cdot\cdot}{F}}-\overset{\cdot\cdot}{\underset{\cdot}{Kr}}-\overset{\cdot\cdot}{\underset{\cdot\cdot}{F}}:$$

c. N: 5 valence electrons
O: 6 valence electrons
+1: -1 valence electron
total: 10 electrons

$$\left[N-O \right]^{+} \longrightarrow \left[:N\equiv O: \right]^{+}$$

d. Br: 7 valence electrons
2O: 2 x 6 valence electrons
-1: 1 valence electron
total: 20 electrons

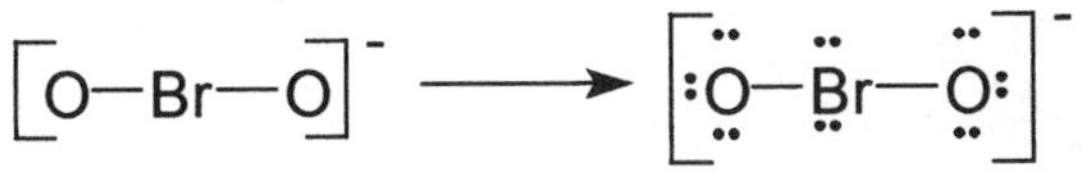

$$\left[O-Br-O \right]^{-} \longrightarrow \left[:\overset{\cdot\cdot}{\underset{\cdot\cdot}{O}}-\overset{\cdot\cdot}{Br}-\overset{\cdot\cdot}{\underset{\cdot\cdot}{O}}: \right]^{-}$$

> **4. *Refer to Section 7.1 and Examples 7.1, 7.2, and 7.4.***

Add up the total number of valence electrons. Draw the skeletal structure, then add the electrons (remember that each bond represents two electrons).

a. P: 5 valence electrons
   4Cl: 4 x 7 valence electrons
   +1: -1 valence electron
   total: 32 electrons

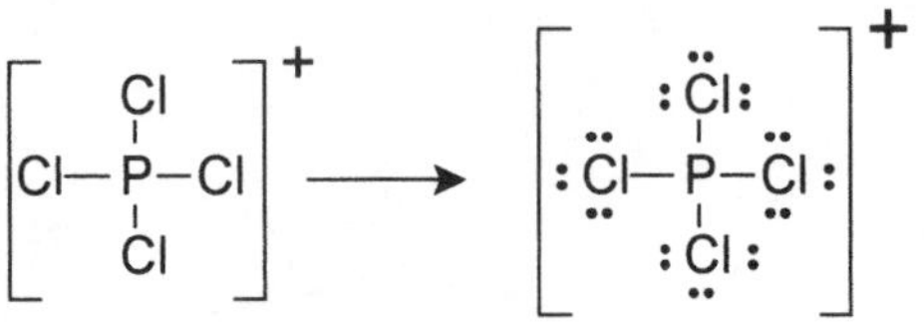

b. Br: 7 valence electrons
   2F: 2 x 7 valence electrons
   +1: -1 valence electron
   total: 20 electrons

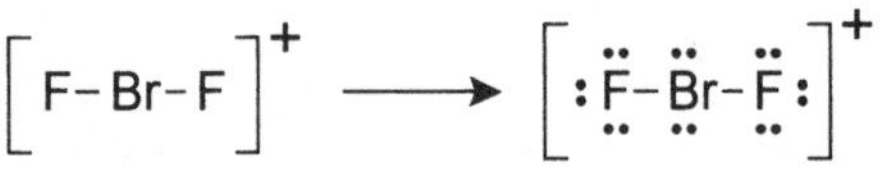

c. 3I: 3 x 7 valence electrons
   -1: 1 valence electron
   total: 22 electrons

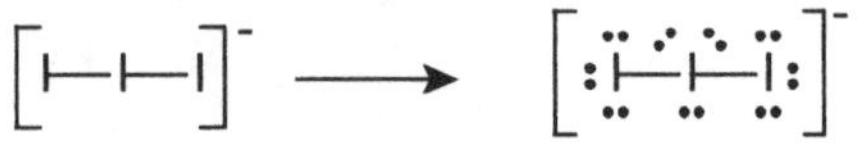

d. Se: 6 valence electrons
   6Br: 6 x 7 valence electrons
   total: 48 electrons

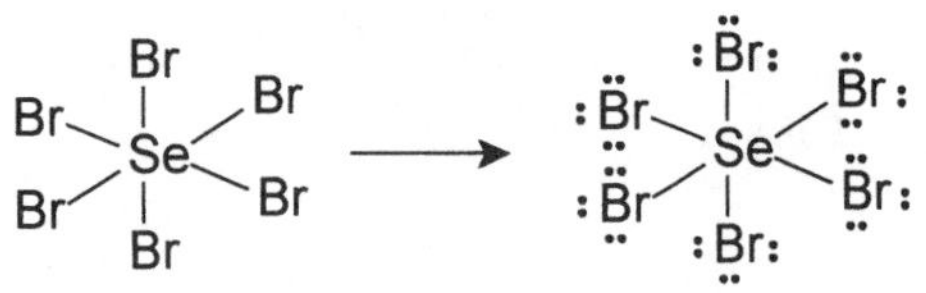

6. *Refer to Section 7.1 and Examples 7.1, 7.2, and 7.4.*

Add up the total number of valence electrons. Draw the skeletal structure, then add the electrons (remember that each bond represents two electrons). Recall the exceptions to the octet rule (for H, I and B).

a. 2C: 2 x 4 valence electrons
    -2: 2 x 1 valence electrons
    total: 10 electrons

$$\left[C{-}C \right]^{2-} \longrightarrow \left[:C{\equiv}C: \right]^{2-}$$

b. N: 5 valence electrons
    F: 7 valence electrons
    O: 6 valence electrons
    total: 18 electrons

$$O{-}N{-}F \longrightarrow :\ddot{O}{=}\ddot{N}{-}\ddot{F}:$$

c. Br: 7 valence electrons
    4F: 4 x 7 valence electrons
    +1: -1 valence electron
    total: 34 electrons

d. N: 5 valence electrons
    3I: 3 x 7 valence electrons
    total: 26 electrons

8. *Refer to Section 7.1 and Example 7.1.*

Add up the total number of valence electrons. Draw the skeleton structure, then add the electrons (remember that each bond represents two electrons). Subtract from the number of valence electrons those used for the skeleton structure. Then give each atom an octet (except H) by assigning to it an unshared pair or a multiple bond.

C: 4 valence electrons
O: 6 valence electrons
H: 1 valence electron
+1: -1 valence electron
total: 10 electrons

$$[H-C-O]^+$$ Valence electrons left: $10 - 4 = 6$

In order to give C and O octets, 5 unshared pairs of electrons (10 electrons) are needed. Since there are only 6 electrons, the octets must be filled using multiple bonds between C and O (recall that H has all the electrons it needs around it).

$$[H-C\equiv O:]^+$$

Note that a double bond between the C and O atoms would not allow C to have an octet.

---

10. Refer to Section 7.1 and Problem 8 above.

a. 2C: 2 x 4 valence electrons
    4H: 4 x 1 valence electrons
    O: 6 valence electrons
    total: 18 electrons

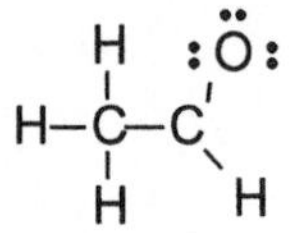

Note that this structure does not have complete octets around the carbon atoms. To complete the octets, move a pair of electrons from the oxygen to form a (double) bond between the carbon and oxygen.

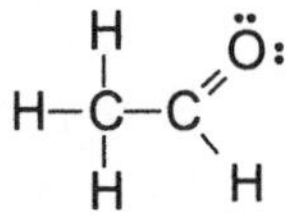

b. S: 6 valence electrons
    2H: 2 x 1 valence electrons
    3O: 3 x 6 valence electrons
    total: 26 electrons

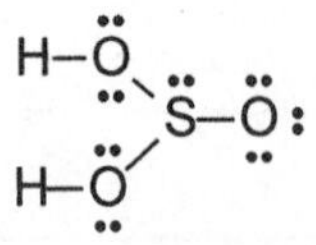

c. 2C: 2 x 4 valence electrons
    2F: 2 x 7 valence electrons
    2Cl: 2 x 7 valence electrons
    total: 36 electrons

118

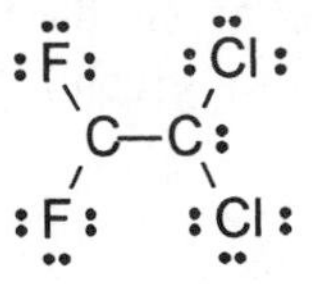

:F: :Cl:
 \   /
  C—C:
 /   \
:F: :Cl:

Note that this structure does not have complete octets around both carbon atoms. To complete the octets, move a pair of electrons from one carbon to form a (double) bond between the two carbon atoms.

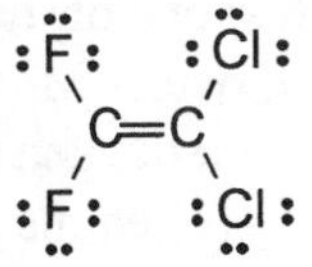

:F: :Cl:
 \   /
  C=C
 /   \
:F: :Cl:

---

12. Refer to Section 7.1 and Example 7.1.

2C: 2 x 4 valence electrons
5H: 5 x 1 valence electrons
2O: 2 x 6 valence electrons
N: 5 valence electrons
total: 30 electrons

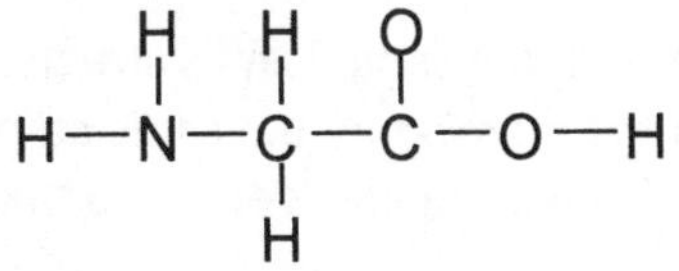

```
    H   H   O
    |   |   ||
H—N—C—C—O—H
        |
        H
```

From the information given, the structure must be that on the left.

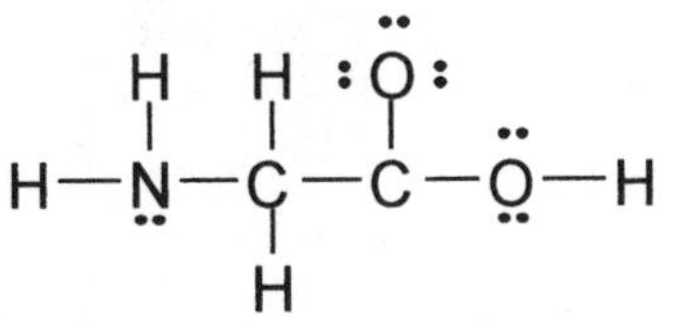

```
    H   H  :O:
    |   |   |
H—N—C—C—O—H
    |   |
        H
```

Note that this structure does not have a complete octet around one of the carbon atoms. To complete the octet, move a pair of electrons from the terminal oxygen to form a C=O double bond.

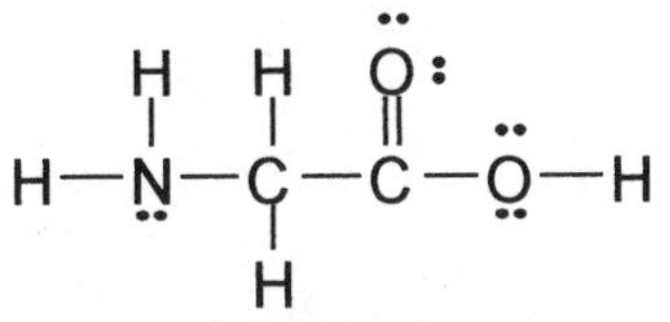

```
    H   H   O:
    |   |   ||
H—N—C—C—O—H
    |   |
        H
```

This structure provides octets for all the atoms and provides that all formal charges are zero.

---

14. Refer to Section 7.1.

3C: 3 x 4 valence electrons
2H: 2 x 1 valence electrons
O: 6 valence electrons
total: 24 electrons

119

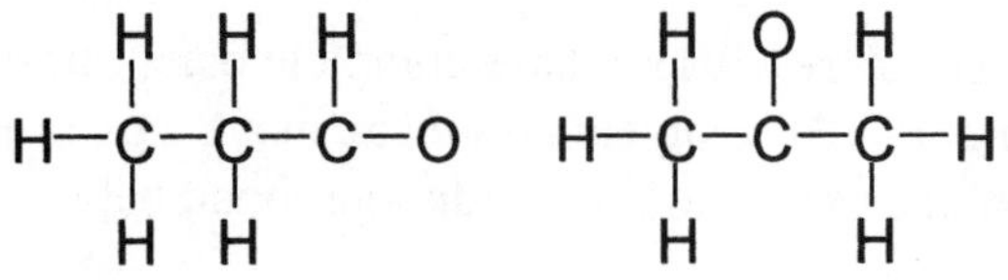

H H H
H—C—C—C—O
H H

H O H
H—C—C—C—H
H H

From the information given, the two structures must be those given to the left.

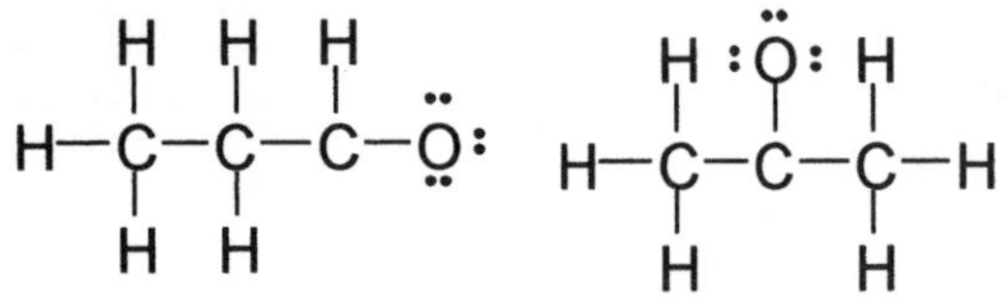

H H H
H—C—C—C—Ö:
H H

H :Ö: H
H—C—C—C—H
H H

In each of the structures, one of the C's does not have a full octet. Move a pair of electrons from one oxygen to form a (double) bond between the carbon and the oxygen.

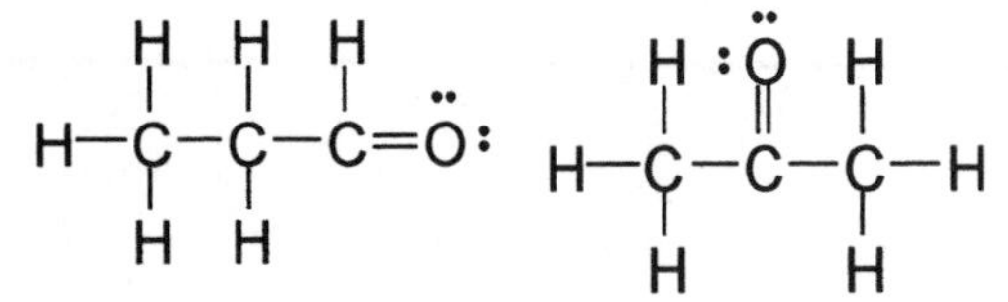

H H H
H—C—C—C=Ö:
H H

H :Ö H
‖
H—C—C—C—H
H H

---

16. Refer to Section 7.1.

Draw the Lewis structure of the ion, then draw a molecule with the same Lewis structure. This can be facilitated by increasing the atomic number of the atom(s) by the same amount as the negative charge (recall that the negative charge results from extra electrons, and that molecules have equal numbers of electrons and protons.

a. O: 6 valence electrons
    H: 1 valence electron
    -1: 1 valence electron
    total: 8 electrons

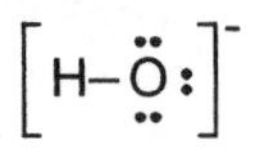

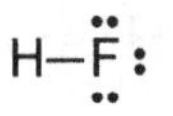

$\left[\text{H—Ö:} \right]^-$

H—F̈:

b. 2O: 2 x 6 valence electrons
    -2: 2 x 1 valence electrons
    total: 14 electrons

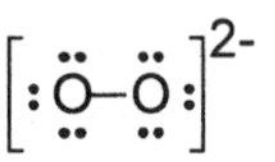

$\left[\text{:Ö—Ö:} \right]^{2-}$

:F̈—F̈:

c. C: 4 valence electrons
    N: 5 valence electrons
    -1: 1 valence electron
    total: 10 electrons

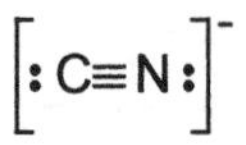

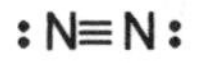

$\left[\text{:C≡N:} \right]^-$

:N≡N:

d. S: 6 valence electrons
    4O: 4 x 6 valence electrons
    -2: 2 x 1 valence electrons
    total: 32 electrons

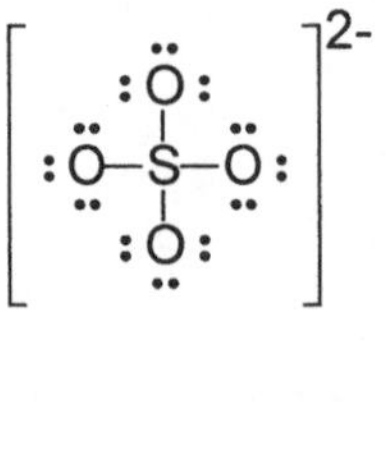

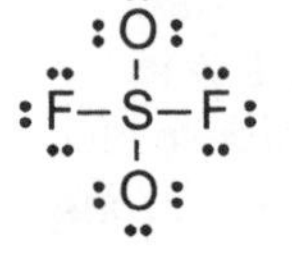

$\left[\begin{matrix} \text{:Ö:} \\ \text{:O—S—O:} \\ \text{:O:} \end{matrix} \right]^{2-}$

:Ö:
:F̈—S—F̈:
:O:

18. *Refer to Section 7.1 and Examples 7.1, 7.2, and 7.4.*

Add up the total number of valence electrons. Draw the skeletal structure, then add the electrons (remember that each bond represents two electrons). Recall the exceptions to the octet rule (for H, I and B).

a. 2P: 2 x 5 valence electrons
7O: 7 x 6 valence electrons
-4: 4 x 1 valence electrons
total: 56 electrons

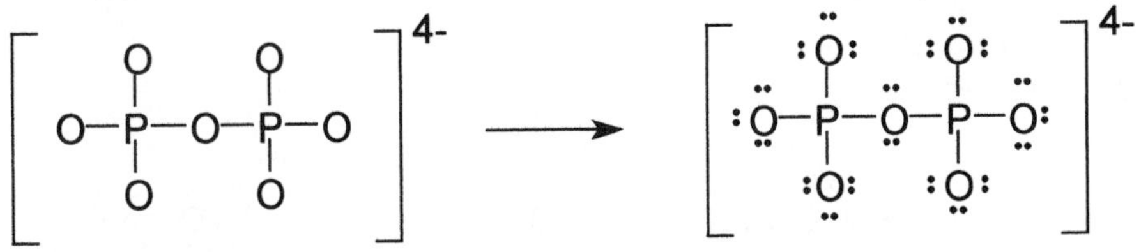

b. H: 1 valence electron
O: 6 valence electrons
Br: 7 valence electrons
total: 14 electrons

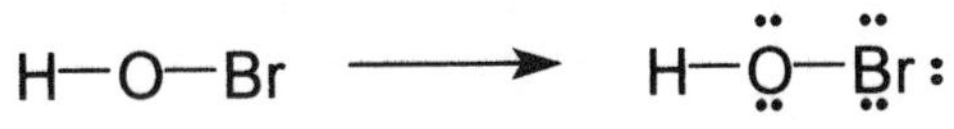

c. N: 5 valence electrons
F: 7 valence electrons
2Br: 2 x 7 valence electrons
total: 26 electrons

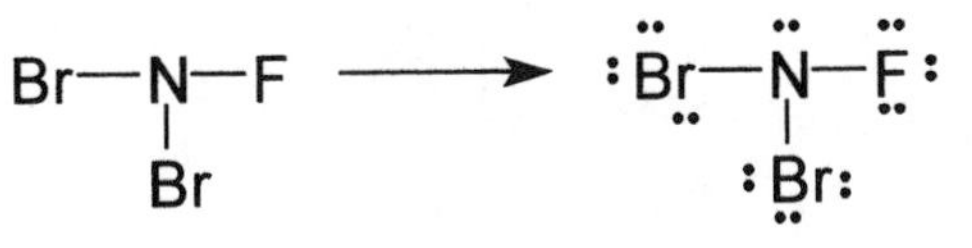

d. I: 7 valence electrons
4F: 4 x 7 valence electrons
-1: 1 valence electron
total: 36 electrons

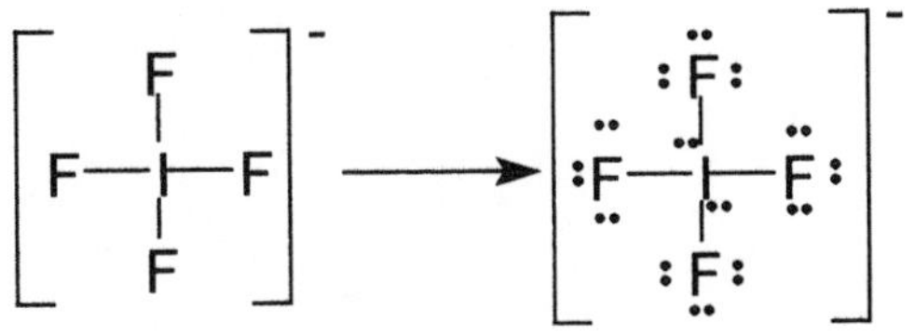

20. Refer to Section 7.1.

Parts b-d have odd numbers of electrons. Consequently, the final structures will have an unpaired electron.

a. Be: 2 valence electrons
2H: 2 x 1 valence electrons
total: 4 electrons

H–Be–H

The skeletal structure is the final Lewis structure.

b. C: 4 valence electrons
O: 6 valence electrons
-1: 1 valence electron
total: 11 electrons

$\left[\text{C–O}\right]^-$ $\left[:\overset{\cdot}{\text{C}}=\overset{\cdot\cdot}{\text{O}}:\right]^-$

c. S: 6 valence electrons
O: 2 x 6 valence electrons
-1: 1 valence electron
total: 19 electrons

$\left[\text{O–S–O}\right]^-$ $\left[:\overset{\cdot\cdot}{\underset{\cdot\cdot}{\text{O}}}-\overset{\cdot\cdot}{\underset{\cdot}{\text{S}}}-\overset{\cdot\cdot}{\underset{\cdot\cdot}{\text{O}}}:\right]^-$

d. C: 4 valence electrons
3H: 3 x 1 valence electrons
total: 7 electrons

$\underset{\text{H}}{\overset{|}{\text{H–C–H}}}$ $\underset{\text{H}}{\overset{\cdot}{\text{H–C–H}}}$

---

22. Refer to Section 7.1 and Example 7.3.

Write the Lewis structure for the compound, then draw resonance forms by changing the positions of electron pairs. Remember that the skeletal structure cannot change, and the new structures must also abide by the rules for Lewis structures.

a. $\left[:\overset{\cdot\cdot}{\underset{\cdot\cdot}{\text{O}}}-\overset{\cdot\cdot}{\text{N}}=\overset{\cdot\cdot}{\text{O}}:\right]^-$ ⟷ $\left[:\overset{\cdot\cdot}{\text{O}}=\overset{\cdot\cdot}{\text{N}}-\overset{\cdot\cdot}{\underset{\cdot\cdot}{\text{O}}}:\right]^-$

b. $:\overset{\cdot\cdot}{\underset{\cdot\cdot}{\text{N}}}-\text{N}≡\text{O}:$ ⟷ $:\overset{\cdot\cdot}{\text{N}}=\text{N}=\overset{\cdot\cdot}{\text{O}}:$ ⟷ $:\text{N}≡\text{N}-\overset{\cdot\cdot}{\underset{\cdot\cdot}{\text{O}}}:$

c. $\left[\underset{\text{H–C}=\overset{\cdot\cdot}{\underset{\cdot\cdot}{\text{O}}}}{\overset{:\overset{\cdot\cdot}{\text{O}}:}{\overset{|}{}}}\right]^-$ ⟷ $\left[\underset{\text{H–C}-\overset{\cdot\cdot}{\underset{\cdot\cdot}{\text{O}}}:}{\overset{:\text{O}:}{\overset{||}{}}}\right]^-$

122

24. Refer to Section 7.1 and Example 7.3.

a. Add up the total number of valence electrons. Draw the skeleton structure, then add the electrons (remember that each bond represents two electrons). Subtract from the number of valence electrons those used for the skeleton structure. Then give each atom an octet (except H) by assigning to it an unshared pair or a multiple bond.

2C: 2 x 4 valence electrons
4O: 4 x 6 valence electrons
-2: 2 x 1 valence electrons
total: 34 electrons

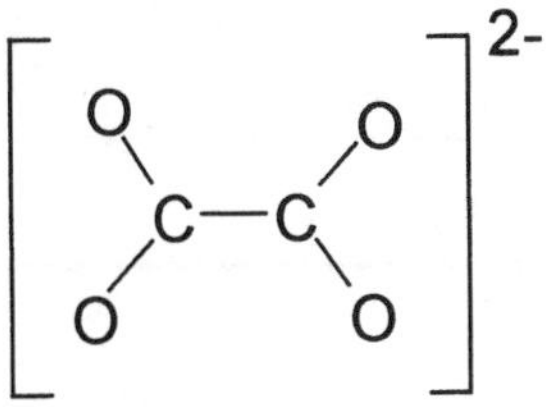

Valence electrons left: 34 – 10 = 24

Try to fill the octet of each atom using unshared pairs. There are 12 unshared pairs available.

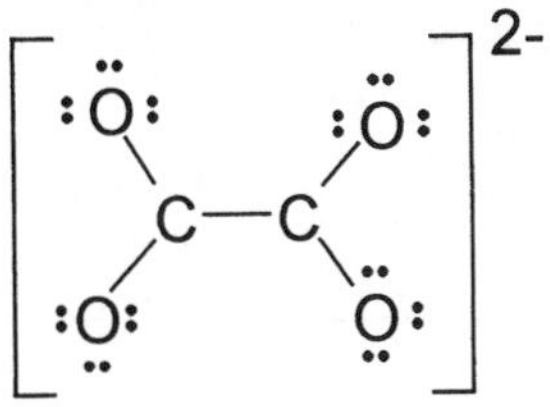

Notice that there are not enough unshared pairs to give carbon an octet.

To complete the octets, move a pair of electrons from each of two oxygens to form (double) bonds between the carbon and oxygen.

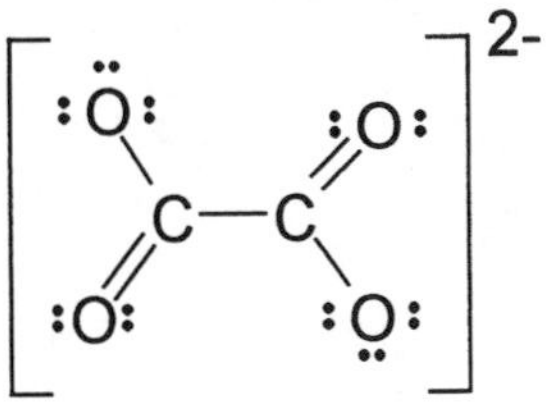

b.   The three resonance structures are drawn by moving the double bonds and the unshared pairs so that each atom still has an octet.

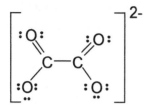

c.   Resonance forms differ only in distribution of electrons, not in arrangement of atoms. Thus, the structure given in c is **not** a resonance form.

---

26. *Refer to Section 7.1 and Example 7.3.*

Write the Lewis structure for borazine, then draw resonance forms by changing the positions of electron pairs. Remember that the skeletal structure cannot change, and the new structures must also abide by the rules for Lewis structures.

3N: 3 x 5 valence electrons
3B: 3 x 3 valence electrons
6H: 6 x 1 valence electrons
total: 30 electrons

---

28. *Refer to Section 7.1 and Table 7.2.*

Draw the Lewis structure. Then apply the formula: $C_f = e_{valence} - (e_{unshared} + \frac{1}{2}(e_{bonding}))$
[formal charge = valence electrons - (unshared electrons + ½(bonding electrons))]
Note that this is the formula given in the text as: $C_f = X - (Y + Z/2)$.

a.   N: 5 valence electrons
2H: 2 x 1 valence electrons
-1: 1 valence electron
total: 8 electrons
$C_f = 5 - (4 + \frac{1}{2}(4)) = -1$

$$\left[H - \overset{..}{\underset{..}{N}} - H \right]^{-}$$

b. Br: 7 valence electrons
    3F: 3 x 7 valence electrons
    total: 28 electrons
    $C_f = 7 - (4 + \frac{1}{2}(6)) = 0$

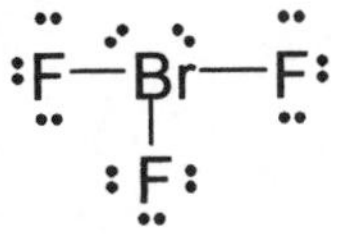

c. H: 1 valence electron
    O: 6 valence electrons
    Cl: 7 valence electrons
    total: 14 electrons
    $C_f = 6 - (4 + \frac{1}{2}(4)) = 0$

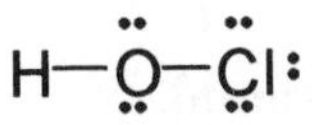

---

30. *Refer to Section 7.2, Examples 7.5 and 7.6, Figure 7.5, and Table 7.3.*

In both structures, all 3 oxygen atoms have a formal charge of -1. $C_f = 6 - (6 + \frac{1}{2}(2)) = -1$. Therefore, the formal charges on the sulfur atoms will determine the better structure.

Structure I
    Central S: $C_f = 6 - (2 + \frac{1}{2}(6)) = +1$
    Chain S: $C_f = 6 - (4 + \frac{1}{2}(4)) = 0$

Structure II
    Central S: $C_f = 6 - (0 + \frac{1}{2}(8)) = +2$
    Terminal S: $C_f = 6 - (6 + \frac{1}{2}(2)) = -1$

Structure I is the better choice of structures since the formal charges on those sulfur atoms are as close to zero as possible.

---

32. *Refer to Section 7.2, Examples 7.5 and 7.6, Figure 7.5, and Table 7.3.*

Draw the Lewis structure of the compound and determine the number of bonded groups and the number of electron pairs. Then use Table 7.3 to assign the geometry.

a. $SO_2 \Rightarrow$:Ö–S̈=Ö:

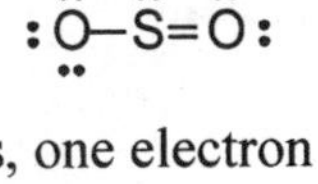

    2 bonded groups, one electron pair, thus AX_2E and **bent**.

b. $BeCl_2 \Rightarrow$ $:\overset{..}{\underset{..}{Cl}}-Be-\overset{..}{\underset{..}{Cl}}:$

2 bonded groups, no electron pairs, thus AX_2 and **linear**.

c. $SeCl_4 \Rightarrow$ $:\overset{..}{\underset{..}{Cl}}-Se-\overset{..}{\underset{..}{Cl}}:$ with $:\overset{..}{Cl}:$ above and $:\overset{}{\underset{..}{Cl}}:$ below

4 bonded groups, one electron pair, thus AX_4E and **seesaw**.

d. $PCl_5 \Rightarrow$ Lewis structure with P center bonded to five Cl atoms

5 bonded groups, no electron pairs, thus AX_5 and **triangular bipyramid**.

---

34. *Refer to Section 7.2, Examples 7.5 and 7.6, Figures 7.4 and 7.5, and Table 7.3.*

Draw the Lewis structure of the compound and determine the number of bonded groups and the number of electron pairs. Then use Table 7.3 to assign the geometry.

a. $NNO \Rightarrow$ $:\overset{..}{N}=N=\overset{..}{O}:$

2 bonded groups, no electron pairs, thus AX_2 and **linear**.

b. $ONCl \Rightarrow$ $:\overset{..}{O}=\overset{..}{N}-\overset{..}{\underset{..}{Cl}}:$

2 bonded groups, one electron pair, thus AX_2E and **bent**.

c. $NH_4^+ \Rightarrow$ $\left[\begin{array}{c} H \\ | \\ H-N-H \\ | \\ H \end{array}\right]^+$

4 bonded groups, no electron pairs, thus AX_4 and **tetrahedron**.

d. $O_3 \Rightarrow$ $:\overset{..}{\underset{..}{O}}-\overset{..}{O}=\overset{..}{O}:$

2 bonded groups, one electron pair, thus AX_2E and **bent**.

Draw the Lewis structure of the compound and determine the number of bonded groups and the number of electron pairs. Then use Table 7.3 to assign the geometry.

a. ClF_2^- $\Rightarrow$ $\left[:\ddot{F}-\ddot{Cl}-\ddot{F}: \right]^-$

2 bonded groups, three electron pairs, thus AX_2E_3 and **linear**.

b. SeF_5Br $\Rightarrow$

6 bonded groups, no electron pairs, thus AX_6 and **octahedral**.

c. SO_3^{2-} $\Rightarrow$ $\left[:\ddot{O}-\ddot{S}=\ddot{O}: \atop \quad :\ddot{O}: \right]^{2-}$

3 bonded groups, one electron pair, thus AX_3E and **triangular pyramid**.

d. BrO_2^- $\Rightarrow$ $\left[:\ddot{O}=\ddot{Br}-\ddot{O}: \right]^-$

2 bonded groups, two electron pairs, thus AX_2E_2 and **bent**.

Draw the Lewis structure, determine the geometry and, from that, the ideal bond angles.

a. $\ddot{O}=C=\ddot{O}$ AX_2, linear   **180°**

b. $H-B-H$ (with H below) AX_3, triangular planar   **120°**

c.

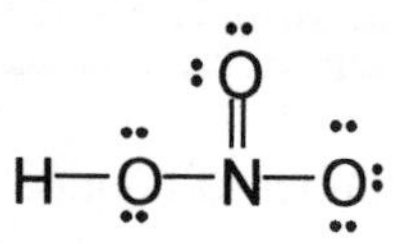

O(left): AX_2E_2, bent   **109.5°**
N: AX_3, triangular planar   **120°**

d.

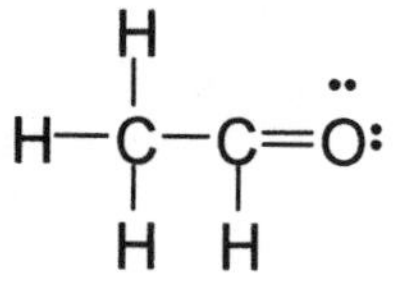

C(left): AX_4, tetrahedron   **109.5°**
C(right): AX_3, triangular   **120°**
planar

---

40. Refer to Section 7.2, Figure 7.5, and Table 7.3.

a.

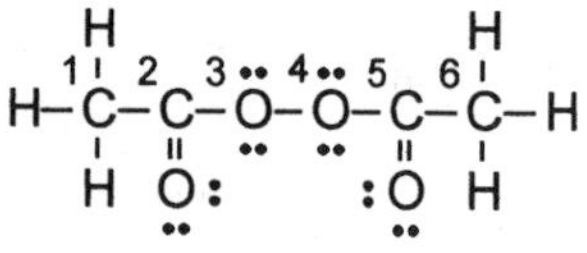

b. C^1, C^6: AX_4, tetrahedron   **109.5°**
    C^2, C^5: AX_3, triangular planar   **120°**
    O^3, O^4: AX_2E_2, bent   **109.5°**

---

42. Refer to Section 7.2.

Draw the Lewis structure for the molecule. Then determine the geometry of the indicated atom and from that, the bond angles.

1	AX_3	bent	**120°**
2	AX_2E_2	bent	**109.5°**
3	AX_3	triangular planar	**120°**

---

44. Refer to Sections 6.8 and 7.3, and Example 7.7.

Consider the electronegativity of the bonded atoms to determine if there are dipoles. If there is a dipole, consider if it is canceled by a dipole in the opposite direction. Then consider the geometry of the molecule to determine if there is a net dipole.
The molecules with bonded atoms which are not symmetrical about the central atom are **(a)** SO_2, and **(c)** $SeCl_4$. These molecules have net dipoles.

---

46. Refer to Section 7.2 and Example 7.7.

First determine if the central atom has an octet. Then consider the electronegativity of the atoms to determine if there are dipoles. Then consider the geometry of the molecule to determine if there is a net dipole.

128

All 4 molecules have octets about the central atom.

a. There is only one dipole, between the central N and the O. Thus, there is a net dipole and **the molecule is a dipole**.

b. There are 2 dipoles, between N-O and N-Cl. Since this molecule is bent, there is a net dipole, and **the molecule is a dipole**.

c. There are 4 dipoles, between each of the N-H bonds. Since the molecule is tetrahedron, the 4 dipoles cancel each other out; there is no net dipole, so **the molecule is not a dipole**.

d. Since all the atoms are identical, there are no dipoles and one would thus assume there is no net dipole. However, a dipole depends only on an unsymmetrical distribution of electrons. Since this molecule is bent, that criterion is met, and **the molecule is a dipole**.

---

48. Refer to Section 7.3.

Draw the Lewis structures of the molecules. Then consider the electronegativity of the atoms to determine if there are dipoles. Then consider the geometry of the molecule to determine if there is a net dipole.

cis

The *cis* structure has a dipole along each N-F bond. The two dipoles don't cancel, so this molecule has a net dipole and is **polar**.

trans

The *trans* structure also has a dipole along each N-F bond, but these two dipoles do cancel, so this molecule does not have a net dipole and is **not polar**.

---

50. Refer to Section 7.4, Example 7.8, and Problem 32 (above).

Recall that the total number of groups (bonded atoms and electron pairs) around the central atom is equal to the number of orbitals that hybridized. Furthermore, the sum of the superscripts in the hybrid orbital notation gives the total number of hybrid orbitals.

a. SO_2 AX_2E 3 groups sp^2

b. $BeCl_2$ AX_2 2 groups sp

c. $SeCl_4$ AX_4E 5 groups sp^3d

d. PCl_5 AX_5 5 groups sp^3d

Recall that the total number of groups (bonded atoms and electron pairs) around the central atom is equal to the number of orbitals that hybridized. Furthermore, the sum of the superscripts in the hybrid orbital notation gives the total number of hybrid orbitals.

a. NNO AX_2 2 groups sp

b. ONCl AX_2E 3 groups sp^2

c. NH_4^+ AX_4 4 groups sp^3

d. O_3 AX_2E 3 groups sp^2

Recall that the total number of groups (bonded atoms and electron pairs) around the central atom is equal to the number of orbitals that hybridized. Furthermore, the sum of the superscripts in the hybrid orbital notation gives the total number of hybrid orbitals.

a. ClF_2^- AX_2E_3 5 groups sp^3d

b. SeF_5Br AX_6 6 groups sp^3d^2

c. SO_3^{2-} AX_3E 4 groups sp^3

d. BrO_2^- AX_2E_2 4 groups sp^3

Draw the Lewis structure for each of the molecules. Then determine the number of electron pairs (whether in a bond or as a lone pair) and the hybridization.

a. AX_4E_2 6 e⁻ pairs sp^3d^2

b.

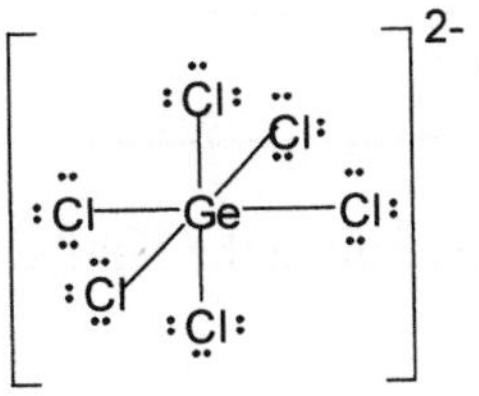

$$\left[\begin{array}{c} :\overset{\cdot\cdot}{Cl}: \quad \overset{\cdot}{Cl}: \\ :\overset{\cdot\cdot}{Cl}-Ge-Cl: \\ :\overset{\cdot}{Cl} \quad :\overset{\cdot\cdot}{Cl}: \end{array}\right]^{2-}$$

AX_6 6 e$^-$ pairs sp^3d^2

c.

$$\left[\begin{array}{c} :\overset{\cdot\cdot}{Cl}: \\ :\overset{\cdot\cdot}{Cl}-Sb-\overset{\cdot\cdot}{Cl}: \\ :\overset{\cdot\cdot}{Cl}: \end{array}\right]^{-}$$

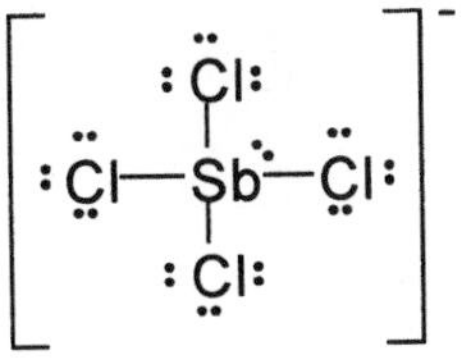

AX_4E 5 e$^-$ pairs sp^3d

---

58. Refer to Section 7.4 and Problem 26 (above).

The Lewis structure for borazine was determined in problem 26. Each B and N is bonded to three other atoms and each has no lone pairs (AX_3). Thus, the hybridization is sp^2.

---

60. Refer to Section 7.4 and Example 7.9.

Draw the Lewis structure for each of the molecules. Then determine the number of electron pairs (shared and unshared) and the hybridization.

a.

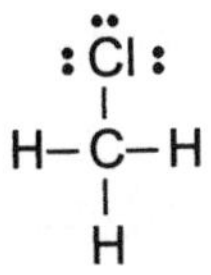

$$\begin{array}{c} :\overset{\cdot\cdot}{Cl}: \\ | \\ H-C-H \\ | \\ H \end{array}$$

AX_4 4 e$^-$ pairs sp^3

b.

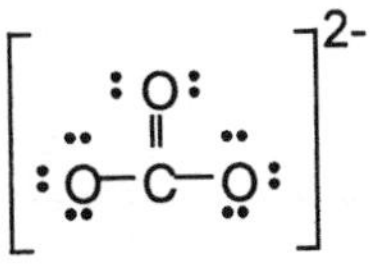

$$\left[\begin{array}{c} :\overset{\cdot\cdot}{O}: \\ \| \\ :\overset{\cdot\cdot}{O}-C-\overset{\cdot\cdot}{O}: \end{array}\right]^{2-}$$

AX_3 3 e$^-$ pairs sp^2

c.

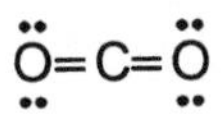

$$\overset{\cdot\cdot}{O}=C=\overset{\cdot\cdot}{O}$$

AX_2 2 e$^-$ pairs sp

d.

$$\begin{array}{c} :O: \\ \| \\ H-C-\overset{\cdot\cdot}{O}-H \end{array}$$

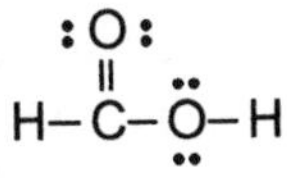

AX_3 3 e$^-$ pairs sp^2

131

Draw the Lewis structure for each of the molecules. Then determine the number of electron pairs (shared and unshared) and the hybridization.

a.

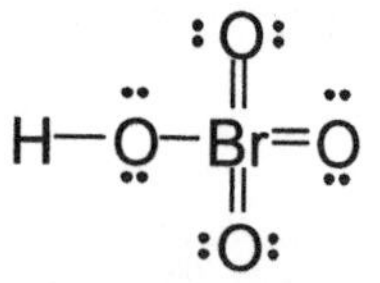

$$:\overset{\cdot\cdot}{O}:$$
$$H-\overset{\cdot\cdot}{\underset{\cdot\cdot}{O}}-Br=\overset{\cdot\cdot}{O}$$
$$:\overset{\cdot\cdot}{O}:$$

AX_4 4 e⁻ pairs sp^3

b.

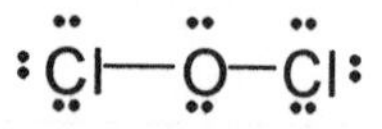

$$:\overset{\cdot\cdot}{\underset{\cdot\cdot}{Cl}}-\overset{\cdot\cdot}{\underset{\cdot\cdot}{O}}-\overset{\cdot\cdot}{\underset{\cdot\cdot}{Cl}}:$$

AX_2E_2 4 e⁻ pairs sp^3

c.

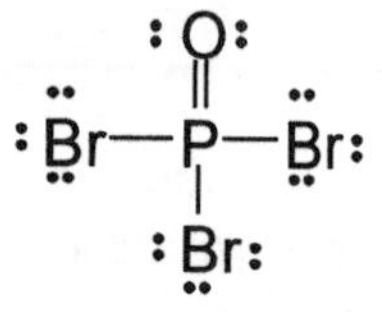

$$:\overset{\cdot\cdot}{O}:$$
$$:\overset{\cdot\cdot}{Br}-\overset{\|}{P}-\overset{\cdot\cdot}{Br}:$$
$$:\overset{\cdot\cdot}{Br}:$$

AX_4 4 e⁻ pairs sp^3

Recall that each single bond is a sigma (σ) bond and each double bond is composed of a sigma (σ) and a pi (π) bond.

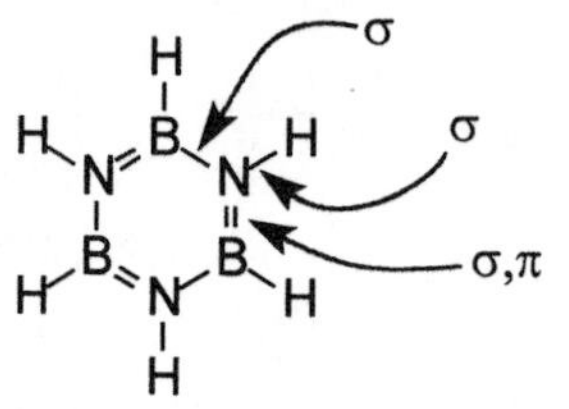

9 single bonds (9 σ) and 3 double bonds (3 σ, and 3 π) adds up to 12 σ-bonds and 3 π-bonds.

a.

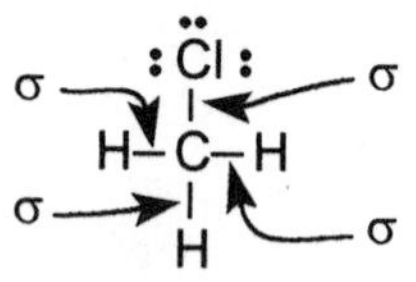

4 σ-bonds

b.

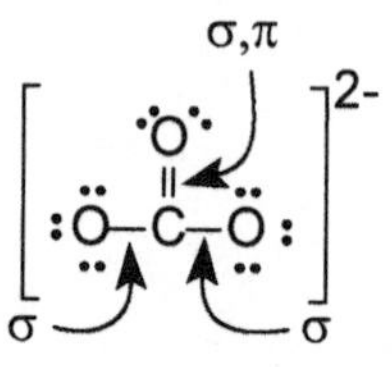

3 σ-bonds, 1 π-bond

132

c.

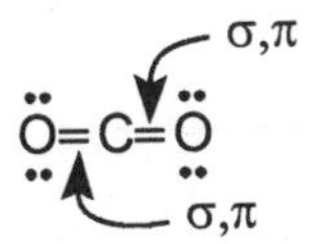

2 σ-bonds, 2 π-bonds

d.

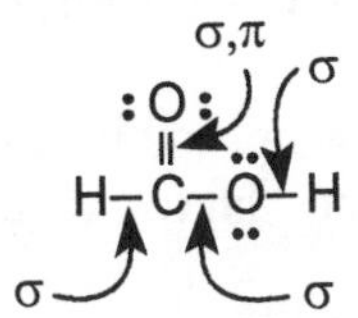

4 σ-bonds, 1 π-bond

---

68. *Refer to Section 7.1.*

Write the skeleton structure from the information given. If there are no Cr-Cr bonds, then there must an O between the two Cr atoms. Since there are no O-O bonds, all O must be bonded to Cr. To be symmetrical, put three O on each Cr.

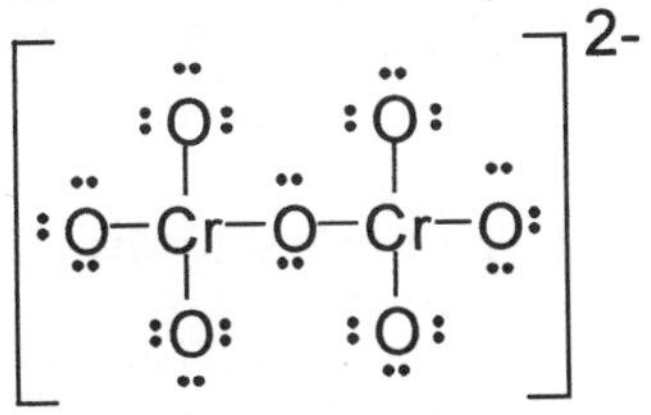

---

70. *Refer to Sections 7.2, 7.3, and 7.4, Tables 7.3 and 7.4, and Figure 7.8.*

For the formula AX_mE_n, m is the number of atoms around Central Atom A and n is the number of electron pairs around A. The geometry is determined by the "Species" formula, as indicated in Table 7.3 and Figure 7.8. The hybridization is also determined by the number of electron pairs (bonding and unshaired, m+n) as shown in Table 7.4. Finally, recall the atoms, X, must be symmetrically disposed around the central atom, A, for the molecule to be nonpolar.

Species	Atoms Around Central Atom A	Unshared Pairs Around A	Geometry	Hybridization	Polarity
AX_2E_2	2	2	bent	sp^3	polar
AX_3	3	0	triangular planar	sp^2	nonpolar
AX_4E_2	4	2	square planar	sp^3d^2	nonpolar
AX_5	5	0	triangular bipyramid	sp^3d	nonpolar

133

As a rule of thumb, the least electronegative atom will be central atom. This can be confirmed by drawing Lewis structures and determining which is the best (obeys all the rules and has the least formal charges).

a. For HCN, recall that H only has single bonds, thus there are two possibilities for the Lewis Structure that obey the octet rule. When H has one bond, the formal charge is 0.

H—N≡C: In this structure, the most electronegative atom, N, has a +1 formal charge and C has a –1 formal charge.

H—C≡N: In this structure, both N and C have formal charges of 0, Thus, it is the best Lewis structure and **C** is the central atom.

b. For NOCl, there are also two possibilities.

:N=O—Cl: Formal charges: Cl: 0; O: +1; N: -1.

:O=N—Cl: Formal charges: Cl:0; O: 0; N: 0. Thus, this is the best Lewis structure and **N** is the central atom.

Draw Lewis structures of each molecule. Electron pairs in unshared pairs exert greater repulsion than electron pairs in bonds. Thus, the structures with unshared pairs will have the bond angles smaller than 109.5°.

$$\begin{array}{ccc} & H & \\ & | & \\ H-Si-H & H-\ddot{P}-H & H-\ddot{S}: \\ & | & | \\ & H & H \\ & & H \end{array}$$

PH_3 and H_2S have at least one unshared pair and will have bond angles less than 109.5°.

Draw Lewis structures of each molecule. Electron pairs in unshared pairs exert greater repulsion than electron pairs in bonds. Thus, the structures with unshared pairs will have the bond angles smaller than 120°.

:Cl—Sn—Cl: One lone pair around Sn. Thus, the bond angle will be **less than 120°**.

:Cl—B—Cl:
   :Cl: No lone pairs around B. Thus, the bond angle will equal 120°.

:O—S—O: One lone pair around S. Thus, the bond angle will be **less than 120°**.

78. *Refer to Chapter 5 and Sections 7.2 - 7.3.*

By calculating the moles of Cl and the total moles of F in the products, one can determine the mole ratio of Cl to F and thus the formula of ClF_x. Then one can proceed to address the questions of geometry and polarity.

$$n = \frac{PV}{RT} = \frac{(3.00\text{ atm})(0.457\text{ L})}{(0.0821\text{ L}\cdot\text{atm/mol.}\cdot\text{K})(348\text{ K})} = 0.0480\text{ mol. Cl}$$

$$5.60\text{ g }UF_6 \times \frac{1\text{ mol. }UF_6}{352\text{ g }UF_6} \times \frac{6\text{ mol. F}}{1\text{ mol. }UF_6} = 0.0960\text{ mol. F}$$

moles Cl = 0.0480 moles.
moles F = 0.0480 + 0.0960 = 0.1440 moles.

$$\frac{0.1440\text{ mol. F}}{0.0480\text{ mol. Cl}} = 3\text{ mol. F/1 mol. Cl}$$

Thus, $x = 3$ ClF_3

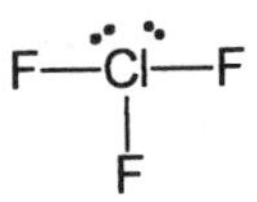

Geometry: AX_3E_2 **T-shaped.**

This molecule is **polar**.

Bond angles are about **90°** and **180°**.

Cl: **sp^{3}d** hybridized

3 σ bonds, **0 π** bonds.

79. *Refer to Sections 7.1, 7.2, and 7.4.*

Lewis Structure:

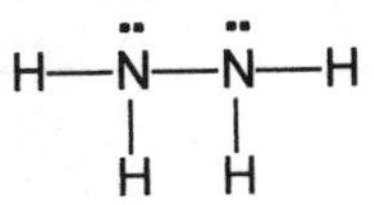

Each N is AX_3E, and thus would be **triangular pyramid**. Due to the lone pairs of electrons on the nitrogens, the bond angles will be slightly less than 109.5°. As drawn, the dipoles are not balanced and thus the molecule is polar.

When one considers rotation about the N-N bond, the dipoles cancel and the molecule is nonpolar.

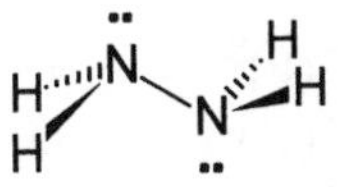

There are 6 groups around the central iodine atom.

The hybridization is sp^3d^2.

The geometry is octahedron.

Geometry:	AX_4: tetrahedron	AX_4: tetrahedron
Hybridization:	sp^3	sp^3
$C_f(S)$:	$6 - (0 + ½(8)) = 2$	$6 - (0 + ½(12)) = 0$
$C_f(O)$ (with single bond):	$6 - (6 + ½(2)) = -1$	$6 - (6 + ½(2)) = -1$
$C_f(O)$ (with double bond):		$6 - (4 + ½(4)) = 0$

Since the atoms in the second structure have charges closer to zero, it is the better structure.

$C_f(P)$:	$5 - (0 + 4) = +1$	$5 - (0 + 5) = 0$
$C_f(O)$:	$6 - (6 + 1) = -1$	$6 - (4 + 2) = 0$
$C_f(Cl)$:	$7 - (6 + 1) = 0$	$7 - (6 + 1) = 0$

Since the second structure has no formal charges, it is the better structure.

Chapter 8: Thermochemistry

---

2. *Refer to Section 8.1 and Example 8.1.*

$q = mc\Delta t$

$1.33 \text{ J} = (5.00 \text{ g})(0.129 \text{ J/g°C})(\Delta t)$

$\Delta t = 2.06°C$

---

4. *Refer to Section 8.1 and Example 8.1.*

Convert the temperature to Celsius, calculate the change in temperature, then calculate q.

$$t(°C) = \frac{375 - 32}{1.8} = 191°C$$

$\Delta t = t_f - t_i = 191°C - 23.00°C = 168°C$

$q = mc\Delta t$

$q = (473 \text{ g})(0.902 \text{ J/g°C})(168°C)$

$q = 71700 \text{ J} = 71.7 \text{ kJ}$

---

6. *Refer to Problem 1.16, Section 8.1, and Example 8.1.*

a. In the solution process, heat is absorbed. Thus, it is **not** exothermic, it is endothermic.

b. Heat is absorbed from the water. Thus, heat exits the water and the process for the water is exothermic. Thus, $q_{H_2O} = -27.6 \text{ kJ}$.

c. $250.0 \text{ mL} \times \dfrac{1.00 \text{ g}}{1 \text{ mL}} = 250.0 \text{ g}$

$-27.6 \text{ kJ} \times \dfrac{1000 \text{ J}}{1 \text{ kJ}} = -27600 \text{ J}$

$q = mc\Delta t$

$-27600 \text{ J} = (250.0 \text{ g})(4.18 \text{ J/g°C}) \Delta t$

$\Delta t = -26.4°C$

$\Delta t = t_f - t_i$

$-26.4°C = t_f - 30.0°C$

$t_f = 3.6°C$

d. $t_{°F} = 1.8t_{°C} + 32$

$t_i = 30.0°C$ $\qquad\qquad\qquad$ $t_f = 3.6°C$

$t_{°F} = 1.8(30.0°C) + 32$ $\qquad\qquad$ $t_{°F} = 1.8(3.6°C) + 32$

$t_i = 86.0°F$ $\qquad\qquad\qquad\quad$ $t_f = 38.5°F$

---

8. *Refer to Section 8.3 and Example 8.3.*

a. $q_{reaction} = -C_{cal} \times \Delta t$

$q_{reaction} = -(10.34 \text{ kJ/°C})(39.7°C - 23.5°C)$

$q_{reaction} = -168 \text{ kJ}$

$q_{cal} = +168 \text{ kJ}$

b. from part a: $q_{reaction} = -168$ kJ per 5.00 mL ethyl ether

c. $1 \text{ mol. ethyl ether} \times \dfrac{74.12 \text{ g ethyl ether}}{1 \text{ mol. ethyl ether}} \times \dfrac{1 \text{ mL}}{0.714 \text{ g}} \times \dfrac{-168 \text{ kJ}}{5.00 \text{ mL ethyl ether}}$

$= -3.49 \times 10^3 \text{ kJ}$

---

10. *Refer to Section 8.3.*

Calculate $q_{reaction}$ for 5.00 g caffeine using 1 mol. caffeine per 4.96×10^3 kJ as a conversion factor. Then use $q_{reaction}$ to calculate C_{cal}.

Since heat is evolved, q is negative. $q_{reaction} = -4.96 \times 10^3$ kJ

$5.00 \text{ g caffeine} \times \dfrac{1 \text{ mol. caffeine}}{194.20 \text{ g caffeine}} \times \dfrac{-4.96 \times 10^3 \text{ kJ}}{1 \text{ mol. caffeine}} = -128 \text{ kJ}$

$q_{reaction} = -C_{cal} \times \Delta t$

$-128 \text{ kJ} = -C_{cal} \times 11.37°C$

$C_{cal} = 11.2 \text{ kJ/°C} = 1.12 \times 10^4 \text{ J/°C}$

---

12. *Refer to Section 8.2 and Example 8.3.*

$0.750 \text{ g acetylene} \times \dfrac{48.2 \text{ kJ}}{1 \text{ g acetylene}} = 36.2 \text{ kJ}$

Since 36.2 kJ of energy is released, $q_{reaction} = -36.2$ kJ

$q_{reaction} = -C_{cal} \times \Delta t$

$-36.2 \text{ kJ} = -(1.117 \text{ kJ/°C})(\Delta t)$

$\Delta t = 32.4°C = t_{final} - t_{initial} = 54.5°C - t_{initial}$

$t_{initial} = 22.1°C$

$$q_{reaction} = -C_{cal} \times \Delta t = -(9.832 \text{ kJ/}°\text{C})(28.4°\text{C} - 25.1°\text{C}) = -32.4 \text{ kJ}$$

$$-32.4 \text{ kJ} \times \frac{1 \text{ mol. } C_{10}H_8}{-5.15 \times 10^3 \text{ kJ}} \times \frac{128.16 \text{ g}}{1 \text{ mol. } C_{10}H_8} \times \frac{1000 \text{ mg}}{1 \text{ g}} = 8.06 \times 10^2 \text{ mg}$$

16. Refer to Sections 8.3 and 8.4, Figure 8.4, and Example 8.4.

a. $CaO(s) + 3C(s) \rightarrow CO(g) + CaC_2(s)$ $\Delta H = +464.8 \text{ kJ}$

b.   Heat is absorbed, therefore the reaction is **endothermic**.

c.

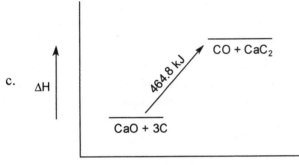

reaction path

The enthalpy of the products is higher than the enthalpy of the reactants.

d.   The formation of one mole of CaC_2 consumes 464.8 kJ, therefore:

$$1.00 \text{ g } CaC_2 \times \frac{1 \text{ mol. } CaC_2}{64.098 \text{ g } CaC_2} \times \frac{464.8 \text{ kJ}}{1 \text{ mol. } CaC_2} = 7.25 \text{ kJ}$$

e.   From the balanced equation, we know there are 3 mol. C per mol. CaC_2, thus:

$$20.00 \text{ kJ} \times \frac{1 \text{ mol. } CaC_2}{464.8 \text{ kJ}} \times \frac{3 \text{ mol. C}}{1 \text{ mol. } CaC_2} \times \frac{12.01 \text{ g C}}{1 \text{ mol. C}} = 1.550 \text{ g C}$$

18. Refer to Section 8.4 and Example 8.4.

a. $$1 \text{ mol. } CO_2 \times \frac{2801 \text{ kJ}}{6 \text{ mol. } CO_2} = 466.8 \text{ kJ}$$

$\Delta H° = 466.8 \text{ kJ}$

b. $$15.00 \text{ g glucose} \times \frac{1 \text{ mol. glucose}}{180.16 \text{ g glucose}} \times \frac{2801 \text{ kJ}}{1 \text{ mol. glucose}} = 233.2 \text{ kJ}$$

233.2 kJ of energy is liberated upon burning 15.00 grams of glucose.

a. $2 \, mol. \, C_3H_5(NO_3)_3 \times \dfrac{227 \, g \, C_3H_5(NO_3)_3}{1 \, mol.} \times \dfrac{-6.26 \, kJ}{1 \, g \, C_3H_5(NO_3)_3} = -2.84 \times 10^3 \, kJ$

$2C_3H_5(NO_3)_3(l) \rightarrow 5H_2O(g) + 3N_2(g) + 6CO_2(g) + \frac{1}{2} O_2(g) \qquad \Delta H = -2.84 \times 10^3 \, kJ$

b. In the balanced thermochemical equation, there are 14.5 total moles of products for 2 moles of $C_3H_5(NO_3)_3$. Use this information as a conversion factor.

$4.65 \, mol. \, product \times \dfrac{2 \, mol. \, C_3H_5(NO_3)_3}{14.5 \, mol. \, product} \times \dfrac{-2.84 \times 10^3 \, kJ}{2 \, mol. \, C_3H_5(NO_3)_3} = -911 \, kJ$

Calculate the amount of heat needed to raise the temperature of the water 3°C. Recall that the heat absorbed by the water will be equal to the heat released by the fat. Finally, use the enthalpy of the reaction to calculate the amount of fat needed to produce the required heat.

$C_{57}H_{104}O_6(s) + 80O_2(g) \rightarrow 57CO_2(g) + 52H_2O(l) \qquad \Delta H = -3.022 \times 10^4 \, kJ/mol.$

$100.0 \, mL \, H_2O = 100.0 \, g \, H_2O$

$q_{H_2O} = (100.0 \, g)(4.18 \, J/g°C)(25.00°C - 22.00°C) = 1.25 \times 10^3 \, J$

$q_{fat} = -q_{H_2O} = -1.25 \times 10^3 \, J$

$-1.25 \times 10^3 \, J \times \dfrac{1 \, kJ}{1000 \, J} \times \dfrac{1 \, mol. \, fat}{-3.022 \times 10^4 \, kJ} \times \dfrac{885.4 \, g \, fat}{1 \, mol. \, fat} = 3.67 \times 10^{-2} \, g \, fat$

Calculate the amount of heat evolved by each process and compare them.

$C_6H_6(l) \rightarrow C_6H_6(s) \qquad \Delta H = -9.84 \, kJ/mol.$

$100.0 \, g \, C_6H_6 \times \dfrac{1 \, mol. \, C_6H_6}{78.114 \, g \, C_6H_6} \times \dfrac{-9.84 \, kJ}{1 \, mol. \, C_6H_6} = -12.6 \, kJ$

$Br_2(l) \rightarrow Br_2(s) \qquad \Delta H = -10.8 \, kJ/mol.$

$100.0 \, g \, Br_2 \times \dfrac{1 \, mol. \, Br_2}{159.808 \, g \, Br_2} \times \dfrac{-10.8 \, kJ}{1 \, mol. \, Br_2} = -6.76 \, kJ$

Thus, freezing 100.0 g of benzene evolves more heat than freezing an equal mass of bromine.

26. Refer to Section 8.4 and Tables 8.1 and 8.2.

Calculate the heat evolved in condensing benzene gas to liquid, then the amount evolved in cooling the liquid to 25.00°C.

$$C_6H_6 \ (g, 80.00°C) \rightarrow C_6H_6 \ (l, 80.00°C) \qquad \qquad \Delta H = -30.8 \text{ kJ/mol.}$$

$$\Delta H_{vap} = 100.00 \text{ g} \times \frac{1 \text{ mol. } C_6H_6}{78.1 \text{ g } C_6H_6} \times \frac{-30.8 \text{ kJ}}{1 \text{ mol.}} = -39.4 \text{ kJ}$$

$$C_6H_6 \ (l, 80.00°C) \rightarrow C_6H_6 \ (l, 25.00°C)$$

$$q = mc\Delta t = (100.00 \text{ g})(1.72 \text{ J/g°C})(25.00°C - 80.00°C) = -9460 \text{ J} = -9.46 \text{ kJ}$$

$$\Delta H = \Delta H \text{ (condensation)} + q \text{ (cooling)}$$

$$\Delta H = (-39.4 \text{ kJ}) + (-9.46 \text{ kJ}) = -48.9 \text{ kJ}$$

28. Refer to Section 8.4.

a. Combine the reactions given to get the desired, overall reaction. The ΔH for the reaction will be the sum of the combined reactions.

$SiO_2(s) + 2C(s) \rightarrow Si(s) + 2CO(g)$	$\Delta H = +689.9 \text{ kJ}$
$Si(s) + 2Cl_2(g) \rightarrow SiCl_4(g)$	$\Delta H = -657.0 \text{ kJ}$
$\underline{SiCl_4(g) + 2Mg(s) \rightarrow 2MgCl_2(s) + Si(s)}$	$\underline{\Delta H = -625.6 \text{ kJ}}$
$SiO_2(s) + 2C(s) + 2Cl_2(g) + 2Mg(s) \rightarrow 2CO(g) + 2MgCl_2(s) + Si(s)$	$\Delta H = -592.7 \text{ kJ}$

b. $\Delta H = -592.7 \text{ kJ}$

c. **Yes**, the overall reaction is exothermic. We know this because the sign of ΔH is negative.

30. Refer to Section 8.4.

Combine the reactions given to get the desired, overall reaction. The ΔH for the reaction will be the sum of the combined reactions. Remember that when a reaction is reversed, the sign for ΔH changes and if the reaction is multiplied through by some factor, ΔH must be multiplied by that same factor.

$2HNO_3(l) \rightarrow N_2O_5(g) + H_2O(l)$	$\Delta H = +73.7 \text{ kJ}$
$\underline{2[\tfrac{1}{2}N_2(g) + \tfrac{3}{2}O_2(g) + \tfrac{1}{2}H_2(g) \rightarrow HNO_3(l)]}$	$\underline{\Delta H = 2(-174.1 \text{ kJ})}$
$2HNO_3(l) + N_2(g) + 3O_2(g) + H_2(g) \rightarrow 2HNO_3(l) + N_2O_5(g) + H_2O(l)$	
$N_2(g) + 3O_2(g) + H_2(g) \rightarrow N_2O_5(g) + H_2O(l)$	$\Delta H = -274.5 \text{ kJ}$

$$N_2(g) + 3O_2(g) + H_2(g) \rightarrow N_2O_5(g) + H_2O(l) \qquad \Delta H = -274.5 \text{ kJ}$$
$$\tfrac{1}{2}[2H_2O(l) \rightarrow 2H_2(g) + O_2(g)] \qquad \underline{\Delta H = \tfrac{1}{2}(+571.6 \text{ kJ})}$$
$$N_2(g) + {}^5\!/_2O_2(g) \rightarrow N_2O_5(g) \qquad \Delta H = 11.3 \text{ kJ}$$

---

32. Refer to Section 8.5, Table 8.3, and Example 8.7.

---

a. $2C(graphite) + H_2(g) \rightarrow C_2H_2(g)$ $\qquad \Delta H = \Delta H_f^\circ = +226.7 \text{ kJ}$

b. $\tfrac{1}{2}N_2(g) + O_2(g) \rightarrow NO_2(g)$ $\qquad \Delta H = \Delta H_f^\circ = +33.2 \text{ kJ}$

c. $Pb(s) + Br_2(l) \rightarrow PbBr_2(s)$ $\qquad \Delta H = \Delta H_f^\circ = -278.7 \text{ kJ}$

d. $\tfrac{1}{4}P_4(s) + {}^5\!/_2Cl_2(g) \rightarrow PCl_5(g)$ $\qquad \Delta H = \Delta H_f^\circ = -374.9 \text{ kJ}$

---

34. Refer to Section 8.5 and Example 8.7.

---

a. $\Delta H_{reaction} = \Sigma \, \Delta H_f^\circ \text{ (products)} - \Sigma \, \Delta H_f^\circ \text{ (reactants)}$
Recall that ΔH_f° for elements in their standard states is zero.
$314.6 \text{ kJ} = [2 \text{mol}.\Delta H_f^\circ \, Cu(s) + 1 \text{mol}.\Delta H_f^\circ \, O_2(g)] - [2 \text{mol}.\Delta H_f^\circ \, CuO(s)]$
$314.6 \text{ kJ} = -2 \text{mol}.\Delta H_f^\circ \, CuO(s)$
$-157.3 \text{ kJ} = \Delta H_f^\circ \, CuO(s)$

b. Use the answer in part a. as a conversion factor.
$$\Delta H^\circ = 13.58 \text{ g CuO} \times \frac{1 \text{ mol.CuO}}{79.55 \text{ g CuO}} \times \frac{-157.3 \text{ kJ}}{1 \text{ mol.CuO}} = -26.85 \text{ kJ}$$

---

36. Refer to Section 8.5 and Chapter 5.

Calculate the amount of energy released by one mole of NH_3, then determine the number of moles of NH_3 in one liter at the given conditions. Finally, calculate the amount of energy released by the one liter of ammonia.

$$NH_3(g) + {}^5\!/_4O_2(g) \rightarrow NO(g) + {}^3\!/_2H_2O(g)$$

$\Delta H_{reaction} = \Sigma \, \Delta H_f^\circ \text{ (products)} - \Sigma \, \Delta H_f^\circ \text{ (reactants)}$
$\quad = [(1 \text{ mol.})(90.2 \text{ kJ/mol.}) + ({}^3\!/_2 \text{ mol.})(-241.8 \text{ kJ/mol.})]$
$\qquad - [(1 \text{ mol.})(-46.1 \text{ kJ/mol.}) + ({}^5\!/_4 \text{ mol.})(0.0 \text{ kJ/mol.})]$
$\quad = -226.4 \text{ kJ}$

$$PV = nRT \Rightarrow n = \frac{PV}{RT} = \frac{(1.01 \text{ atm})(1.00 \text{ L})}{(0.0821 \text{ L} \cdot \text{atm/mol.} \cdot \text{K})(296 \text{ K})} = 0.0416 \text{ mol.}$$

$$\Delta H = \frac{-226.4 \text{ kJ}}{1 \text{ mol.}} \times 0.0416 \text{ mol.} = -9.42 \text{ kJ}$$

Thus **9.42 kJ** of heat are evolved.

38. Refer to Section 8.5, Example 8.8, and Table 8.3.

a. $\Delta H_{reaction} = \Sigma \, \Delta H_f^\circ \, (products) - \Sigma \, \Delta H_f^\circ \, (reactants)$

$\Delta H_{reaction} = [\Delta H_f^\circ \, Zn^{2+}(aq) + \Delta H_f^\circ \, H_2(g)] - [\Delta H_f^\circ \, Zn(s) + 2\Delta H_f^\circ \, H^+(aq)]$

$\Delta H_{reaction} = [(1 \text{ mol.})(-153.9 \text{ kJ/mol.}) + (1 \text{ mol.})(0.0 \text{ kJ/mol.})]$
$\quad\quad\quad - [(1 \text{ mol.})(0.0 \text{ kJ/mol.}) + (2 \text{ mol.})(0.0 \text{ kJ/mol.})]$

$\Delta H_{reaction} = -153.9 \text{ kJ}$

b. $\Delta H_{reaction} = \Sigma \, \Delta H_f^\circ \, (products) - \Sigma \, \Delta H_f^\circ \, (reactants)$

$\Delta H_{reaction} = [2\Delta H_f^\circ \, SO_2(g) + 2\Delta H_f^\circ \, H_2O(g)] - [2\Delta H_f^\circ \, H_2S(g) + 3\Delta H_f^\circ \, O_2(g)]$

$\Delta H_{reaction} = [(2 \text{ mol.})(-296.8 \text{ kJ/mol.}) + (2 \text{ mol.})(-241.8 \text{ kJ/mol.})]$
$\quad\quad\quad - [(2 \text{ mol.})(-20.6 \text{ kJ/mol.}) + (3 \text{ mol.})(0.0 \text{ kJ/mol.})]$

$\Delta H_{reaction} = -1036.0 \text{ kJ}$

c. $\Delta H_{reaction} = \Sigma \, \Delta H_f^\circ \, (products) - \Sigma \, \Delta H_f^\circ \, (reactants)$

$\Delta H_{reaction} = [3\Delta H_f^\circ \, Ni^{+2}(aq) + 2\Delta H_f^\circ \, NO(g) + 4\Delta H_f^\circ \, H_2O(l)] -$
$\quad\quad\quad [3\Delta H_f^\circ \, Ni(s) + 2\Delta H_f^\circ \, NO_3^-(aq) + 8\Delta H_f^\circ \, H^+(aq)]$

$\Delta H_{reaction} = [(3 \text{ mol.})(-54.0 \text{ kJ/mol.}) + (2 \text{ mol.})(+90.2 \text{ kJ/mol.}) + (4 \text{ mol.})(-285.8 \text{ kJ/mol.})]$
$\quad\quad\quad - [(3 \text{ mol.})(0.0 \text{ kJ/mol.}) + (2 \text{ mol.})(-205.0 \text{ kJ/mol.}) + (8 \text{ mol.})(0.0 \text{ kJ/mol.})]$

$\Delta H_{reaction} = -714.8 \text{ kJ}$

40. Refer to Section 8.5.

a. $CuO(s) + CO(g) \rightarrow Cu(s) + CO_2(g)$

$\Delta H_{reaction} = \Sigma \, \Delta H_f^\circ \, (products) - \Sigma \, \Delta H_f^\circ \, (reactants)$

$\Delta H_{reaction} = [\Delta H_f^\circ \, Cu(s) + \Delta H_f^\circ \, CO_2(g)] - [\Delta H_f^\circ \, CuO(s) + \Delta H_f^\circ \, CO(g)]$

$\Delta H_{reaction} = [(1 \text{ mol.})(0.0 \text{ kJ/mol.}) + (1 \text{ mol.})(-393.5 \text{ kJ/mol.})]$
$\quad\quad\quad - [(1 \text{ mol.})(-157.3 \text{ kJ/mol.}) + (1 \text{ mol.})(-110.5 \text{ kJ/mol.})]$

$\Delta H_{reaction} = -125.7 \text{ kJ}$

b. $CH_3OH(l) \rightarrow CH_4(g) + \frac{1}{2}O_2(g)$

$\Delta H_{reaction} = \Sigma \, \Delta H_f^\circ \, (products) - \Sigma \, \Delta H_f^\circ \, (reactants)$

$\Delta H_{reaction} = [\Delta H_f^\circ \, CH_4(g) + \frac{1}{2}\Delta H_f^\circ \, O_2(g)] - [\Delta H_f^\circ \, CH_3OH(l)]$

$\Delta H_{reaction} = [(1 \text{ mol.})(-74.8 \text{ kJ/mol.}) + (\frac{1}{2} \text{ mol.})(0.0 \text{ kJ/mol.})] - [(1 \text{ mol.})(-238.7 \text{ kJ/mol.})]$

$\Delta H_{reaction} = +163.9 \text{ kJ}$

42. Refer to Section 8.5 and Example 8.9.

a. $CaCO_3(s) + 2NH_3(g) \rightarrow CaCN_2(s) + 3H_2O(l)$ $\quad\quad\quad \Delta H = 90.1 \text{ kJ/mol.}$

b. $\Delta H_{reaction} = \Sigma \, \Delta H_f^\circ \, (products) - \Sigma \, \Delta H_f^\circ \, (reactants)$

$\Delta H_{reaction} = [\Delta H_f^\circ \, CaCN_2(s) + 3\Delta H_f^\circ \, H_2O(l)] - [\Delta H_f^\circ \, CaCO_3(s) + 2\Delta H_f^\circ \, NH_3(g)]$

$90.1 \text{ kJ} = [(1 \text{ mol.})(\Delta H_f^\circ \, CaCN_2(s)) + (3 \text{ mol.})(-285.8 \text{ kJ/mol.})]$
$\quad\quad\quad - [(1 \text{ mol.})(-1206.9 \text{ kJ/mol.}) + (2 \text{ mol.})(-46.1 \text{ kJ/mol.})]]$

$\Delta H_f^\circ \, CaCN_2(s) = -351.6 \text{ kJ/mol.}$

44. Refer to Section 8.5, Table 8.3, and Example 8.9.

a. $4C_3H_5(NO_3)_3(l) \rightarrow 10H_2O(l) + 6N_2(g) + 12CO_2(g) + O_2(g)$ $\Delta H_{reaction} = -2.290 \times 10^4$ kJ

b. $\Delta H_{reaction} = \Sigma \Delta H_f^{\circ}{}_{(products)} - \Sigma \Delta H_f^{\circ}{}_{(reactants)}$
$\Delta H_{reaction} = [10\Delta H_f^{\circ} H_2O(l) + 6\Delta H_f^{\circ} N_2(g) + 12\Delta H_f^{\circ} CO_2(g) + \Delta H_f^{\circ} O_2(g)]$
$\qquad - [4\Delta H_f^{\circ} C_3H_5(NO_3)_3(l)]$
-2.290×10^4 kJ $= [(10 \text{ mol.})(-285.8 \text{ kJ/mol.}) + (6 \text{ mol.})(0.0 \text{ kJ/mol.})$
$\qquad + (12 \text{ mol.})(-393.5 \text{ kJ/mol.}) + (1 \text{ mol.})(0.0 \text{ kJ/mol.})]$
$\qquad - [(4 \text{ mol.})(\Delta H_f^{\circ} C_3H_5(NO_3)_3(l)]$
-1.532×10^4 kJ $= -[(4 \text{ mol.})(\Delta H_f^{\circ} C_3H_5(NO_3)_3(l)]$

$\Delta H_f^{\circ} C_3H_5(NO_3)_3(l) = +3830$ kJ/mol.

46. Refer to Section 8.5.

Write a balanced thermochemical equation. Use ΔH_f° values to calculate $\Delta H_{reaction}$. Then use the ideal gas law to calculate the number of moles. Finally, use $\Delta H_{reaction}$ as a conversion factor to calculate heat.

$2NH_3(g) + 3N_2O(g) \rightarrow 3H_2O(l) + 4N_2(g)$

$\Delta H_{reaction} = \Sigma \Delta H_f^{\circ}{}_{(products)} - \Sigma \Delta H_f^{\circ}{}_{(reactants)}$
$\Delta H_{reaction} = [3\Delta H_f^{\circ} H_2O(l) + 4\Delta H_f^{\circ} N_2(g)] - [2\Delta H_f^{\circ} NH_3(g) + 3\Delta H_f^{\circ} N_2O(g)]$
$\Delta H_{reaction} = [(3 \text{ mol.})(-285.8 \text{ kJ/mol.}) + (4 \text{ mol.})(0.0 \text{ kJ/mol.})]$
$\qquad - [(2 \text{ mol.})(-46.1 \text{ kJ/mol.}) + (3 \text{ mol.})(82.05 \text{ kJ/mol.})]$
$\Delta H_{reaction} = -1011.4$ kJ

$$n = \frac{PV}{RT} = \frac{(0.943 \text{ atm})(0.346 \text{ L})}{(0.0821 \text{ L} \cdot \text{atm/mol.} \cdot \text{K})(298 \text{ K})} = 0.0133 \text{ mol. N}_2$$

$$0.0133 \text{ mol. N}_2 \times \frac{-1011.4 \text{ kJ}}{4 \text{ mol. N}_2} = -3.36 \text{ kJ}$$

48. Refer to Section 8.7 and Example 8.9.

$$12.2 \text{ kJ} \times \frac{1 \text{ L} \cdot \text{atm}}{0.1013 \text{ kJ}} = 120 \text{ L} \cdot \text{atm}$$

50. Refer to Section 8.7 and Example 8.9.

    a. $\Delta E = q + w$

        $-72 \text{ J} = q - 54 \text{ J}$

        $q = -18 \text{ J}$

    b. $\Delta E = q + w$

        $\Delta E = -38 \text{ J} + 102 \text{ J}$

        $\Delta E = 64 \text{ J}$

52. Refer to Sections 8.5 and 8.7 and Chapter 5.

$H_2O(l) \rightarrow H_2O(g)$

    a. $\Delta H_{vap} = 40.7 \text{ kJ}$ *(from Table 8.2)*

    b. $\Delta PV = PV_{products} - PV_{reactants}$ (assume $P = 1.00$ atm)

$$V_{reactants} = 1 \text{ mol. } H_2O \times \frac{18.02 \text{ g } H_2O}{1 \text{ mol. } H_2O} \times \frac{1 \text{ mL } H_2O}{1.00 \text{ g } H_2O} \times \frac{1 \text{ L}}{1000 \text{ mL}} = 0.01802 \text{ L } H_2O$$

$PV_{reactants} = 1.00 \text{ atm} \times 0.01802 \text{ L} = 0.01802 \text{ L·atm}$

$PV_{products} = nRT = (1.00 \text{ mol.})(0.0821 \text{ L·atm/mol.·K})(373 \text{ K}) = 30.6 \text{ L·atm}$

$\Delta PV = 30.6 \text{ L·atm} - 0.0821 \text{ L·atm} = 30.6 \text{ L·atm}$
    (Note that PV for the liquid is so small relative to that of the gas it could be ignored.)

$$30.6 \text{ L·atm} \times \frac{0.1013 \text{ kJ}}{1 \text{ L·atm}} = 3.10 \text{ kJ}$$

    c. $\Delta H = \Delta E + \Delta(PV)$

        $40.7 \text{ kJ} = \Delta E + 3.10 \text{ kJ}$

        $\Delta E = 37.6 \text{ kJ}$

54. Refer to Sections 8.5 and 8.7.

    a. $C_2H_2(g) + {}^5/_2 O_2(g) \rightarrow 2CO_2(g) + H_2O(l)$

        $\Delta H° = [2\Delta H_f° \ CO_2(g) + \Delta H_f° \ H_2O(l)] - [\Delta H_f° \ C_2H_2(g) + {}^5/_2 \Delta H_f° \ O_2(g)]$

        $\Delta H° = [(2 \text{ mol.})(-393.5 \text{ kJ/mol.}) + (1 \text{ mol.})(-285.8 \text{ kJ/mol.})]$

                $- [(1 \text{ mol.})(+226.7 \text{ kJ/mol.}) + ({}^5/_2 \text{ mol.})(0.0 \text{ kJ/mol.})]$

        $\Delta H° = -1229.5 \text{ kJ}$

b. $\Delta H = \Delta E + \Delta(PV)$
$\Delta(PV) = \Delta nRT$

$\qquad \Delta n = n_{\text{products}} - n_{\text{reactants}} = 2 - (1 + 2.5) = -1.5$ *(this applies to the gases only)*

$\Delta(PV) = -1.5 \text{ mol.}(0.0821 \text{ L·atm/mol.·K})(298 \text{ K}) = -36.7 \text{ L·atm}$

$$\Delta PV = -36.7 \text{ L·atm} \times \frac{0.1013 \text{ kJ}}{1 \text{ L·atm}} = -3.72 \text{ kJ}$$

$-1299.5 \text{ kJ} = \Delta E + (-3.72 \text{ kJ})$

$\Delta E = -1295.8 \text{ kJ}$

---

56. Refer to Sections 8.2 and 8.4.

a. $1 \text{ therm} \times \dfrac{1 \times 10^5 \text{ BTU}}{1 \text{ therm}} \times \dfrac{1 \text{ J}}{9.48 \times 10^{-4} \text{ BTU}} = 1.05 \times 10^8 \text{ J}$

b. $C_3H_8(g) + 5O_2(g) \rightarrow 3CO_2(g) + 4H_2O(g)$

$\Delta H_{\text{reaction}} = \Sigma \, \Delta H_f^{\circ}{}_{\text{(products)}} - \Sigma \, \Delta H_f^{\circ}{}_{\text{(reactants)}}$

$\Delta H_{\text{reaction}} = [3\Delta H_f^{\circ} \, CO_2(g) + 4\Delta H_f^{\circ} \, H_2O(g)] - [\Delta H_f^{\circ} \, C_3H_8(g) + 5\Delta H_f^{\circ} \, O_2(g)]$

$\Delta H_{\text{reaction}} = [(3 \text{ mol.})(-393.5 \text{ kJ/mol.}) + (4 \text{ mol.})(-241.8 \text{ kJ/mol.})]$
$\qquad\qquad - [(1 \text{ mol.})(-10.3.8 \text{ kJ/mol.}) + (5 \text{ mol.})(0.0 \text{ kJ/mol.})]$

$\Delta H_{\text{reaction}} = -2043.9 \text{ kJ}$

$$1.00 \text{ mol. } C_3H_8 \times \frac{2043.9 \text{ kJ}}{1 \text{ mol.} C_3H_8} \times \frac{1000 \text{ J}}{1 \text{ kJ}} \times \frac{1 \text{ therm}}{1.05 \times 10^8 \text{ J}} = 0.0195 \text{ therms}$$

Note: this answer is consistent with the answer in the first printing of the text. Combustion reactions typically produce liquid water instead of steam. If the reaction is treated as a typical combustion reaction, the answer is 0.0211 therms.

---

58. Refer to Section 8.5 and Example 8.7.

Rearrange the equations given so that when they are added they give the equation for the formation of N_2H_4 (hydrazine). Remember to mathematically do to the ΔH value whatever you do to the chemical equation.

$N_2(g) + 2H_2O(g) \rightarrow N_2H_4(l) + O_2(g)$ $\qquad\qquad \Delta H = +534.2 \text{ kJ}$

$\underline{2H_2(g) + O_2(g) \rightarrow 2H_2O(g)} \qquad\qquad\qquad\quad \underline{\Delta H = -483.6 \text{ kJ}}$

$N_2(g) + 2H_2(g) \rightarrow N_2H_4(l)$ $\qquad\qquad\qquad \Delta H = +534.2 \text{ kJ} - 483.6 \text{ kJ} = 50.6 \text{ kJ}$

60. Refer to Section 8.1, Table 1.3, and Chemistry Beyond the Classroom.

$$1 \text{ lb fat} \times \frac{453.6 \text{ g}}{1 \text{ lb}} \times \frac{32 \text{ kJ}}{1.00 \text{ g fat}} \times \frac{1 \text{ kcal}}{4.184 \text{ kJ}} \times \frac{0.5 \text{ hr}}{225 \text{ kcal}} = 7.7 \text{ hr}$$

62. Refer to Sections 8.1 and 8.2.

Calculate the volume and mass of the brass cube. The final temperature of the water will also be the final temperature of the brass cube. Also, the heat lost by the cube will equal the heat gained by the water. Set up the equations for each, set them equal to one another, and solve for t_f.

$$(22.00 \text{ mm})^3 \times \frac{(1 \text{ cm})^3}{(10 \text{ mm})^3} \times \frac{8.25 \text{ g}}{1 \text{ cm}^3} = 87.9 \text{ g brass}$$

$q_{brass} = -q_{water}$

$87.9 \text{ g} \times 0.362 \text{ J/g°C} \times (t_f - 95.0°\text{C}) = -[20.0 \text{ g} \times 4.18 \text{ J/g°C} \times (t_f - 22.0°\text{C})]$

$(31.8 \text{ J/°C})(t_f - 95.0°\text{C}) = -(83.6 \text{ J/°C})(t_f - 22.0°\text{C})$

$31.8 \text{ J/°C}(t_f) - 3021 \text{ J} = -83.6 \text{ J/°C}(t_f) - 1839 \text{ J}$

$115.4 \text{ J/°C}(t_f) = 4860 \text{ J}$

$t_f = 42.1°\text{C}$

64. Refer to Chapter 6 and Section 8.3.

$q_{reaction} = -mc\Delta t$

$q_{reaction} = -(350 \text{ g})(4.18 \text{ J/g·°C})(99.0°\text{C} - 23.0°\text{C})$

$q_{reaction} = -1.11 \times 10^5 \text{ J}$

$$12.5 \text{ cm} \times \frac{1 \text{ m}}{100 \text{ cm}} = 0.125 \text{ m}$$

$$E_{(per\ photon)} = \frac{hc}{\lambda} = \frac{6.626 \times 10^{-34} \text{ J·s})(2.998 \times 10^8 \text{ m/s})}{0.125 \text{ m}} = 1.59 \times 10^{-24} \text{ J}$$

$$E_{(per\ mole)} = \frac{1.59 \times 10^{-24} \text{ J}}{1 \text{ photon}} \times \frac{6.022 \times 10^{23} \text{ photons}}{1 \text{ mol.}} = 0.957 \text{ J/mol.}$$

$$1.11 \times 10^5 \text{ J} \times \frac{1 \text{ mol.}}{0.957 \text{ J}} = 1.16 \times 10^5 \text{ moles of photons}$$

Write the equations. Remember that changes which require the input of energy (such as melting and vaporization) will have a positive (+) ΔH, while those that release energy (such as freezing and condensing) will have a negative (-) ΔH.

a. $Hg(s) \rightarrow Hg(l)$ $\Delta H = +2.33$ kJ/mol.

b. $Br_2(l) \rightarrow Br_2(g)$ $\Delta H = +29.6$ kJ/mol.

c. $C_6H_6(l) \rightarrow C_6H_6(s)$ $\Delta H = -9.84$ kJ/mol.

d. $Hg(g) \rightarrow Hg(l)$ $\Delta H = -59.4$ kJ/mol.

e. This phase change is not in the table and must be calculated with Hess's Law.

$$\begin{array}{ll} C_{10}H_8(s) \rightarrow C_{10}H_8(l) & \Delta H = +43.3 \text{ kJ/mol.} \\ C_{10}H_8(l) \rightarrow C_{10}H_8(g) & \Delta H = +19.3 \text{ kJ/mol.} \\ \hline C_{10}H_8(s) \rightarrow C_{10}H_8(g) & \Delta H = +62.6 \text{ kJ/mol.} \end{array}$$

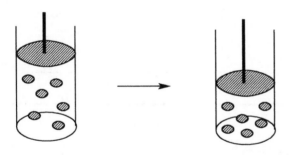

$q_A = - q_B$

$m_A c_A \Delta t_A = - m_B c_B \Delta t_B$

$m_A = m_B$

$\Delta t_A = 80°C - 100°C = -20°C$

$\Delta t_B = 80°C - 50°C = +30°C$

Substitute values in and solve for c_A

$m_A c_A(-20°C) = - m_A c_B(+30°C)$

$$c_A = \frac{-m_A c_B (30°C)}{m_A(-20°C)} = \frac{30 c_B}{20}$$

c_A is 1.5 times bigger than c_B.

---

72. *Refer to Section 8.1.*

a. **True**, this statement is the definition of specific heat, c.

b. **False**, heat flow, q, has units of J, whereas specific heat, c, has units of J/g°C.

c. **False**, since different substances have different values of specific heat, the temperature change that these substances will experience will be different.

$q = 20 \text{ J} = mc\Delta t$

If q and m are constant, and c is variable depending upon the substance, then Δt will also vary depending upon the substance.

d. **False**, temperature can be measured °C. Heat is measured with units of energy such as Joules.

---

74. *Refer to Section 8.3.*

a. Mass of soda $= 6 \text{ cans} \times \dfrac{12.0 \text{ oz}}{1 \text{ can}} \times \dfrac{1 \text{ lb}}{16 \text{ oz}} \times \dfrac{454 \text{ g}}{1 \text{ lb}} = 2.04 \times 10^3 \text{ g}$

$q = -mc\Delta t$

$q_{soda} = -(2.04 \times 10^3 \text{ g})(4.10 \text{ J/g·°C})(5.0°C - 25.0°C)$
$q_{soda} = 1.68 \times 10^5 \text{ J}$

$q_{Al} = -(6 \text{ cans} \times 12.5 \text{ g/can})(0.902 \text{ J/g·°C})(5.0°C - 25.0°C)$
$q_{Al} = 1.35 \times 10^3 \text{ J}$

$q_{total} = q_{soda} + q_{Al} = 1.68 \times 10^5 \text{ J} + 1.35 \times 10^3 \text{ J} = 1.69 \times 10^5 \text{ J}$

b. $\Delta H_{fus} = 6.00 \text{ kJ/mol.} = 6.00 \times 10^3 \text{ J/mol.}$

$1.69 \times 10^5 \text{ J} \times \dfrac{1 \text{ mol. ice}}{6.00 \times 10^3 \text{ J}} \times \dfrac{18.02 \text{ g ice}}{1 \text{ mol. ice}} = 508 \text{ g ice}$

Some reasonable assumptions to make are:

    density of tea = density of H_2O = 1.00 g/mL

    density of ice = 1.0 g/mL (although, since ice floats, the density is actually less than 1)

    room temperature = 25°C

    volume of the glass = 480 mL (16 fl. oz.)

Then, let x = mass of ice and $(480 - x)$ = mass of tea.

q_{ice} = heat to melt ice + heat to warm ice to 25°C

$$q_{melt} = x \text{ g } H_2O \times \frac{1 \text{ mol. } H_2O}{18.02 \text{ g } H_2O} \times \frac{6.00 \text{ kJ}}{1 \text{ mol } H_2O} \times \frac{1000 \text{ J}}{1 \text{ kJ}} = 333x \text{ J}$$

$q_{warming} = mc\Delta t$

(note that the sign is positive since the water is warming, and thus gaining heat)

$q_{warming}$ = $(x$ g$)(4.18$ J/g·°C$)(25$°C - 0°C$)$

$q_{warming}$ = $105x$ J

q_{ice} = $333x$ J + $105x$ J = $438x$ J

q_{tea} = $-(480$ g - x g$)(4.18$ J/g·°C$)(0$°C - 25°C$)$

q_{tea} = $50200 - 105x$

q_{ice} = q_{tea}

$438x$ J = 50200 J - $105x$ J

$543x$ J = 50200 J

$x = 92.4$ g (of ice)

$(480 - x = 480 - 92.4 = 388$ g tea$)$

Since the densities of the tea and water (from the ice) equal 1.00 g/mL, density = mass and:

$$\text{fraction to be left empty} = \frac{92.4 \text{ mL}}{480 \text{ mL}} \times 100\% = 19.3\%$$

a. $\Delta H_{reaction} = \Sigma \Delta H_f^\circ \text{ (products)} - \Sigma \Delta H_f^\circ \text{ (reactants)}$

          = [(1 mol.)(-1675.7 kJ/mol.) + (2 mol.)(0.0 kJ/mol.)]

          - [(2 mol.)(0.0 kJ/mol.) + (1 mol.)(-824.2 kJ/mol.)]

          = -851.5 kJ

b.   First calculate the mass of Al_2O_3 and Fe for the reaction of 1 mole of Al_2O_3. Then calculate t_f, using $\Delta H_{reaction} = q_{Al_2O_3} + q_{Fe}$ and $q = -mc(\Delta t)$

$$1 \text{ mol. } Al_2O_3 \times \frac{101.96 \text{ g } Al_2O_3}{1 \text{ mol. } Al_2O_3} = 101.96 \text{ g } Al_2O_3$$

$$2 \text{ mol. } Fe \times \frac{55.85 \text{ g } Fe}{1 \text{ mol. } Fe} = 111.7 \text{ g } Fe$$

$\Delta H_{reaction} = -851.5 \text{ kJ} = -8.515 \times 10^5 \text{ J}$
$\quad = -(101.96 \text{ g } Al_2O_3)(0.77 \text{ J/g·°C})(t_f - 25°C) - (111.7 \text{ g } Fe)(0.45 \text{ J/g·°C})(t_f - 25°C)$
$\quad = -78.5t_f + 1963 - 50.3t_f + 1257$
$-8.515 \times 10^5 \text{ J} = -128.8t_f + 3220$
$-8.547 \times 10^5 \text{ J} = -128.8t_f$
$t_f = 6.6 \times 10^3 °C$

c. $111.7 \text{ g } Fe \times \dfrac{270 \text{ J}}{1 \text{ g}} = 30200 \text{ J} = 30.2 \text{ kJ}$

Yes, the $6.6 \times 10^3 °C$ is much higher than the melting point of Fe, and the 851.5 kJ produced in the reaction (part a) is more than the 30.2 kJ needed to melt the Fe, so the reaction will definitely produce molten iron.

---

77. *Refer to Section 7.3.*

$q_{reaction} = -C_{cal}\Delta T$
$\quad = -(22.51 \text{ kJ/°C})(1.67°C)$
$\quad = -37.6 \text{ kJ}$

$$-37.6 \text{ kJ} \times \frac{1 \text{ mol. sucrose}}{-5.64 \times 10^3 \text{ kJ}} \times \frac{342.30 \text{ g sucrose}}{1 \text{ mol. sucrose}} = 2.28 \text{ g sucrose}$$

$$\text{mass \% sucrose} = \frac{2.28 \text{ g}}{3.000 \text{ g}} \times 100\% = 76.0\%$$

Chapter 9: Liquids and Solids

2. *Refer to Section 9.1, Example 9.1, and Chapter 5.*

$$P = 325 \text{ mm Hg} \times \frac{1 \text{ atm}}{760 \text{ mm Hg}} = 0.428 \text{ atm}$$

$$T = 80°\text{C} + 273 = 353 \text{ K}$$

a. $\dfrac{P_1 V_1}{n_1 T_1} = \dfrac{P_2 V_2}{n_2 T_2}$

Since neither volume nor moles change, V and n are constant and the equation becomes:

$$\frac{P_1}{T_1} = \frac{P_2}{T_2} \quad \text{and} \quad P_2 = \frac{P_1 T_2}{T_1}$$

at 50°C: $\quad P_2 = \dfrac{(0.428 \text{ atm})(50 + 273)}{353 \text{ K}} = 0.392 \text{ atm}$

$$0.392 \text{ atm} \times \frac{760 \text{ mm Hg}}{1 \text{ atm}} = 298 \text{ mm Hg}$$

at 60°C: $\quad P_2 = \dfrac{(0.428 \text{ atm})(60 + 273)}{353 \text{ K}} = 0.404 \text{ atm}$

$$0.404 \text{ atm} \times \frac{760 \text{ mm Hg}}{1 \text{ atm}} = 307 \text{ mm Hg}$$

b. At 50°C, the calculated vapor pressure is **greater than** the equilibrium vapor pressure of benzene.
At 60°C, the calculated vapor pressure is **less than** the equilibrium vapor pressure of benzene.

c. The pressure exerted by the benzene vapor will never exceed the vapor pressure, therefore, at 50°C, $P = 269$ mm Hg, and at 60°C, $P = 307$ mm Hg.

4. *Refer to Section 9.1, Example 9.1, and Chapter 5.*

a. $\quad P = 0.466 \text{ mm Hg} \times \dfrac{1 \text{ atm}}{760 \text{ mm Hg}} = 6.13 \times 10^{-4} \text{ atm}$

$$30°\text{C} + 273 = 303 \text{ K}$$

$PV = nRT$

$(6.13 \times 10^{-4} \text{ atm})(1.00 \text{ L}) = (n)(0.0821 \text{ L·atm/mol·K})(303 \text{ K})$

$n = 2.46 \times 10^{-5} \text{ mol. I}_2$

$$2.46 \times 10^{-5} \text{ mol. I}_2 \times \frac{253.8 \text{ g}}{1 \text{ mol. I}_2} = 6.26 \times 10^{-3} \text{ g I}_2 = 6.26 \text{ mg I}_2$$

b. $$2.0 \text{ mg} \times \frac{1 \text{ g}}{1000 \text{ mg}} \times \frac{1 \text{ mol. I}_2}{253.8 \text{ g}} = 7.9 \times 10^{-6} \text{ mol. I}_2$$

$PV = nRT$

$(P)(1.00 \text{ L}) = (7.9 \times 10^{-6} \text{ mol.})(0.0821 \text{ L·atm/mol·K})(303 \text{ K})$

$P = 2.0 \times 10^{-4} \text{ atm}$

$$P = 2.0 \times 10^{-4} \text{ atm} \times \frac{760 \text{ mm Hg}}{1 \text{ atm}} = 0.15 \text{ mm Hg}$$

c. There is more I$_2$ (10 mg) than will sublime (6.26 mg), thus, there will be some unsublimed I$_2$, and P will equal the vapor pressure of I$_2$, 0.466 mm Hg.

---

6. *Refer to Section 9.1, Example 9.1, and Chapter 5.*

a. Use $PV = nRT$ to determine the moles of p-dichlorobenzene.

$V = 750 \text{ mL flask} = 0.750 \text{ L}$

$T = 20°C = 20 + 273 = 293 \text{ K}$

$R = 0.0821 \text{ L·atm/mol·K}$

$$P = 0.40 \text{ mm Hg} \times \frac{1 \text{ atm}}{760.0 \text{ mm Hg}} = 5.3 \times 10^{-4} \text{ atm}$$

$5.3 \times 10^{-4} \text{ atm} \times 0.750 \text{ L} = n \times 0.0821 \text{ L·atm/mol·K} \times 293 \text{ K}$

$n = 1.7 \times 10^{-5} \text{ moles } p\text{-dichlorobenzene}$

$$1.7 \times 10^{-5} \text{ mol.} \times \frac{147.00 \text{ g C}_6\text{H}_4\text{Cl}_2}{1 \text{ mol. C}_6\text{H}_4\text{Cl}_2} \times \frac{1000 \text{ mg}}{1 \text{ g}} = 2.5 \text{ mg C}_6\text{H}_4\text{Cl}_2$$

b. 5.0 mg – 2.5 mg = 2.5 mg of p-dichlorobenzene remaining.

c. $$2.0 \text{ mg} \times \frac{1 \text{ g}}{1000 \text{ mg}} \times \frac{1 \text{ mol. C}_6\text{H}_4\text{Cl}_2}{147.00 \text{ g}} = 1.36 \times 10^{-5} \text{ mol. C}_6\text{H}_4\text{Cl}_2$$

$5.3 \times 10^{-4} \text{ atm} \times 0.500 \text{ L} = n \times 0.0821 \text{ L·atm/mol·K} \times 293 \text{ K}$

$n = 1.1 \times 10^{-5} \text{ moles } p\text{-dichlorobenzene}$

There is more C$_6$H$_4$Cl$_2$ (1.36 $\times$ 10^{-5} mol.) than will sublime (1.1 $\times$ 10^{-5} mol.), thus, there will be some unsublimed C$_6$H$_4$Cl$_2$, and P will equal the vapor pressure of C$_6$H$_4$Cl$_2$, 0.40 mm Hg.

8. Refer to Section 9.1 and Example 9.2.

a. $\ln\left(\dfrac{P_2}{P_1}\right) = \dfrac{+\Delta H_{vap}}{R}\left(\dfrac{1}{T_1} - \dfrac{1}{T_2}\right)$

$\ln\left(\dfrac{760\,atm}{203\,atm}\right) = \dfrac{+\Delta H_{vap}}{8.31\,J/mol.\cdot K}\left(\dfrac{1}{308\,K} - \dfrac{1}{337.8\,K}\right)$

$1.32 = \dfrac{+\Delta H_{vap}}{8.31\,J/mol.\cdot K}\left(2.86\times10^{-4}\,K^{-1}\right)$

$\Delta H_{vap} = 3.84 \times 10^4$ J/mol.

$\Delta H_{vap} = 38.4$ kJ/mol.

b. $\ln\left(\dfrac{P_2}{P_1}\right) = \dfrac{+\Delta H_{vap}}{R}\left(\dfrac{1}{T_1} - \dfrac{1}{T_2}\right)$

$\ln\left(\dfrac{760\,atm}{P_1}\right) = \dfrac{3.84\times10^4\,J/mol.}{8.31\,J/mol.\cdot K}\left(\dfrac{1}{313.2\,K} - \dfrac{1}{337.8\,K}\right)$

$\ln\left(\dfrac{760\,atm}{P_1}\right) = 4.62\times10^3\,K(2.32\times10^{-4}\,K^{-1})$

$\ln(760\,atm) - \ln P_1 = 1.07$

$5.56 = \ln P_1$

$P_1 = 260$ mm Hg

10. Refer to Section 9.1 and Example 9.2.

$P = 681\,mm\,Hg \times \dfrac{1\,atm}{760\,mm\,Hg} = 0.896\,atm$

$\ln\left(\dfrac{P_2}{P_1}\right) = \dfrac{+\Delta H_{vap}}{R}\left(\dfrac{1}{T_1} - \dfrac{1}{T_2}\right)$

$\ln\left(\dfrac{0.896\,atm}{1.00\,atm}\right) = \dfrac{4.07\times10^4\,J/mol.}{8.31\,J/mol.\cdot K}\left(\dfrac{1}{373\,K} - \dfrac{1}{T_2}\right)$

$-0.110 = (4.90\times10^3\,K)\left(0.00268\,K^{-1} - \dfrac{1}{T_2}\right)$

$$-2.25 \times 10^{-5} \, K^{-1} = 0.00268 \, K^{-1} - \frac{1}{T_2} \Rightarrow \frac{1}{T_2} = 0.00270$$

$T_2 = 370 \, K = 97°C$

12. Refer to Section 9.1 and Example 9.2.

$$\ln\left(\frac{P_2}{P_1}\right) = \frac{+\Delta H_{vap}}{R}\left(\frac{1}{T_1} - \frac{1}{T_2}\right)$$

$$\ln\left(\frac{1.75 \, atm}{1.00 \, atm}\right) = \frac{40{,}700 \, J/mol.}{8.31 \, J/mol.\cdot K}\left(\frac{1}{373} - \frac{1}{T_2}\right)$$

$$0.56 = 4900\left(\frac{1}{373} - \frac{1}{T_2}\right)$$

$$\frac{1}{T_2} = \frac{1}{373} - \frac{0.56}{4900} = 0.0026$$

$T_2 = 390 \, K = 117°C$

14. Refer to Section 9.2 and Figure 9.5.

a. **Liquid and vapor in equilibrium**

b. 0.5 atm = 380 mm Hg; **Vapor**

c. **Liquid**

16. Refer to Section 9.2.

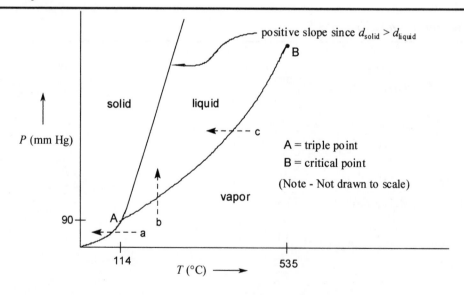

a. Iodine vapor at 80 mm Hg condenses to the **solid** when cooled sufficiently. Since this is below the triple point of 90 mm Hg, the liquid cannot form.

b. Iodine vapor at 125°C condenses to the **liquid** when enough pressure is applied. Since the temperature is above that of the triple point, condensation will be to the liquid state.

c. Iodine vapor at 700 mm Hg condenses to the **liquid** when cooled above the triple point temperature. The pressure is above that of the triple point, so condensation will be to the liquid state.

---

18. *Refer to Section 9.2.*

---

a.

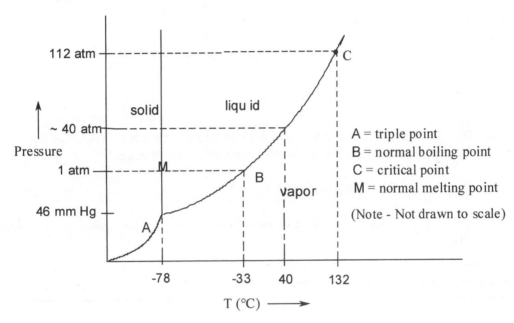

A = triple point
B = normal boiling point
C = critical point
M = normal melting point

(Note - Not drawn to scale)

b. As shown in the phase diagram above, the pressure will be about 40 atm. The value one gets will depend on the curve draw for the vapor-liquid boundary, but must be between 1 atm (normal boiling point) and 112 atm (critical point).

20. *Refer to Section 9.2.*

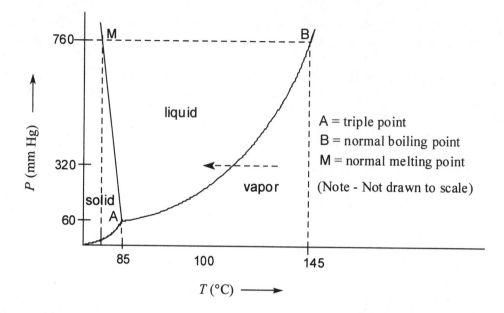

a. The phase diagram is drawn above with the appropriate labels. Note that the slope is negative since mp decreases slightly with increasing pressure.

b. The estimated boiling point is 145 °C.

c. The phase change will be from vapor to liquid, as noted with dashed arrow in the phase diagram.

22. *Refer to Section 9.3 and Example 9.4.*

The only intermolecular force present for each of the elements listed is dispersion. Thus, boiling point will correspond with the strength of the dispersion forces, which depends on two factors: (1) the number of electrons and (2) ease of electron dispersion. The second factor is the same for each of the listed elements, thus, the determining factor is the number of electrons. Xe has the most, He has the least.

He < Ne < Ar < Xe

24. *Refer to Section 9.3, Examples 9.4 and 9.6, and Chapter 7.*

All molecules have dispersion forces. One can consider each molecule to be roughly spherical. Those that have an even distribution of charge over the surface (b and c) will not have a dipole, while those that have an uneven distribution (a and d) will have a dipole.

a. PH₃: **Dispersion and dipole**. The electrons are not symmetrically distributed, so this molecule has a dipole. Each H bears a slight negative charge and the P bears a slight positive charge. Recall from Chapter 7 that PH₃ is triangular pyramid in shape.

b. N₂: **Dispersion**. The electrons are symmetrically distributed, so this molecule does not have a dipole. Each N has identical, zero, charge.

c. CH₄: **Dispersion**. The electrons are symmetrically distributed, so this molecule does not have a dipole. Each H bears an identical, slight positive charge, thus, the surface of the "sphere" has an even distribution of charge.

d. H₂O: **Dispersion and dipole**. The electrons are not symmetrically distributed, so this molecule has a dipole. The O bears a slight negative charge, while the H's each bear a slight positive charge. Recall from Chapter 7 that water is bent.

---

26. *Refer to Section 9.3, Example 9.5, and Chapter 7.*

In order for a molecule to have H-bonding, a hydrogen must be bonded to one of the following atoms: N, O, or F. Draw the molecule and look for the appropriate bond.

a.   This molecule has no F-H bond, thus, there is **no H-bonding.**

b.   This molecule does have O-H bonds, thus, there is **H-bonding.**

c.   This molecule has N-H bonds, thus, there is **H-bonding.**

d.   This molecule has no O-H bond, thus, there is **no H-bonding.**

---

28. *Refer to Sections 9.3 and 9.4, and Example 9.4.*

a.   NaBr is an ionic compound, which typically have high melting points.
Br₂ has dispersion forces only, which results in Br₂ having the lower melting point.

b.   C₂H₅OH has dispersion forces, dipole forces, and H-bonding, while
C₄H₁₀ has only dispersion forces. Thus, C₄H₁₀ will have the lower boiling point.

c. H_2O has dispersion forces, dipole forces, and H-bonding, while H_2Te has only dispersion and dipole forces, resulting a lower boiling point for H_2Te.

d. CH_3CO_2H and $C_6H_5CO_2H$ both have dispersion forces, dipole forces, and H-bonding. However, $C_6H_5CO_2H$ is a larger molecule with more electrons, consequently, it has greater dispersion forces and a higher boiling point.

---

30. *Refer to Section 9.3.*

a. **IMF's.** This is a phase change from solid to gas, phase changes only involve overcoming intermolecular forces.

b. **IMF's.** This is a phase change from liquid to gas, phase changes only involve overcoming intermolecular forces.

c. **Breaking Covalent Bonds.** H_2O decomposes to H_2 and O_2, thus, covalent bonds must be broken.

d. **Breaking Covalent Bonds.** This is a chemical reaction. Chemical reactions involve the making and/or breaking of bonds.

---

32. *Refer to Section 9.3 and Example 9.4.*

a. **LiCl or CCl_4.** LiCl is an ionic compound, which typically have high melting and boiling points, while CCl_4 has only dispersion forces, and thus, the lower boiling point.

b. **CH_3OH or CH_3F.** Both of these compounds have both dispersion and dipole forces. CH_3OH, however, also has hydrogen bonding, and has a higher boiling point.

c. **H_2O or SO_2.** Both of these compounds have both dispersion and dipole forces. H_2O, however, also has hydrogen bonding, and has a higher boiling point.

d. **N_2 or Cl_2.** Both of these molecules have only dispersion forces. N_2, however, has fewer electrons and thus, weaker dispersion forces and a lower boiling point.

---

34. *Refer to Section 9.3.*

a. Melting H_2O requires breaking IMF's, the strongest of which is **H-bonding**.

b. Subliming Br_2 requires breaking IMF's. The only intermolecular force between Br_2 molecules is **dispersion forces**.

c. Boiling $CHCl_3$ requires breaking IMF's, the strongest of which is **dipole forces**.

d. Vaporizing benzene (C_6H_6) requires breaking **dispersion forces**, the only IMF's present between benzene molecules (see 32b).

36. *Refer to Section 9.4, Example 9.7, and Table 9.5.*

   a. The water insolubility indicates network covalent, molecular or metallic. The low melting point narrows it to a **molecular** or a **low melting metallic solid**.

   b. Conduction of electricity when melted indicates the compound is **ionic**.

   c. The lack of solubility in water indicates the solid is not ionic. Conductivity in the solid state indicates it is **metallic**.

38. *Refer to Section 9.4, Example 9.7, and Table 9.5.*

   a. Three types of solids are generally insoluble in water: **metallic, network covalent and nonpolar molecular solids**.

   b. **Network covalent and ionic compounds** generally have high melting points. Many of the metallic solids have high melting points, but several (notably Hg) have quite low melting points.

   c. **Metals and some network covalent** compounds conduct electricity as solids (most network covalent compounds do not conduct electricity, but a few (such as graphite) do. Ionic compounds only do so in solution or in a melt.

40. *Refer to Section 9.4, Example 9.7, Table 9.5, and Figure 9.11.*

   a. **Metallic** Note position in periodic table.

   b. **Molecular.** Compounds of two nonmetals will be molecular.

   c. **Network Covalent.** Diamond is made only of carbon, but is a 3-dimensional network of -C-C-C- bonds.

   d. **Ionic.** This molecule is composed of two ions, NH_4^+ and CO_3^{2-}.

   e. **Molecular.** F_2 is not a metal (note position on periodic table), not a network covalent molecule (very low b.p.) and cannot be ionic since it is composed of a single element.

42. *Refer to Section 9.4.*

Note that many answers are possible given the large number of molecules that contain O.

   a. C_2H_5OH, ethanol

   b. $CaCO_3$ (found in TUMS®)

   c. SiO_2

   d. O_2

44. Refer to Section 9.4.

a. Graphite is a network covalent compound, so the C atoms are held together in an extended network of covalent bonds.

b. Silicon carbide is a network covalent compound, so the Si and C atoms are held together in an extended network of covalent bonds.

c. $FeCl_2$ is an ionic solid. The structural units are composed of individual Fe^{2+} and Cl^- ions.

d. Acetylene (C_2H_2) is a molecular solid, composed of individual C_2H_2 molecules held together by dispersion forces.

46. Refer to Section 9.5 and Example 9.8.

Determine which of the equations relating the sides of the unit cell to the given atomic radius is valid. The geometry corresponding to the equation is the geometry of the unit cell.

$$2\,r = s = 2(0.162) = 0.324 \neq 0.458$$

(not simple cubic)

$$4\,r = s\sqrt{3} \qquad \frac{4(0.162)}{\sqrt{3}} = s = 0.374 \neq 0.458$$

(not body-centered cubic)

$$4\,r = s\sqrt{2} \qquad \frac{4(0.162)}{\sqrt{2}} = s = 0.458$$

Thus, the unit cell is **face-centered cubic**.

48. Refer to Section 9.5 and Example 9.8.

A unit cell is a cube. For a cube, volume = s^3
$$s = \sqrt[3]{0.0278} = 0.303 \text{ nm}$$
For body-centered cubic structure,

$$4\,r = s\sqrt{3} \qquad r = \frac{0.303\sqrt{3}}{4} = 0.131 \text{ nm}$$

50. Refer to Section 9.5, Figures 9.17 and 9.18, and Example 9.8.

K^+ : $r = 0.133$ nm
I^- : $r = 0.216$ nm

a. one side of a cube

K$^+$ and I$^-$ ions touch along an edge of a cell

s = 0.216 nm + 2(0.133 nm) + 0.216 nm = 0.698 nm

b. face diagonal

$$d = s\sqrt{2}$$
$$d = 0.698(1.41)$$
$$d = 0.987 \text{ nm}$$

---

52. Refer to Section 9.5.

Cs$^+$: $r = 0.169$ nm

Cl$^-$: $r = 0.181$ nm

Figure 9.18 indicates that CsCl forms a BCC structure.

a. Body diagonal $= 4r = 2(r_{Cr^+}) + 2(r_{Cl^-})$

$\qquad\qquad\qquad = 2(0.169 \text{ nm}) + 2(0.181 \text{ nm})$

$\qquad\qquad\qquad = 0.338 \text{ nm} + 0.362 \text{ nm}$

$\qquad\qquad\qquad = 0.700 \text{ nm}$

b. $4r = s\sqrt{3}$

$\quad 0.700 \text{ nm} = s(1.73)$

$\quad s = 0.404 \text{ nm}$

---

54. Refer to Section 9.5 and Figure 9.18.

Note that you are seeing only two dimensions. Visualize the unit cell in three dimensions with ions coming up towards you, and going back away from you.

The unit cell has in the center one whole cesium ion. At each corner, $^1/_8$ of a chloride ion is inside the cell. There are 8 corners, thus, the number of chloride ions is 8 x $^1/_8$ = 1 chloride ion. Thus, there is 1 Cs$^+$ ion and 1 Cl$^-$ ion.

---

56. Refer to Section 9.5 and Table 9.6.

The formula of packing efficiency is

$$\text{packing efficiency} = \frac{\text{volume of the atom in the cell}}{\text{volume of the unit cell}} \times 100$$

a. For a **simple cubic** unit cell:

Volume of the unit cell: $2r = s \qquad (2r)^3 = s^3 = 8r^3$

$$\text{Volume of 1 atom} = \frac{4\pi\, r^3}{3}$$

$$\text{packing efficiency} = \frac{\left(\dfrac{4\pi\, r^3}{3}\right)}{8\,r^3} \times 100 = \frac{\left(\dfrac{\pi}{3}\right)}{2} \times 100 = 52.4\%$$

b. For a **face-centered cubic** unit cell:

Volume of the unit cell: $4\,r = s\sqrt{2}$ $\qquad \left(\dfrac{4\,r}{\sqrt{2}}\right)^3 = s^3 = \dfrac{32\,r^3}{\sqrt{2}}$

Volume of 4 atoms $= \dfrac{16\pi\, r^3}{3}$

$$\text{packing efficiency} = \frac{\left(\dfrac{16\pi\, r^3}{3}\right)}{\left(\dfrac{32\,r^3}{\sqrt{2}}\right)} \times 100 = \frac{\left(\dfrac{\pi}{3}\right)}{\left(\dfrac{2}{\sqrt{2}}\right)} \times 100 = 74.0\%$$

c. For a **body-centered cubic** unit cell:

Volume of the unit cell: $4\,r = s\sqrt{3}$ $\qquad \left(\dfrac{4\,r}{\sqrt{3}}\right)^3 = s^3 = \dfrac{64\,r^3}{3\sqrt{3}}$

Volume of 2 atoms $= \dfrac{8\pi\, r^3}{3}$

$$\text{packing efficiency} = \frac{\left(\dfrac{8\pi\, r^3}{3}\right)}{\left(\dfrac{64\,r^3}{3\sqrt{3}}\right)} \times 100 = \frac{\pi}{\left(\dfrac{8}{\sqrt{3}}\right)} \times 100 = 68.0\%$$

To check your answers for accuracy, compare these values to the values for % empty space found in Table 9.6

For simple cubic : 100% - 52.4% = 47.6% empty space √
For body-centered cubic: 100% - 68.0% = 32.0% empty space √
For face-centered cubic: 100% - 74.0% = 26.0% empty space √

---

58. *Refer to Section 9.1 and Chapter 5.*

$T = 20°C + 273 = 293$ K

amount of HCOOH vaporized = 10.00 g – 7.50 g = 2.50 g

$$n = 2.50 \text{ g HCOOH} \times \frac{1 \text{ mol. HCOOH}}{46.03 \text{ g}} = 5.43 \times 10^{-2} \text{ mol. HCOOH}$$

$PV = nRT$
$P(30.0 \text{ L}) = (5.43 \times 10^{-2} \text{ mol.})(0.0821 \text{ L·atm/mol.·K})(293 \text{ K})$
$P = 4.35 \times 10^{-2}$ atm

a.

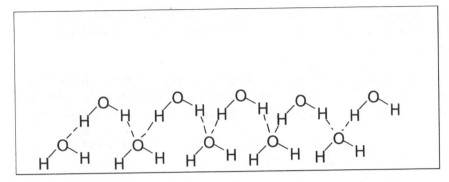

b.

c.

a. A covalent bond is the sharing of 2 electrons between two atoms within a molecule (an **intra**molecular interaction), while hydrogen bonding is the sharing of 2 electrons between a hydrogen (on an electronegative atom) and a second electronegative atom (usually on another molecule). Hydrogen bonding is an **inter**molecular interaction.

b. Boiling point is defined as the temperature at which the vapor pressure equals ambient pressure, while normal boiling point is the temperature at which the vapor pressure equals 1 atm.

c. The triple point is the temperature and pressure at which all three phases (solid, liquid and gas) coexist in equilibrium. The critical point is the condition beyond which a liquid can no longer exist.

d. The vapor pressure curve is a plot of temperature versus pressure at which vapor and liquid are in equilibrium. The phase diagram also includes plots for solid-liquid and solid-gas equilibria. Thus, a vapor pressure curve is a subset of a phase diagram.

e. Volume changes have no effect on vapor pressure for a given liquid. Temperature changes, however, do affect vapor pressure.

---

64. *Refer to Section 9.1 and Figure 9.2.*

a. **A,** the liquid with the lowest boiling point will have the weakest intermolecular forces.

b. **A,** the "normal" boiling point, at 760 mm Hg, from the chart for A is about 15°C.

c. **≈ 33°C,** line B crosses the 500 mm Hg line around 33°C.

d. **gas,** from the graph, 25°C is higher than the normal boiling point of A, so it is a gas at 25°C.

e. **≈ 200 mmHg,** the line for liquid C crosses the 40°C line at 200 mm Hg.

---

65. *Refer to Section 9.5.*

First calculate the volume of one cell. Using density as a conversion factor, calculate the mass per cell and mass per atom of Fe. Using the mass of a single Fe atom and the mass of a mole of Fe (atomic mass), calculate the number of atoms in a mole.

For a body centered cube: $4r = s\sqrt{3}$. Thus

$$s = \frac{4(0.124\,\text{nm})}{\sqrt{3}} = 0.286\,\text{nm}$$

$$s = 0.286\,\text{nm} \times \frac{1\,\text{m}}{1 \times 10^9\,\text{nm}} \times \frac{100\,\text{cm}}{1\,\text{m}} = 2.86 \times 10^{-8}\,\text{cm}$$

$$V = s^3 = (2.86 \times 10^{-8}\,\text{cm})^3 = 2.34 \times 10^{-23}\,\text{cm}^3$$

$$\frac{2.34 \times 10^{-23}\,\text{cm}^3}{1\,\text{cell}} \times \frac{7.86\,\text{g}}{1\,\text{cm}^3} \times \frac{1\,\text{cell}}{2\,\text{atoms}} = 9.19 \times 10^{-23}\,\text{g/atom}$$

$$\frac{55.847\,\text{g}}{1\,\text{mole}} \times \frac{1\,\text{atom}}{9.19 \times 10^{-23}\,\text{g}} = 6.08 \times 10^{23}\,\text{atoms/mole}$$

First, determine the limiting reagent in the reaction. Then calculate the moles of water that would be produced, and the pressure that would result if all the water existed as vapor. Compare that to the vapor pressure of water at 27°C.

a. $0.400 \, \text{g H}_2 \times \dfrac{1 \, \text{mol. H}_2}{2.016 \, \text{g}} \times \dfrac{2 \, \text{mol. H}_2\text{O}}{2 \, \text{mol. H}_2} = 0.198 \, \text{mol. H}_2\text{O}$

$3.20 \, \text{g O}_2 \times \dfrac{1 \, \text{mol. O}_2}{32.00 \, \text{g}} \times \dfrac{2 \, \text{mol. H}_2\text{O}}{1 \, \text{mol. O}_2} = 0.200 \, \text{mol. H}_2\text{O}$

Thus O_2 and H_2 are present in nearly stoichiometric quantities. Consequently, the only gas in the flask will be H_2O. Our calculations will be based on H_2.

$P = \dfrac{nRT}{V} = \dfrac{(0.198 \, \text{mol.})(0.0821 \, \text{L} \cdot \text{atm/mol.} \cdot \text{K})(300 \, \text{K})}{10.0 \, \text{L}} = 0.488 \, \text{atm}$

$P_{H_2O \, (27°C)} = 26.74 \, \text{mm Hg} \times \dfrac{1 \, \text{atm}}{760 \, \text{mm Hg}} = 0.0352 \, \text{atm}$

$P > P_{H_2O}$ Since the total pressure cannot exceed the vapor pressure of the water at 27°C, much of the water will exist as liquid, and **both the liquid and vapor phases** will be present.

b. The final pressure in the flask will be 0.0352 atm (26.7 mm Hg), the vapor pressure of the H_2O at 27°C.

c. $3.2 \, \text{g H}_2 \times \dfrac{1 \, \text{mol. H}_2}{2.016 \, \text{g}} \times \dfrac{2 \, \text{mol. H}_2\text{O}}{2 \, \text{mol. H}_2} = 1.6 \, \text{mol. H}_2\text{O}$

$3.2 \, \text{g O}_2 \times \dfrac{1 \, \text{mol. O}_2}{32.00 \, \text{g}} \times \dfrac{2 \, \text{mol. H}_2\text{O}}{1 \, \text{mol. O}_2} = 0.20 \, \text{mol. H}_2\text{O}$

Thus, O_2 is the limiting reactant and 1.4 mol. H_2 will be left over. The total pressure will be P_{H_2O} (0.0352 atm, see part a) plus P_{H_2}.

$P = \dfrac{nRT}{V} = \dfrac{(1.4 \, \text{mol.})(0.0821 \, \text{L} \cdot \text{atm/mol.} \cdot \text{K})(300 \, \text{K})}{10.0 \, \text{L}} = 3.4 \, \text{atm}$

$P_{total} = 3.4 \, \text{atm} + 0.0352 \, \text{atm} = 3.4 \, \text{atm}$

Calculate the moles of trichloroethane and the pressure that would result if all the material vaporized.

$$1\,\text{cup} \times \frac{1\,\text{qt.}}{4\,\text{cups}} \times \frac{1\,\text{L}}{1.057\,\text{qt.}} \times \frac{1000\,\text{mL}}{1\,\text{L}} \times \frac{1.325\,\text{g}}{1\,\text{mL}} \times \frac{1\,\text{mol.}}{133.39\,\text{g}} = 2.35\,\text{mol.}\,C_2H_3Cl_3$$

$$18\,\text{ft}^3 \times \frac{28.32\,\text{L}}{1\,\text{ft}^3} = 510\,\text{L}$$

$$(39°\text{F} - 32) \times \frac{5}{9} = 3.9°\text{C} = 280\,\text{K}$$

$$P = \frac{nRT}{V} = \frac{(2.35\,\text{mol.})(0.0821\,\text{L}\cdot\text{atm/mol.}\cdot\text{K})(280\,\text{K})}{510\,\text{L}} = 0.11\,\text{atm}$$

Now we must calculate the vapor pressure of trichloroethane at 3.9°C and compare the vapor pressure to the pressure calculated above. Calculate ΔH_{vap} and from that, the vapor pressure at 3.9°C.

$$\ln\left(\frac{P_2}{P_1}\right) = \frac{+\Delta H_{vap}}{R}\left(\frac{1}{T_1} - \frac{1}{T_2}\right)$$

$$\ln\left(\frac{760}{100}\right) = \frac{+\Delta H_{vap}}{8.31\,\text{J/mol.}\cdot\text{K}}\left(\frac{1}{293\,\text{K}} - \frac{1}{347.3\,\text{K}}\right)$$

$$2.03 = \frac{+\Delta H_{vap}}{8.31\,\text{J/mol.}\cdot\text{K}}\left(5.34 \times 10^{-4}\,\text{K}^{-1}\right)$$

$\Delta H_{vap} = 3.17 \times 10^4$ J/mol. (Now calculate vapor pressure at 3.9°C, (277 K))

$$\ln\left(\frac{760}{P_1}\right) = \frac{3.17 \times 10^4\,\text{J/mol.}}{8.31\,\text{J/mol.}\cdot\text{K}}\left(\frac{1}{277\,\text{K}} - \frac{1}{347.3\,\text{K}}\right) = 2.79$$

$$\frac{760\,\text{mm Hg}}{P_1} = e^{2.79} = 16.3$$

$$P_1 = 46.6\,\text{mm Hg} \times \frac{1\,\text{atm}}{760\,\text{mm Hg}} = 0.0613\,\text{atm}$$

Since the pressure that would result from complete vaporization is greater than the vapor pressure of trichloroethane, not all the $C_2H_3Cl_3$ would evaporate. To calculate the percentage that remains as liquid, we need to determine the amount that would evaporate to produce a pressure of 0.0613 atm. The rest would remain as liquid.

$$n = \frac{PV}{RT} = \frac{(0.0613\,\text{atm})(510\,\text{L})}{(0.0821\,\text{L}\cdot\text{atm/mol.}\cdot\text{K})(277\,\text{K})} = 1.37\,\text{mol.}\ \ \text{(would vaporize)}$$

2.35 mol. - 1.37 mol. = 0.98 mol. (would remain)

$$\text{Percent remaining} = \frac{0.98}{2.35} \times 100\% = 42\%$$

---

68. *Refer to Section 9.2.*

The pressure exerted by a skater is:

$$\text{Pressure} = \frac{120\,\text{lbs}}{0.10\,\text{in}^2} \times \frac{1\,\text{atm}}{15\,\text{lbs/in}^2} = 80\,\text{atm}$$

Statement #2 under the heading melting point states "An increase in pressure of 134 atm is required to lower the melting point of ice by 1°C." Thus, unless the ice is only a half degree below freezing, the concept that pressure melts the ice is implausible.

Another possible scenario is heat conduction, given that metals are good conductors of heat. This explanation also falls short, however, since skating is frequently enjoyed when ambient temperatures are well below the freezing point of water.

Friction also fails as an explanation since skating is relatively frictionless.

The most likely explanation for the ease at which a skater glides over ice is a low coefficient of friction between ice and steel.

---

69. *Refer to Section 9.5.*

LiCl is a face-centered cube, thus,
$$4r = s\sqrt{2}$$

The length of one side $= s = 2(r_{\text{cation}} + r_{\text{anion}})$

The diagonal $= 4r_{\text{anion}} = s\sqrt{2}$

Substituting:
$$4r_{\text{anion}} = 2(r_{\text{cation}} + r_{\text{anion}})\sqrt{2}$$

$$4r_{\text{anion}} = 2\sqrt{2}\,r_{\text{cation}} + 2\sqrt{2}\,r_{\text{anion}}$$

$$\frac{4r_{\text{anion}}}{2\sqrt{2}\,r_{\text{anion}}} = \frac{2\sqrt{2}\,r_{\text{cation}}}{2\sqrt{2}\,r_{\text{anion}}} + \frac{2\sqrt{2}\,r_{\text{anion}}}{2\sqrt{2}\,r_{\text{anion}}} = \frac{r_{\text{cation}}}{r_{\text{anion}}} + 1$$

$$\frac{r_{\text{cation}}}{r_{\text{anion}}} = \frac{4}{2\sqrt{2}} - 1 = 0.414$$

The critical temperature of N_2 is -147°C, thus N_2 exists entirely as a gas at both 20°C and 10°C. Propane, on the other hand, has a critical temperature of 97°C and thus is present as an equilibrium between the vapor and liquid phases. The change in pressure observed with the N_2 is due to a "contraction" of the gas, or a reduction in the thermal motion of the N_2 molecules, while the pressure change with the propane is due to a change in vapor pressure. Changes in vapor pressure are much more sensitive to temperature changes than are changes in gas pressure.

Chapter 10: Solutions

---

2. *Refer to Section 10.1 and Example 10.2.*

a. The mass of acetone, C_3H_6O, and the mass of ethanol, C_2H_6O, must be calculated.

$$35.0 \, \text{mL} \times \frac{0.790 \, \text{g} \, C_3H_6O}{\text{mL} \, C_3H_6O} = 27.6 \, \text{g} \, C_3H_6O$$

$$50.0 \, \text{mL} \times \frac{0.789 \, \text{g} \, C_2H_6O}{\text{mL} \, C_2H_6O} = 39.4 \, \text{g} \, C_2H_6O$$

$$\text{mass percent of } C_3H_6O = \frac{\text{mass}_{C_3H_6O}}{\text{mass}_{C_2H_6O} + \text{mass}_{C_3H_6O}} \times 100\%$$

$$\frac{27.6 \, \text{g} \, C_3H_6O}{39.4 \, \text{g} \, C_3H_6O + 27.6 \, \text{g} \, C_3H_6O} \times 100\% = 41.2\%$$

b. $$\text{volume percent of } C_2H_6O = \frac{\text{volume}_{C_2H_6O}}{\text{volume}_{C_2H_6O} + \text{volume}_{C_3H_6O}} \times 100\%$$

$$\frac{50.0 \, \text{mL} \, C_2H_6O}{50.0 \, \text{mL} \, C_2H_6O + 35.0 \, \text{mL} \, C_3H_6O} \times 100\% = 58.8\%$$

c. The moles of acetone, C_3H_6O, and the moles of ethanol, C_2H_6O, must be calculated.

$$35.0 \, \text{mL} \times \frac{0.790 \, \text{g} \, C_3H_6O}{\text{mL} \, C_3H_6O} \times \frac{1 \, \text{mol.} \, C_3H_6O}{58.08 \, \text{g} \, C_3H_6O} = 0.476 \, \text{mol.} \, C_3H_6O$$

$$50.0 \, \text{mL} \times \frac{0.789 \, \text{g} \, C_2H_6O}{\text{mL} \, C_2H_6O} \times \frac{1 \, \text{mol.} \, C_2H_6O}{47.07 \, \text{g} \, C_2H_6O} = 0.856 \, \text{mol.} \, C_2H_6O$$

$$\text{mole fraction of } C_3H_6O = \frac{\text{mole}_{C_3H_6O}}{\text{mole}_{C_2H_6O} + \text{mole}_{C_3H_6O}}$$

$$\frac{0.476 \, \text{mol.} \, C_3H_6O}{0.856 \, \text{mol.} \, C_2H_6O + 0.476 \, \text{mol.} \, C_3H_6O} = 0.357$$

---

4. *Refer to Section 10.1 and Example 10.4.*

Recall that 0.90% by mass means 0.90 g NaCl per 100 g solution. Assume 100 g solution for simplification of calculations.

$$0.90 \, \text{g} \, \text{NaCl} \times \frac{1 \, \text{mol.} \, \text{NaCl}}{58.44 \, \text{g} \, \text{NaCl}} = 0.015 \, \text{mol.} \, \text{NaCl}$$

Since the density is 1.00 g/mL; 100 g solution = 100 mL

$$M = \frac{\text{moles solute}}{\text{L solution}} = \frac{0.015\,\text{mol. NaCl}}{0.100\,\text{L}} = 0.15\,M$$

---

6. Refer to Section 10.1.

$$\text{ppm} = \frac{\text{mass of solute}}{\text{mass of solution}} \times 10^6$$

$$0.250\,\text{ppm} = \frac{\text{mass (g) of Pb}}{1.00\,\text{g blood}} \times 10^6$$

mass of Pb = 2.50×10^{-7} g Pb

$$2.50 \times 10^{-7}\,\text{g Pb} \times \frac{1\,\text{mol. Pb}}{207.2\,\text{g Pb}} = 1.21 \times 10^{-9}\,\text{mol. Pb}$$

---

8. Refer to Section 10.1.

	Mass of Solute	Volume of Solution	Molarity
a.	**38.76 g**	1.370 L	0.08415 *M*
b.	12.01 g	**0.617 L**	0.0579 *M*
c.	26.44 g	2.750 L	**0.02860 *M***

a. $1.370\,\text{L} \times \dfrac{0.08415\,\text{mol. Ba(ClO}_4)_2}{\text{L}} \times \dfrac{336.2\,\text{g Ba(ClO}_4)_2}{1\,\text{mol. Ba(ClO}_4)_2} = 38.76\,\text{g Ba(ClO}_4)_2$

b. $12.10\,\text{g Ba(ClO}_4)_2 \times \dfrac{1\,\text{mol. Ba(ClO}_4)_2}{336.2\,\text{g Ba(ClO}_4)_2} \times \dfrac{\text{L}}{0.0579\,\text{mol. Ba(ClO}_4)_2} = 0.617\,\text{L}$

c. $\dfrac{26.44\,\text{g Ba(ClO}_4)_2}{2.750\,\text{L solution}} \times \dfrac{1\,\text{mol. Ba(ClO}_4)_2}{336.2\,\text{g Ba(ClO}_4)_2} = 0.02860\,M\,\text{Ba(ClO}_4)_2$

---

10. Refer to Section 10.1 and Examples 10.4 and 10.5.

	Molality	Mass Percent of Solvent	Ppm Solute	Mole Fraction of Solvent
a.	2.577	**86.58%**	**1.340 x 10⁵**	**0.9556**
b.	**20.4**	45.0	**5.50 x 10⁵**	**0.731**
c.	**0.07977**	99.5232%	4768	**0.9986**
d.	**12.6**	**57.0%**	**4.30 x 10⁵**	0.815

Since these are aqueous solutions, the solvent is water. The solute is urea, $CO(NH_2)_2$. Molar mass of urea: $(12.01) + 16.00 + 2(14.01 + 2(1.008)) = 60.06$ g/mol.

a. 2.577 m indicates that there are 2.577 mol. of solute per 1000 g of solvent and the calculations are based on that ratio.

$$2.577 \text{ mol.} \times \frac{60.06 \text{ g urea}}{1 \text{ mol. urea}} = 154.8 \text{ g urea}$$

$$1 \text{ kg} \times \frac{1000 \text{ g}}{1 \text{ kg}} \times \frac{1 \text{ mol. } H_2O}{18.02 \text{ g } H_2O} = 55.49 \text{ mol. } H_2O$$

$$\text{mass \%} = \frac{1000 \text{ g}}{(1000 \text{ g} + 154.8 \text{ g})} \times 100\% = \frac{1000 \text{ g}}{1155 \text{ g}} = \mathbf{86.58\%}$$

$$\text{ppm solute} = \frac{154.8 \text{ g}}{(1000 \text{ g} + 154.8 \text{ g})} \times 10^6 = \mathbf{1.340 \times 10^5}$$

$$X_{H_2O} = \frac{55.49 \text{ mol.}}{(55.49 \text{ mol.} + 2.577 \text{ mol.})} = \mathbf{0.9556}$$

b. 45.0 mass % of solvent means that 45.0 g of solvent are present for each 100.0 g of solution. Calculations are then based on this ratio.

$$45.0 \text{ g} \times \frac{1 \text{ mol. } H_2O}{18.02 \text{ g}} = 2.50 \text{ mol. } H_2O$$

mass of solute = 100.0 g (total) – 45.0 g (water) = 55.0 g $CO(NH_2)_2$

$$55.0 \text{ g urea} \times \frac{1 \text{ mol. urea}}{60.06 \text{ g urea}} = 0.916 \text{ mol. urea}$$

$$\text{molality} = \frac{0.916 \text{ mol. urea}}{0.0450 \text{ kg } H_2O} = \mathbf{20.4 \, m} \text{ (remember to convert mass of solvent to kg)}$$

$$\text{ppm solute} = \frac{55.0 \text{ g urea}}{100 \text{ g}} \times 10^6 = \mathbf{5.50 \times 10^5}$$

$$X_{H_2O} = \frac{2.50 \text{ mol.}}{(2.50 \text{ mol.} + 0.916 \text{ mol.})} = \mathbf{0.731}$$

c. $$\text{ppm} = \frac{\text{mass of solute}}{\text{mass of solution}} \times 10^6 = 4768 \text{ g}$$

If we assume a total mass of 10^6 g, then the mass of solute = 4768 g by definition. Any assumption for mass is valid here, this one was chosen for simplicity.

$$4768 \text{ g urea} \times \frac{1 \text{ mol. urea}}{60.06 \text{ g urea}} = 79.39 \text{ mol. urea}$$

$$10^6 \text{ g} - 4768 \text{ g} = 995232 \text{ g}; \quad 995232 \text{ g H}_2\text{O} \times \frac{1 \text{ mol.}}{18.0152 \text{ g}} = 55244.0 \text{ mol. H}_2\text{O}$$

$$X_{H_2O} = \frac{55244.0 \text{ mol.}}{(55244.0 \text{ mol.} + 79.39 \text{ mol.})} = \textbf{0.9986}$$

$$\text{mass \%} = \frac{9.95232 \times 10^5 \text{ g}}{10^6 \text{ g}} \times 100\% = \textbf{99.5232\%}$$

$$\text{molality} = \frac{79.39 \text{ mol.}}{995.232 \text{ kg}} = \textbf{0.07977 } \boldsymbol{m}$$

d. $X_{solvent} = 0.815$ indicates that there are 0.815 moles H_2O per 1 mole of solvent and solute combined. Consequently, there must be 0.185 mol. (1-0.815) urea.

$$0.185 \text{ mol. urea} \times \frac{60.06 \text{ g urea}}{1 \text{ mol. urea}} = 11.1 \text{ g urea}$$

$$0.815 \text{ mol. H}_2\text{O} \times \frac{18.02 \text{ g}}{1 \text{ mol.}} = 14.7 \text{ g H}_2\text{O}$$

$$\frac{0.185 \text{ mol. urea}}{0.0147 \text{ kg H}_2\text{O}} = \textbf{12.6 } \boldsymbol{m} \quad \text{(remember to convert mass of solvent to kg)}$$

$$\text{Mass \%} = \frac{\text{mass of solvent}}{\text{total mass}} \times 100\% = \frac{14.7 \text{ g}}{(11.1 \text{ g} + 14.7 \text{ g})} \times 100\% = \textbf{57.0\%}$$

$$\text{ppm} = \frac{\text{mass of solute}}{\text{mass of solution}} \times 10^6 = \frac{11.1 \text{ g urea}}{(11.1 \text{ g} + 14.7 \text{ g})} \times 10^6 = \textbf{4.30} \times \textbf{10}^5$$

---

12. Refer to Section 10.1 and Example 10.1.

a. A 0.750 M solution requires 0.750 mol. $Ba(OH)_2$ per 1 L of solution. Thus, calculate the mass of 0.750 mol. $Ba(OH)_2$ and dissolve that in sufficient water to make exactly one liter of solution.

$$0.750 \text{ mol. Ba(OH)}_2 \times \frac{171.32 \text{ g Ba(OH)}_2}{1 \text{ mol. Ba(OH)}_2} = 128 \text{ g Ba(OH)}_2$$

a. $[Ba(OH)_2]_c(V)_c = [Ba(OH)_2]_d(V)_d$
    $(6.00\ M)(V)_c = (0.750\ M)(1.00 \text{ L})$
    $(V)_c = 0.125 \text{ L}$ Thus, one would obtain 0.125 L of 6.00 M Ba(OH)$_2$ and add sufficient water to make one liter of solution.

174

14. *Refer to Section 10.1 and Example 10.1.*

a. $0.7850 \, \text{L} \times \dfrac{1.262 \, \text{mol. K}_2\text{S}}{\text{L}} \times \dfrac{110.27 \, \text{g K}_2\text{S}}{1 \, \text{mol. K}_2\text{S}} = 109.2 \, \text{g K}_2\text{S}$

b. $\dfrac{109.2 \, \text{g K}_2\text{S}}{2.000 \, \text{L}} \times \dfrac{1 \, \text{mol. K}_2\text{S}}{110.27 \, \text{g K}_2\text{S}} = 0.4951 \, M \, \text{K}_2\text{S}$

Remember that $K_2\text{S}$ dissociates in solution $K_2\text{S}(s) \rightarrow 2K^+(aq) + S^{2-}(aq)$

$\dfrac{0.4951 \, \text{mol. K}_2\text{S}}{\text{L solution}} \times \dfrac{2 \, \text{mol. K}^+}{1 \, \text{mol. K}_2\text{S}} = 0.9903 \, M \, \text{K}^+$

$\dfrac{0.4951 \, \text{mol. K}_2\text{S}}{\text{L solution}} \times \dfrac{1 \, \text{mol. S}^{2-}}{1 \, \text{mol. K}_2\text{S}} = 0.4951 \, M \, \text{S}^{2-}$

16. *Refer to Section 10.1 and Examples 10.1 and 10.4.*

29.89% NH_3 means that there are 29.89 g NH_3 per 100.00 g solution. Consequently, there must be (100.00 − 29.89 =) 70.11 g H_2O.

a. Molarity: Convert grams NH_3 to moles, and grams of solution to volume.

$29.89 \, \text{g NH}_3 \times \dfrac{1 \, \text{mol. NH}_3}{17.03 \, \text{g NH}_3} = 1.755 \, \text{mol. NH}_3$

$100.00 \, \text{g solution} \times \dfrac{1 \, \text{mL}}{0.8960 \, \text{g}} \times \dfrac{1 \, \text{L}}{1000 \, \text{mL}} = 0.1116 \, \text{L solution}$

$M = \dfrac{1.755 \, \text{mol. H}_3\text{PO}_4}{0.1116 \, \text{L}} = 15.73 \, M$

b. $M_cV_c = M_dV_d$
$(15.73 \, M)(0.2500 \, \text{L}) = M_d(3.00 \, \text{L})$
$M_d = 1.31 \, M$

18. *Refer to Section 10.1 and Example 10.4.*

	Density g/mL	Molarity (*M*)	Molality (*m*)	Mass % of Solute
a.	1.06	0.886	**0.940**	**11.0%**
b.	1.15	**2.27**	**2.66**	26.0%
c.	1.23	**2.70**	3.11	**29.1%**

a. 0.866 M indicates that there are 0.866 moles $(NH_4)_2SO_4$ per 1 L of solvent. For simplicity, our calculations will be based on a volume of 1 L. Molality requires moles of solute and kg of solvent, thus we must calculate mass of H_2O.

$$0.866 \text{ mol. } (NH_4)_2SO_4 \times \frac{132.15 \text{ g } (NH_4)_2SO_4}{1 \text{ mol. } (NH_4)_2SO_4} = 117 \text{ g } (NH_4)_2SO_4$$

$$1 \text{ L solution} \times \frac{1000 \text{ mL}}{1 \text{ L}} \times \frac{1.06 \text{ g}}{1 \text{ mL}} = 1060 \text{ g solution}$$

$$1060 \text{ g solution} - 117 \text{ g } (NH_4)_2SO_4 = 943 \text{ g } H_2O = 0.943 \text{ kg } H_2O$$

$$\frac{0.886 \text{ mol. } (NH_4)_2SO_4}{0.943 \text{ kg } H_2O} = \mathbf{0.940 \, m}$$

$$\text{Mass \%} = \frac{\text{mass of solute}}{\text{total mass}} \times 100\% = \frac{117 \text{ g}}{(117 \text{ g} + 943 \text{ g})} \times 100\% = \mathbf{11.0\%}$$

b. 26.0% indicates that there are 26.0 g $(NH_4)_2SO_4$ per 100 g of solution (and 74.0 g H_2O). For simplicity, our calculations will be based on these masses.

$$26.0 \text{ g } (NH_4)_2SO_4 \times \frac{1 \text{ mol. } (NH_4)_2SO_4}{132.15 \text{ g } (NH_4)_2SO_4} = 0.197 \text{ mol. } (NH_4)_2SO_4$$

$$\frac{0.197 \text{ mol. } (NH_4)_2SO_4}{0.0740 \text{ kg } H_2O} = \mathbf{2.66 \, m}$$

$$100 \text{ g solution} \times \frac{1 \text{ cm}^3}{1.15 \text{ g}} \times \frac{1 \text{ mL}}{1 \text{ cm}^3} \times \frac{1 \text{ L}}{1000 \text{ mL}} = 0.0870 \text{ L solution}$$

$$M = \frac{0.197 \text{ mol. } (NH_4)_2SO_4}{0.0870 \text{ L}} = \mathbf{2.27 \, M}$$

c. 3.11 m indicates that there are 3.11 mol. $(NH_4)_2SO_4$ per 1 kg of solvent. Again, our calculations will be based on these amounts.

$$3.11 \text{ mol. } (NH_4)_2SO_4 \times \frac{132.15 \text{ g } (NH_4)_2SO_4}{1 \text{ mol. } (NH_4)_2SO_4} = 411 \text{ g } (NH_4)_2SO_4$$

$$(1000 + 411) \text{ g solution} \times \frac{1 \text{ cm}^3}{1.23 \text{ g}} \times \frac{1 \text{ mL}}{1 \text{ cm}^3} \times \frac{1 \text{ L}}{1000 \text{ mL}} = 1.15 \text{ L solution}$$

$$M = \frac{3.11 \text{ mol. } (NH_4)_2SO_4}{1.15 \text{ L}} = \mathbf{2.70 \, M}$$

$$\text{Mass \%} = \frac{\text{mass of solute}}{\text{total mass}} \times 100\% = \frac{411 \text{ g}}{(1000 \text{ g} + 411 \text{ g})} \times 100\% = \mathbf{29.1\%}$$

The compound which exhibits intermolecular forces most similar to water will be the more soluble in water (like dissolves like). Recall that water has dispersion, dipole, and H-bonding forces.

a. CH_3Cl: dispersion forces
   CH_3OH: dispersion, dipole, and H-bonding forces
   CH_3OH would be more soluble since it shares H-bonding with water.

b. NI_3:   dispersion and dipole forces
   KI: ionic
   KI would be more soluble because it is ionic and ionic compounds generally exhibit high solubility in water.

c. $LiCl$:   ionic
   C_2H_5Cl: dispersion and dipole forces
   $LiCl$ would be more soluble because it is ionic and ionic compounds generally exhibit high solubility in water.

d. NH_3:   dispersion, dipole, and H-bonding forces
   CH_4:   dispersion and dipole forces
   NH_3 would be more soluble because of the H-bonding.

a. For the solution process, $NaOH(s) \rightarrow Na^+(aq) + OH^-(aq)$
   $\Delta H = [(1 \text{ mol.})(-240.1 kJ/mol.) + (1 \text{ mol.})(-230.0 kJ/mol.)] - [(1 \text{ mol.})(-425.6 kJ/mol.)]$
   $\Delta H = -44.5 kJ$

b. Since the solution process is exothermic ($\Delta H < 0$), an increase in temperature will decrease the solubility of NaOH.

a. $\dfrac{1.0 \times 10^{-3} M}{\text{atm}} \times \dfrac{1 \text{ atm}}{760.0 \text{ mm Hg}} = \dfrac{1.3 \times 10^{-6} M}{\text{mm Hg}}$

b. $C_g = kP_g = \dfrac{1.3 \times 10^{-6} M}{\text{mm Hg}} \times 693 \text{ mm Hg} = 9.1 \times 10^{-4} M$ argon

c. $25 \text{ L} \times 693 \text{ mm Hg} \times \dfrac{1.3 \times 10^{-6} \dfrac{\text{mol.}}{\text{L}}}{\text{mm Hg}} \times \dfrac{39.95 \text{ g Ar}}{1 \text{ mol. Ar}} = 0.90 \text{ g Ar}$

26. Refer to Section 10.2 and Example 10.5.

a. 21 mol. % O_2 is identical to $X = 0.21$.

$$P_{O_2} = X_{O_2} \cdot P_{tot} = (0.21)(1.00 \text{ atm}) = 0.21 \text{ atm}$$

$$0.21 \text{ atm} \times \frac{3.30 \times 10^{-4} M}{1 \text{ atm}} = 6.9 \times 10^{-5} M$$

$$1.00 \text{ L} \times \frac{6.9 \times 10^{-5} \text{ mol. } O_2}{1 \text{ L}} \times \frac{32.00 \text{ g } O_2}{1 \text{ mol. } O_2} = 2.2 \times 10^{-3} \text{ g } O_2$$

b. $$0.21 \text{ atm} \times \frac{2.85 \times 10^{-4} M}{1 \text{ atm}} = 6.0 \times 10^{-5} M$$

$$1.00 \text{ L} \times \frac{6.0 \times 10^{-5} \text{ mol. } O_2}{1 \text{ L}} \times \frac{32.00 \text{ g } O_2}{1 \text{ mol. } O_2} = 1.9 \times 10^{-3} \text{ g } O_2$$

c. 2.2×10^{-3} g - 1.9×10^{-3} g = 0.3×10^{-3} g = 3×10^{-4} g

$$\frac{3 \times 10^{-4} \text{ g}}{2.2 \times 10^{-3} \text{ g}} \times 100\% = 10\% \qquad \text{(note that this answer has only one significant figure)}$$

28. Refer to Sections 10.1 and 10.3, and Example 10.6.

a. $$\Delta P = (X_{C_2H_6O_2})(P^\circ_{H_2O}) = (0.288)(657.6 \text{ mm Hg}) = 189 \text{ mm Hg}$$

$$\Delta P = P^\circ_{H_2O} - P_{H_2O} \quad \Rightarrow \quad 189 \text{ mm Hg} = 657.6 \text{ mm Hg} - P_{H_2O}$$

$$P_{H_2O} = 469 \text{ mm Hg}$$

b. 39.0 mass % means 39.0 g $C_2H_6O_2$ per 100 g solution, while the remainder (61.0 g) must be water.

$$39.0 \text{ g } C_2H_6O_2 \times \frac{1 \text{ mol. } C_2H_6O_2}{62.07 \text{ g } C_2H_6O_2} = 0.628 \text{ mol. } C_2H_6O_2$$

$$61.0 \text{ g } H_2O \times \frac{1 \text{ mol. } H_2O}{18.02 \text{ g } H_2O} = 3.39 \text{ mol. } H_2O$$

$$X_{C_2H_6O_2} = \frac{0.628 \text{ mol. } C_2H_6O_2}{(3.39 + 0.628) \text{ mol.}} = 0.156$$

$$\Delta P = (X_{C_2H_6O_2})(P^\circ_{H_2O}) = (0.156)(657.6 \text{ mm Hg}) = 103 \text{ mm Hg}$$

$$P_{H_2O} = 657.6 - 103 = 555 \text{ mm Hg}$$

c. 2.42 *m* means 2.42 moles $C_2H_6O_2$ are present in 1.00 kg solvent (H_2O). Calculate the moles of water, and from that, the mole fraction. Then proceed as above.

$$1.00 \text{ kg } H_2O \times \frac{1000 \text{ g}}{1 \text{ kg}} \times \frac{1 \text{ mol. } H_2O}{18.02 \text{ g } H_2O} = 55.5 \text{ mol. } H_2O$$

$$X_{C_2H_6O_2} = \frac{2.42 \text{ mol. } C_2H_6O_2}{(55.5 + 2.42) \text{ mol.}} = 0.0418$$

$$\Delta P = (X_{C_2H_6O_2})(P_{H_2O}^{\circ}) = (0.0418)(657.6 \text{ mm Hg}) = 27.5 \text{ mm Hg}$$

$$P_{H_2O} = 657.6 - 27.5 = 630.1 \text{ mm Hg}$$

---

30. *Refer to Section 10.3.*

The first step is to calculate the mole fraction of oxalic acid.

$$\Delta P = P_{H_2O}^{\circ} - P_{\text{solution}} = 22.38 \text{ mm Hg} - 21.97 \text{ mm Hg} = 0.41 \text{ mm Hg}$$

$$\Delta P = (X_{H_2C_2O_4})(P_{H_2O}^{\circ}) \implies 0.41 \text{ mm Hg} = (X_{H_2C_2O_4})(22.38 \text{ mm Hg})$$

$$X_{H_2C_2O_4} = 0.018$$

Assuming 1 mole total, this means we have 0.018 mol. $H_2C_2O_4$ and 0.982 mol. water (1.00 - 0.018 = 0.982). The next step is to calculate the masses associated with these quantities, and from that the mass of solution and volume of solution.

$$0.982 \text{ mol. } H_2O \times \frac{18.02 \text{ g } H_2O}{1 \text{ mol. } H_2O} = 17.7 \text{ g } H_2O$$

$$0.018 \text{ mol. } H_2C_2O_4 \times \frac{90.04 \text{ g } H_2C_2O_4}{1 \text{ mol. } H_2C_2O_4} = 1.6 \text{ g } H_2C_2O_4$$

$$(17.7 + 1.6) \text{ g solution} \times \frac{1 \text{ mL}}{1.05 \text{ g}} \times \frac{1 \text{ L}}{1000 \text{ mL}} = 0.0184 \text{ L}$$

Now one can either:

1. Calculate molarity (mol./L) and convert mol./L to grams/L (using molecular mass) or

2. Directly calculate grams of $H_2C_2O_4$ in one liter (as shown below)

$$1.00 \text{ L solution} \times \frac{1.6 \text{ g } H_2C_2O_4}{0.0184 \text{ L solution}} = 87 \text{ g } H_2C_2O_4$$

Thus, to prepare the prescribed solution, one must dissolve 87 g $H_2C_2O_4$ in enough water to make 1.00 L of solution.

32. _Refer to Section 10.3 and Example 10.8._

$$0.250 \text{ g pepsin} \times \frac{1 \text{ mol.}}{3.50 \times 10^4 \text{ g}} = 7.14 \times 10^{-6} \text{ mol. pepsin}$$

$$\frac{7.14 \times 10^{-6} \text{ mol.}}{0.0550 \text{ L}} = 1.30 \times 10^{-4} \ M$$

$$\pi = MRT$$

$$\pi = (1.30 \times 10^{-4} \ M)(0.0821 \text{ L·atm/mol.·K})(303 \text{ K})$$

$$\pi = 0.00323 \text{ atm}$$

$$0.00323 \text{ atm} \times \frac{760 \text{ mm Hg}}{1 \text{ atm}} = 2.45 \text{ mm Hg}$$

34. _Refer to Section 10.3 and Example 10.8._

$$0.118 \text{ mm Hg} \times \frac{1 \text{ atm}}{760 \text{ mm Hg}} = 1.55 \times 10^{-4} \text{ atm}$$

$$\pi = MRT$$

$$1.55 \times 10^{-4} \text{ atm} = M(0.0821 \text{ L·atm/mol.·K})(298 \text{ K})$$

$$M = 6.34 \times 10^{-6} \ M$$

$$0.225 \text{ L} \times \frac{6.34 \times 10^{-6} \text{ mol.}}{1 \text{ L}} = 1.43 \times 10^{-6} \text{ mol. lysozyme}$$

$$\text{molar mass} = \frac{0.020 \text{ g lysozyme}}{1.43 \times 10^{-6} \text{ mol. lysozyme}} = 1.4 \times 10^4 \text{ g/mol.}$$

36. _Refer to Section 10.3._

a. $\Delta T_b = k_b m$

$2.0°\text{C} = (2.75°\text{C}/m)(m)$

$m = 0.73$

$$0.1000 \text{ kg cyclohexane} \times \frac{0.73 \text{ mol. } C_6H_8O_7}{1 \text{ kg cyclohexane}} \times \frac{192.2 \text{ g } C_6H_8O_7}{1 \text{ mol. } C_6H_8O_7} = 14 \text{ g } C_6H_8O_7$$

$\Delta T_f = k_f m$

$1.0°\text{C} = (20.2°\text{C}/m)(m)$

$m = 0.050$

$$0.1000 \text{ kg cyclohexane} \times \frac{0.050 \text{ mol. } C_6H_8O_7}{1 \text{ kg cyclohexane}} \times \frac{192.2 \text{ g}}{1 \text{ mol. } C_6H_8O_7} = 0.96 \text{ g } C_6H_8O_7$$

b. $\Delta T_b = k_b m$

   $2.0°C = (2.75°C/m)(m)$

   $m = 0.73$

   $0.1000 \text{ kg cyclohexane} \times \dfrac{0.73 \text{ mol.}}{1 \text{ kg cyclohexane}} \times \dfrac{194.2 \text{ g } C_8H_{10}N_4O_2}{1 \text{ mol.}} = 14 \text{ g } C_8H_{10}N_4O_2$

   $\Delta T_f = k_f m$

   $1.0°C = (20.2°C/m)(m)$

   $m = 0.050$

   $0.1000 \text{ kg cyclohexane} \times \dfrac{0.050 \text{ mol.}}{1 \text{ kg cyclohexane}} \times \dfrac{194.2 \text{ g } C_8H_{10}N_4O_2}{1 \text{ mol.}} = 0.97 \text{ g } C_8H_{10}N_4O_2$

---

38. Refer to Section 10.3 and Example 10.7, and Table 7.2.

a. $t_{°F} = 1.8 t_{°C} + 32$

   $-20°F = 1.8 t_{°C} + 32 \Rightarrow t_{°C} = -29°C$

   $\Delta T_f = 0 - (-29°C)$

   $\Delta T_f = k_f m$

   $29°C = (1.86°C/m)(m)$

   $m = 16\ m$

   The minimum molality of antifreeze required to protect the car engine to $-20°F$ is **16m**.

b. $250 \text{ mL } H_2O \times \dfrac{1 \text{ g } H_2O}{1 \text{ mL } H_2O} \times \dfrac{1 \text{ kg}}{1000 \text{ g}} = 0.250 \text{ kg } H_2O$

   $0.250 \text{ kg } H_2O \times \dfrac{16 \text{ mol. } C_2H_6O_2}{1 \text{ kg } H_2O} \times \dfrac{62.07 \text{ g } C_2H_6O_2}{1 \text{ mol. } C_2H_6O_2} \times \dfrac{1 \text{ mL}}{1.12 \text{ g } C_2H_6O_2} = 2.2 \times 10^2 \text{ mL } C_2H_6O_2$

   $m = \dfrac{\text{mol. solute } (C_2H_6O_2)}{\text{kg solvent}} = \dfrac{7.22 \text{ mol. } C_2H_6O_2}{0.600 \text{ kg } H_2O} = 12.0\ m$

---

40. Refer to Sections 10.3 and 3.3, and Example 10.7.

Use the freezing point depression to calculate molality, and from that, moles of the compound and the compound's molecular mass.

$\Delta T_f = T_f° - T_f = 178.40°C - 173.44°C = 4.96°C$

$\Delta T_f = k_f m$

$4.96°C = (40.0°C/m)(m)$

$m = 0.124 \; m$

$$50.00 \text{ g camphor} \times \frac{1 \text{ kg}}{1000 \text{ g}} = 0.0500 \text{ kg camphor}$$

$$m = \frac{\text{mol. solute}}{\text{kg solvent}} \implies 0.124 \; m = \frac{x \text{ mol.}}{0.05000 \text{ kg camphor}}$$

$x = 0.00620 \text{ mol.}$

$$\frac{2.50 \text{ g}}{0.00620 \text{ mol.}} = 403 \text{ g/mol.}$$

---

42. *Refer to Sections 10.3 and 3.3.*

Use the freezing point depression to calculate molality, and from that, moles of the compound and the compound's molecular mass. Then calculate empirical formula and molecular formula.

$\Delta T_f = T_f^\circ - T_f = 6.50°C - 0.0°C = 6.5°C$

$\Delta T_f = k_f m$

$6.5°C = (20.2°C/m)(m)$
$m = 0.322 \; m$

$$75.0 \text{ mL cyclohexane} \times \frac{1 \text{ cm}^3}{1 \text{ mL}} \times \frac{0.779 \text{ g}}{1 \text{ cm}^3} \times \frac{1 \text{ kg}}{1000 \text{ g}} = 0.0584 \text{ kg cyclohexane}$$

$$m = \frac{\text{mol. solute}}{\text{kg solvent}} \implies 0.322 \; m = \frac{x \text{ mol.}}{0.0584 \text{ kg cyclohexane}}$$

$x = 0.0188 \text{ mol.}$

$$\frac{3.16 \text{ g}}{0.0188 \text{ mol.}} = 168 \text{ g/mol.}$$

Empirical formula:

$$42.9 \text{ g C} \times \frac{1 \text{ mol. C}}{12.01 \text{ g C}} = 3.57 \text{ mol. C}$$

$$2.4 \text{ g H} \times \frac{1 \text{ mol. H}}{1.008 \text{ g H}} = 2.38 \text{ mol. H}$$

$$16.6 \text{ g N} \times \frac{1 \text{ mol. N}}{14.01 \text{ g N}} = 1.18 \text{ mol. N}$$

$$38.1 \text{ g O} \times \frac{1 \text{ mol. O}}{16.00 \text{ g O}} = 2.38 \text{ mol. O}$$

(3.57 mol. C) / (1.18 mol.) = 3
(2.38 mol. H) / (1.18 mol.) = 2
(1.18 mol. N) / (1.18 mol.) = 1
(2.38 mol. O) / (1.18 mol.) = 2

empirical formula = $C_3H_2NO_2$ (emp. mass = 84.06 g/mol.)

$\frac{168}{84} = 2$, thus multiply the subscripts of empirical formula by two ($C_{3 \times 2}H_{2 \times 2}N_{1 \times 2}O_{2 \times 2}$).

empirical formula = $C_6H_4N_2O_4$ (mol. mass = 168 g/mol.)

---

44. *Refer to Section 10.3 and Problem 42 (above).*

$\Delta T_f = k_f m$ $\qquad\qquad$ $\Delta T_f = 5.50°C - 3.03°C = 2.47°C$

$2.47°C = (5.10°C/m)(m)$
$m = 0.484 \ m$

$$100.0 \text{ mL benzene} \times \frac{0.877 \text{ g benzene}}{1 \text{ mL benzene}} \times \frac{1 \text{ kg}}{1000 \text{ g}} \times \frac{0.484 \text{ mol. caffeine}}{1 \text{ kg benzene}} = 0.0424 \text{ mol. caffeine}$$

$$\text{Molar mass} = \frac{8.25 \text{ g caffeine}}{0.0424 \text{ mol. caffeine}} = 194 \text{ g/mol.}$$

Caffeine is given as having a composition of 49.5% C, 5.2% H, 16.5% O, and 28.9% N.

$$\frac{49.5 \text{ g C}}{12.01 \text{ g/mol. C}} = 4.12 \text{ mol. C} \qquad \frac{4.12 \text{ mol. C}}{1.03} = 4.00 \text{ mol. C}$$

$$\frac{5.2 \text{ g H}}{1.01 \text{ g/mol. H}} = 5.15 \text{ mol. H} \qquad \frac{5.15 \text{ mol. H}}{1.03} = 5.00 \text{ mol. H}$$

$$\frac{28.9 \text{ g N}}{14.01 \text{ g/mol. N}} = 2.06 \text{ mol. N} \qquad \frac{2.06 \text{ mol. N}}{1.03} = 2.00 \text{ mol. N}$$

$$\frac{16.5 \text{ g O}}{16.00 \text{ g/mol. O}} = 1.03 \text{ mol. O} \qquad \frac{1.03 \text{ mol. O}}{1.03} = 1.00 \text{ mol. O}$$

The empirical formula of caffeine is $C_4H_5N_2O$ and the empirical mass is 97.1 g/mol.

$$\frac{\text{molar mass}}{\text{empirical mass}} = \text{multiple for molecular formula}$$

$$\frac{194 \text{ g/mol.}}{97.1 \text{ g/mol.}} = 2$$

The molecular formula of caffeine is $C_{4 \times 2}H_{5 \times 2}N_{2 \times 2}O_2 = \mathbf{C_8H_{10}N_4O_2}$.

---

46. Refer to Section 10.3.

$\pi = MRT$

7.7 atm = (M)(0.0821 L·atm/mol.·K)(298 K)

$M = 0.32$ mol./L

---

48. Refer to Section 10.3 and Example 10.8.

$$4.60 \text{ mm Hg} \times \frac{1 \text{ atm}}{760 \text{ mm Hg}} = 6.05 \times 10^{-3} \text{ atm}$$

$\pi = MRT$

6.05×10^{-3} atm = (M)(0.0821 L·atm/mol.·K)(293 K)

$M = 2.52 \times 10^{-4}$ mol./L

$$0.200 \text{ L} \times \frac{2.52 \times 10^{-4} \text{ mol.}}{1 \text{ L}} = 5.03 \times 10^{-5} \text{ mol.}$$

$$\frac{3.27 \text{ g hemoglobin}}{5.03 \times 10^{-5} \text{ mol.}} = 6.50 \times 10^4 \text{ g/mol.}$$

---

50. Refer to Section 10.4 and Example 10.11.

All molecules are of the same concentration, so determine which molecule produces the greatest number of species in solution (i). The ones with the larger i's will be the ones with the lowest freezing point and the highest boiling point.

a. $Fe(NO_3)_3 \rightarrow Fe^{3+}(aq) + 3NO_3^-(aq)$ $\qquad\qquad i = 4$

b. $C_2H_5OH(aq) \rightarrow C_2H_5OH(aq)$ $\qquad\qquad\qquad i = 1$

c. $Ba(OH)_2 \rightarrow Ba^{2+}(aq) + 2OH^-(aq)$ $\qquad\qquad i = 3$

d. $CaCr_2O_7 \rightarrow Ca^{2+}(aq) + Cr_2O_7^{2-}(aq)$ $\qquad\quad i = 2$

freezing point: $\quad Fe(NO_3)_3 < Ba(OH)_2 < CaCr_2O_7 < C_2H_5OH$
boiling point: $\qquad C_2H_5OH < CaCr_2O_7 < Ba(OH)_2 < Fe(NO_3)_3$

Calculate the moles of sucrose in 1 kg of maple syrup, then calculate the molality.
1 kg syrup is 66% sucrose, thus 660 g sucrose, with the remainder (340 g) being the water.

$$660 \text{ g } C_{12}H_{22}O_{11} \times \frac{1 \text{ mol. } C_{12}H_{22}O_{11}}{342.3 \text{ g } C_{12}H_{22}O_{11}} = 1.93 \text{ mol. } C_{12}H_{22}O_{11}$$

$$\frac{1.93 \text{ mol. } C_{12}H_{22}O_{11}}{0.340 \text{ kg } H_2O} = 5.68 \, m$$

$\Delta T_f = k_f m$

$\Delta T_f = (1.86°C/m)(5.68m)$
$\Delta T_f = 11°C$
$T_f = -11°C$

a. $32.48 \text{ g FeCl}_3 \times \dfrac{1 \text{ mol. FeCl}_3}{162.2 \text{ g FeCl}_3} = 0.2002 \text{ mol. FeCl}_3$

   $\text{Molarity} = \dfrac{0.2002 \text{ mol.}}{0.1000 \text{ L}} = 2.002 \, M$

b. $100.0 \text{ mL solution} \times \dfrac{1.249 \text{ g}}{1 \text{ mL}} = 124.9 \text{ g solution}$

   $124.9 \text{ g solution} - 32.48 \text{ g FeCl}_3 = 92.4 \text{ g } H_2O = 0.0924 \text{ kg } H_2O$

   $\text{Molality} = \dfrac{0.2002 \text{ mol.}}{0.0924 \text{ kg}} = 2.17 \, m$

c. $\pi = iMRT$

   $\pi = (4)(2.002 \text{ mol./L})(0.0821 \text{ L·atm/mol.·K})(298 \text{ K})$

   $\pi = 196 \text{ atm}$

d. $\Delta T_f = i k_f m$

   $\Delta T_f = (4)(1.86°C/m)(2.17 \, m) = 16.1°C$

   $T_f = 0°C - 16.1°C = -16.1°C$

a. $132 \text{ mL chloroform} \times \dfrac{1.49 \text{ g chloroform}}{\text{mL}} \times \dfrac{\text{kg}}{1000 \text{ g}} = 0.197 \text{ kg chloroform}$

$\dfrac{0.146 \text{ mol. nonelectrolyte}}{0.197 \text{ kg chloroform}} = 0.741 \text{ mol./kg}$

$\Delta T = 64.4^\circ C - 61.7^\circ C = 2.7^\circ C$
$\Delta T_b = k_b m$
$2.7^\circ C = k_b \times 0.741 m$
$k_b = 3.6^\circ C/m$

b. $\Delta T = 66.7^\circ C - 61.7^\circ C = 5.0^\circ C$

$427.5 \text{ mL chloroform} \times \dfrac{1.49 \text{ g chloroform}}{\text{mL}} \times \dfrac{\text{kg}}{1000 \text{ g}} = 0.637 \text{ kg chloroform}$

$? \, m \text{ electrolyte} = \dfrac{45.2 \text{ g electrolyte} \left(\dfrac{\text{mol. electrolyte}}{154 \text{ g electrolyte}} \right)}{0.637 \text{ kg chloroform}} = 0.461 \text{ mol./kg}$

$\Delta T_b = i \, k_b m$
$5.0^\circ C = i \, (3.6^\circ C/m)(0.461 m)$
$i = 3$

a. $P_{radon} = X_{radon} P_{tot}$
$P_{radon} = (2.7 \times 10^{-6})(28 \text{ atm}) = 7.6 \times 10^{-5} \text{ atm} = 5.8 \times 10^{-2} \text{ mm Hg}$
$C_g = kP_g$
$C_g = (9.57 \times 10^{-6} \, M/\text{mm Hg})(5.8 \times 10^{-2} \text{ mm Hg}) = 5.5 \times 10^{-7} \, M$

b. Note 1.0 g/mL = 1000 g/L

$\dfrac{5.5 \times 10^{-7} \text{ mol.}}{1 \text{ L sample}} \times \dfrac{1 \text{ L}}{1000 \text{ g}} \times \dfrac{222 \text{ g radon}}{1 \text{ mol. radon}} = \dfrac{1.2 \times 10^{-7} \text{ g radon}}{\text{g sample}}$

$(1.2 \times 10^{-7} \text{ g}) \times 10^6 = 0.12 \text{ ppm}$

60. Refer to Section 10.1.

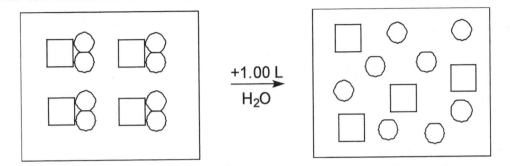

There are 4 moles $CaCl_2$ in 1.00 L, thus the molarity is 4.

$CaCl_2 \rightarrow Ca^{2+} + 2Cl^-$

$$\frac{4.0 \text{ mol. } CaCl_2}{1\,L} \times \frac{1 \text{ mol. } Ca^{2+}}{1 \text{ mol. } CaCl_2} = 4.0\ M\ Ca^{2+}$$

$$\frac{4.0 \text{ mol. } CaCl_2}{1\,L} \times \frac{2 \text{ mol. } Cl^-}{1 \text{ mol. } CaCl_2} = 8.0\ M\ Cl^-$$

62. Refer to Section 10.1.

a. **Solution 2.** Since the molar mass of X > the molar mass of Y, the molarity of X solution < molarity of the Y solution, since there will be fewer moles of X. Therefore, solution 2 will have the higher molarity.

b. **Neither.** Since each solution contains equal amounts of solute and equal amounts of solvent, they will have equal mass percents.

c. **Solution 2.** Since the molar mass of X > the molar mass of Y, the molality of X solution < molality of the Y solution, since there will be fewer moles of X. Therefore, solution 2 will have the higher molality.

d. **Neither.** Since they are both nonelectrolytes, both will have $i = 1$.

e. **Solution 1.** Since the molar mass of X > the molar mass of Y, there will be a larger number of moles of Y than X. Therefore, the solution with the larger mole fraction of solvent will be the solution with the fewest number of moles of solute. Solution 1 has the fewest number of moles of solute.

64. Refer to Section 10.4.

$\Delta T_f = k_f m\ i$
$0.38°C = (1.86°C/m)(0.20\ m)(i)$
$i = 1$ Thus, HF is non-ionized; if it were ionized, $i = 2$.

For dilute solution, $M \approx m$

$\Delta T_f = k_f m \, i$
$0.38°C = (1.86°C/m)(0.10 \, m)(i)$
$i = 2$

Thus, two moles of ions are formed, indicating that the best equation is **b**.

a. A saturated solution is not always a concentrated solution. If the solubility is low, a saturated solution will be a dilute solution.

b. The water solubility of a solid decreases with a drop in temperature if the solubility process is endothermic. If the solubility process is exothermic, the water solubility of a solid increases with a drop in temperature.

c. Molarity and molality are approximately equal if the solutions are dilute. They are not equal if the solutions are concentrated.

d. $CaCl_2$: $i = 3$; KCl: $i = 2$. Thus, the freezing point depression for $0.10 \, m$ $CaCl_2$ is $^3/_2$ that for $0.10 \, m$ KCl.

e. Sucrose: $i = 1$; NaCl: $i = 2$. Thus, the osmotic pressure for sucrose is ½ that for NaCl.

a. Determine if the solution conducts a current. If so, it is an electrolyte.

b. Beer is a solution of CO_2 (among other things). As it warms, the solubility of the CO_2 decreases because gas solubility decreases with increasing temperature.

c. Molality is moles of solute per kg of solvent, while mole fraction is moles of solute per total moles (solvent plus solute). In general, the kg of solvent is considerably less than the moles of solvent or even moles of solvent + solute, thus molality is usually greater than mole fraction.

d. The presence of a (non-volatile) solute raises the boiling point of the solvent because it lowers the vapor pressure of the solvent. (Remember that boiling point is the temperature at which the vapor pressure equals the ambient pressure.) This vapor pressure lowering is a consequence of the relative disorders between the pure solvent and solvent vapor and between solution and solvent vapor.

From the change in boiling point, one can calculate the molality and from that the mass of solute in 1.00 kg of solvent. This also yields the total mass of the solution. To calculate density, one also needs to know the volume of solution. From osmotic pressure, one can calculate molarity. Using the moles of solute and mass of **solution** calculated from molality, one can now calculate density.

$$KNO_3(s) \rightarrow K^+(aq) + NO_3^-(aq) \qquad (i = 2)$$

$$\Delta T_b = 103.0°C - 100.0°C = 3.0°C$$

$$\Delta T_b = k_b mi$$

$$3.0°C = (2)(0.52°C/m)(m)$$

$$m = 2.9\ m$$

$$\frac{2.9\ mol.\ KNO_3}{1.000\ kg\ solvent} \times \frac{101\ g}{1\ mol.\ KNO_3} = 290\ g\ KNO_3/1.000\ kg\ solvent$$

Thus the total mass of solution is: 290 g solute + 1000 g solvent = 1290 g solution.

$$\pi = MRTi$$
$$122\ atm = (M)(0.0821\ L·atm/mol.·K)(298\ K)(2)$$
$$M = 2.49\ M$$

$$\frac{2.49\ mol.\ KNO_3}{1.000\ L\ solution} \times \frac{1290\ g\ solution}{2.9\ mol.\ KNO_3} \times \frac{1\ L}{1000\ mL} = 1.1\ g/mL$$

Molality is moles solute per kilogram of solvent, so convert mass of solute to moles, and liters of solution to kg of solvent.

$$158.2\ g\ KOH \times \frac{1\ mol.\ KOH}{56.11\ g\ KOH} = 2.820\ mol.\ KOH$$

$$1.000\ L\ solution \times \frac{1000\ mL}{1\ L} \times \frac{1.13\ g}{1\ mL} = 1130\ g\ solution$$

1130 g solution - 158.2 g solute = 972 g solvent = 0.972 kg solvent

$$molality = \frac{2.820\ mol.}{0.972\ kg} = 2.90\ m$$

This solution is about 2.9 m, approximately 10 times stronger than the solution the technician needs. Thus, he will need to increase the volume approximately 10 fold (to ~1 L).

If there are 2.820 moles KOH in 1.000 L, then there are 0.2820 moles in the 100 mL sample. Likewise, the 100 mL sample also has only 0.0972 kg of solvent. The problem can then be solved by setting up a ratio.

$$\frac{0.250 \text{ mol. KOH}}{1.000 \text{ kg solvent}} = \frac{0.2820 \text{ mol. KOH}}{(0.0972 \text{ kg} + x \text{ kg) solvent}}$$

$$0.0972 \text{ kg H}_2\text{O} + x \text{ kg H}_2\text{O} = 1.128 \text{ kg H}_2\text{O}$$

$$x = 1.03 \text{ kg H}_2\text{O}$$

Thus, the technician must add 1.03 kg (1.03 L) H_2O to the 100 mL of KOH solution.

---

74. Refer to Section 10.1.

One can prove the generality of this equation in one of two ways:
1. Combine simple relationships (i.e. m = moles solute/ kg solvent) to come up with the complex equation given (as is done in the answer in the back of the text book).
2. Simplify the complex equation to come up with a final simple relationship which we already know to be true (as is done below).

$$m = \frac{M}{d - \frac{(MM)(M)}{1000}} = \frac{\dfrac{\text{moles of solute}}{\text{L of solution}}}{\dfrac{\text{g solution}}{\text{mL of solution}} - \dfrac{\left(\dfrac{\text{g solute}}{\text{moles solute}}\right)\left(\dfrac{\text{moles of solute}}{\text{L of solution}}\right)}{1000}}$$

Multiplying through the $(MM)(M)$ terms simplifies the equation slightly to:

$$m = \frac{\dfrac{\text{moles of solute}}{\text{L of solution}}}{\dfrac{\text{g solution}}{\text{mL of solution}} - \dfrac{\dfrac{\text{g solute}}{\text{L of solution}}}{1000}}$$

Since $\dfrac{\dfrac{\text{g solute}}{\text{L of solution}}}{1000} = \dfrac{\text{g solute}}{(1000)(\text{L of solution})}$ and $(1000)(L) = mL$, we get:

$$m = \frac{\dfrac{\text{moles of solute}}{\text{L of solution}}}{\dfrac{\text{g solution}}{\text{mL of solution}} - \dfrac{\text{g solute}}{\text{mL of solution}}} = \frac{\dfrac{\text{moles of solute}}{\text{L of solution}}}{\dfrac{\text{g solution - g solute}}{\text{mL of solution}}} .$$

Recognizing that (g solution) – (g solute) = (g solvent), one can substitute and further simplify to:

$$m = \frac{\dfrac{\text{moles of solute}}{\text{L of solution}}}{\dfrac{\text{g solvent}}{\text{mL of solution}}} = \left(\frac{\text{moles of solute}}{\text{L of solution}}\right)\left(\frac{\text{mL of solution}}{\text{g solvent}}\right)$$

Recognizing that (1000)(L) = mL, and likewise (1000)(kg) = g, one can substitute and simplify the equation further to:

$$m = \frac{\text{moles of solute}}{\text{L of solution}} \times \frac{(1000)(\text{L of solution})}{(1000)(\text{kg solvent})} = \frac{\text{moles of solute}}{\text{kg solvent}}$$

This is the definition of molality, so the equation is true and therefore valid for any solution.

The second part of the question asks why $m = M$ for dilute aqueous solutions. This is because such solutions have a density of approximately 1. Consequently, L solution = kg solvent.

Substituting into the first equation above, one gets:

$$m = \frac{M}{1 - \dfrac{\left(\dfrac{\text{g solute}}{\text{moles solute}}\right)\left(\dfrac{\text{moles of solute}}{\text{kg of solvent}}\right)}{1000}} = \frac{M}{1 - \dfrac{\text{g solute}}{\text{g of solvent}}}$$

For dilute solutions, grams of solute is so much smaller than grams of solvent that

$$1 - \frac{\text{g solute}}{\text{g of solvent}} \approx 1 \quad \text{and so} \quad m = \frac{M}{1} = M.$$

---

75. *Refer to Section 10.3.*

From freezing point depression one can calculate the molality of the solution. Since the mass of the solvent is given, one can also calculate the moles of solute. From the molar masses of the two solutes and the mass of the mixture, one can then determine the mole ratio.

$\Delta T_f = k_f m$

$0.500°C = (1.86°C/m)(m)$

$m = 0.269\ m$

$$0.269\ m = \frac{\text{mol. solute}}{1.00 \times 10^{-3}\ \text{kg solvent}} \Rightarrow \text{moles of solute} = 2.69 \times 10^{-4}\ \text{mol.}$$

Thus, the total moles of solute (moles sugar + moles X) = 2.69×10^{-4} mol.

If mass of solute = x g, then mass of sugar = $(0.100 - x)$ g.

Combining the above we get

$$x \text{ g} \times \frac{1 \text{ mol. X}}{410 \text{ g}} + (0.100 \text{ g} - x \text{ g}) \times \frac{1 \text{ mol. sugar}}{342 \text{ g}} = 2.69 \times 10^{-4} \text{ mol.}$$

$$\frac{x \text{ mol.}}{410} + \frac{0.100}{342} - \frac{x \text{ mol.}}{342} = 2.69 \times 10^{-4} \text{ mol.} \Rightarrow \frac{x \text{ mol.}}{410} - \frac{x \text{ mol.}}{342} = -2.34 \times 10^{-5} \text{ mol.}$$

$$\frac{(410)(x \text{ mol.})}{410} - \frac{(410)(x \text{ mol.})}{342} = (410)(-2.34 \times 10^{-5} \text{ mol.})$$

x mol. $- 1.20x$ mol. $= -9.59 \times 10^{-3}$ mol.

$-0.20x$ mol. $= -9.59 \times 10^{-3}$ mol.

$x = 0.048$ and

mass percent of $X = \dfrac{0.048 \text{ g}}{0.100 \text{ g}} \times 100\% = 48\%$

---

76. Refer to Chapters 1 and 3.

Calculate the amount of alcohol that enters the blood (recall that to convert a percent to a decimal, one divides by 100). Then, calculate the concentration of alcohol in the blood.

142 g/martini x 2 martinis x 0.30 (30% alcohol) x 0.15 (15% enter bloodstream) = 13 g alcohol.

$$7.0 \text{ L blood} \times \frac{1000 \text{ mL}}{1 \text{ L}} = 7000 \text{ mL}$$

$$\text{Conc. of alcohol in blood} = \frac{13 \text{ g alcohol}}{7000 \text{ mL blood}} = 0.0019 \text{ g/cm}^3$$

Thus, the person is legally intoxicated.

---

77. Refer to Section 10.1 and Chapter 5.

a. $49.92 \text{ g NaOH} \times \dfrac{1 \text{ mol.}}{40.00 \text{ g NaOH}} = 1.248 \text{ mol. NaOH}$

$\dfrac{1.248 \text{ mol. NaOH}}{0.600 \text{ L}} = 2.08 \, M$

b. $1.248 \text{ mol. NaOH} \times \dfrac{1 \text{ mol. OH}^-}{1 \text{ mol. NaOH}} \times \dfrac{3 \text{ mol. H}_2}{2 \text{ mol. OH}^-} = 1.872 \text{ mol. H}_2$

c. Determine the limiting reactant first.

$$41.28 \text{ g Al} \times \frac{1 \text{ mol. Al}}{26.98 \text{ g Al}} \times \frac{3 \text{ mol. H}_2}{2 \text{ mol. Al}} = 2.30 \text{ mol. H}_2$$

The amount of H_2 that could be produced by the Al is less than the actual amount produced, thus, the limiting reactant is NaOH.

$$P = (758.6 \text{ mm Hg} - 23.8 \text{ mm Hg}) \times \frac{1 \text{ atm}}{760 \text{ mm Hg}} = 0.967 \text{ atm}$$

$$V = \frac{nRT}{P} = \frac{(1.872 \text{ mol.})(0.0821 \text{ L} \cdot \text{atm/mol.} \cdot \text{K})(298 \text{ K})}{0.967 \text{ atm}} = 47.4 \text{ L}$$

---

78. Refer to Section 10.3.

The ideal gas law is: $V = \dfrac{nRT}{P}$

Henry's law is: $C_g = kP_g$, where C_g is the concentration of the gas in solution.

If we hold the volume of solution constant, then $n_g = kP_g$, where n_g is the moles of gas.

Substituting gives: $V = \dfrac{(kP) RT}{P} = kRT$

Thus, the volume of gas is directly proportional only to the temperature; pressure is not a factor.

Chapter 11: Rate of Reaction

2. *Refer to Section 11.1.*

$$\text{rate} = \frac{-\Delta[N_2H_4]}{2\Delta t} = \frac{-\Delta[N_2O_4]}{\Delta t} = \frac{\Delta[N_2]}{3\Delta t} = \frac{\Delta[H_2O]}{\Delta t}$$

a. $\text{rate} = \dfrac{-\Delta[N_2O_4]}{\Delta t}$

b. $\text{rate} = = \dfrac{\Delta[N_2]}{3\Delta t}$

4. *Refer to Section 11.1.*

$$\text{rate} = \frac{-\Delta[Br^-]}{5\Delta t} = \frac{-\Delta[BrO_3^-]}{\Delta t} = \frac{-\Delta[H^+]}{6\Delta t} = \frac{\Delta[Br_2]}{3\Delta t} = \frac{\Delta[H_2O]}{3\Delta t}$$

a. $\dfrac{\Delta[Br_2]}{\Delta t} = 0.039 \,\text{mol./L} \cdot \text{s} \Rightarrow \dfrac{\Delta[Br_2]}{3\Delta t} = \dfrac{0.039 \,\text{mol./L} \cdot \text{s}}{3}$

  $\dfrac{\Delta[Br_2]}{3\Delta t} = \dfrac{\Delta[H_2O]}{3\Delta t} \Rightarrow \dfrac{\Delta[H_2O]}{3\Delta t} = \dfrac{0.039 \,\text{mol./L} \cdot \text{s}}{3}$

  $\dfrac{\Delta[H_2O]}{\Delta t} = \dfrac{3(0.039 \,\text{mol./L} \cdot \text{s})}{3} = 0.039 \,\text{mol./L} \cdot \text{s}$

b. $\dfrac{-\Delta[Br^-]}{5\Delta t} = \dfrac{\Delta[Br_2]}{3\Delta t}$

  $\dfrac{-\Delta[Br^-]}{5\Delta t} = \dfrac{0.039 \,\text{mol./L} \cdot \text{s}}{3}$

  $\dfrac{-\Delta[Br^-]}{\Delta t} = \dfrac{5(0.039 \,\text{mol./L} \cdot \text{s})}{3} = 0.065 \,\text{mol./L} \cdot \text{s}$

c. $\dfrac{-\Delta[H^+]}{6\Delta t} = \dfrac{\Delta[Br_2]}{3\Delta t} \Rightarrow \dfrac{-\Delta[H^+]}{6\Delta t} = \dfrac{0.039 \,\text{mol./L} \cdot \text{s}}{3}$

  $\dfrac{-\Delta[H^+]}{\Delta t} = \dfrac{6(0.039 \,\text{mol./L} \cdot \text{s})}{3} = 0.078 \,\text{mol./L} \cdot \text{s}$

6. Refer to Section 11.1.

a. $N_2(g) + 3H_2(g) \rightarrow 2NH_3(g)$

b. $\text{rate} = \dfrac{\Delta[NH_3]}{2\Delta t}$

c. $\text{rate} = \dfrac{0.815\,M - 0.257\,M}{2(15-0)\text{min}} = 1.86 \times 10^{-2}\,M/\text{min}$

8. Refer to Section 11.1 and Figure 11.2.

Time (s)	0	10	20	30	40	50
[X] (M)	0.0038	0.0028	0.0021	0.0016	0.0012	0.00087

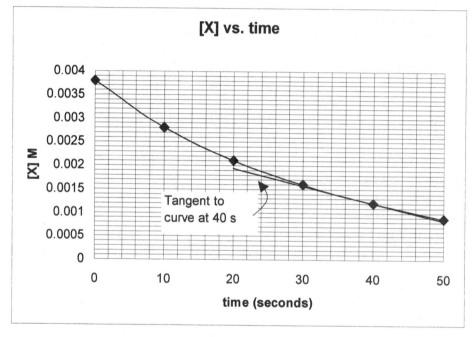

a. The graph shows the change in concentration of X over time.

b. The negative slope of the tangent gives the instantaneous rate ($3.7 \times 10^{-5}\,M/\text{s}$) at that point in time.

$$\text{slope} = \frac{0.00074}{-20} = -3.7 \times 10^{-5}\,M/s$$

c. The average rate over the 10 to 50 second time frame is the change in concentration over that given time.

$$\text{rate}_{avg} = \frac{\Delta[X]}{\Delta t} = \frac{0.00087M - 0.0028M}{50s - 10s} = 4.8 \times 10^{-5}\,M/s$$

d. The instantaneous rate (3.7×10^{-5} M/s) is less than the average rate (4.8×10^{-5} M/s) over that 40 sec interval.

---

10. Refer to Section 11.2.

a. rate = $k_1[X]^2[Y]$
$m = 2$; so the reaction is **2nd order in X**.
$n = 1$; so the reaction is **1st order in Y**.
$m + n = 3$; so the reaction is **3rd order overall**.

b. rate = $k_2[X]$
$m = 1$; so the reaction is **1st order in X**.
Since [Y] is not in the rate law, $n = 0$; so the reaction is **0 order in Y**
$m = 1$; so the reaction is **1st order overall**.

c. rate = $k_3[X]^2[Y]^2$
$m = 2$; so the reaction is **2nd order in X**.
$n = 2$; so the reaction is **2nd order in Y**.
$m + n = 4$; so the reaction is **4th order overall**.

d. rate = k_4
$m = 0$; so the reaction is **0th order in X**.
$n = 0$; so the reaction is **0th order in Y**.
$m + n = 0$; so the reaction is **0th order overall**.

---

12. Refer to Section 11.2.

a. rate = $k_1[X]^2[Y]$
$$k_1 = \frac{rate}{[M]^2[M]} = \frac{mol./L \cdot s}{(mol./L)^2(mol./L)} = \frac{1/s}{(mol./L)^2} = L^2/mol.^2 \cdot s$$

b. rate = $k_2[X]$
$$k_1 = \frac{rate}{[M]} = \frac{mol./L \cdot s}{mol./L} = \frac{1/s}{1} = 1/s = s^{-1}$$

c. rate = $k_3[X]^2[Y]^2$
$$k_1 = \frac{rate}{[M]^2[M]^2} = \frac{mol./L \cdot s}{(mol./L)^2(mol./L)^2} = \frac{1/s}{(mol./L)^3} = L^3/mol.^3 \cdot s$$

d. rate = k_4
k_4 = rate = mol./L·s

---

14. Refer to Section 11.2.

	[X] (M)	[Y] (M)	k (L/mol.·h)	Rate (mol./L·h)
a.	0.100	0.400	1.89	**0.0756**
b.	0.600	**0.300**	0.884	0.159
c.	**0.143**	0.025	13.4	0.0479
d.	0.592	0.233	**0.00812**	0.00112

a. rate = k[X][Y]
   rate = (1.89 L/mol.·h)(0.100 M)(0.400 M) = **0.0756** M/h

b. rate = k[X][Y]
   0.159 M/h = (0.884 L/mol.·h)(0.600 M)[Y]
   [Y] = **0.300** M

c. rate = k[X][Y]
   0.0479 M/h = (13.4 L/mol.·h) [X] (0.025 M)
   [X] = **0.143** M

d. rate = k[X][Y]
   0.00112 M/h = k(0.592 M)(0.233 M)
   k = **0.00812 L/mol.·h**

---

16. *Refer to Section 11.2.*

a. rate = k[NH_3]^0
   rate = k
   rate = 2.5×10^{-4} mol./L·min

b. Rate is independent of [NH_3], thus rate = k.
   rate = 2.5×10^{-4} mol./L·min

c. In a zero order reaction, rate is always equal to the rate constant, k, thus, at any [NH_3], rate = k.

---

18. *Refer to Section 11.2.*

a. rate = k[ICl][H_2]
   4.89×10^{-5} mol./L·s = k(0.100 M)(0.030 M)
   k = 0.016 L/mol.·s

b. rate = k[ICl][H_2]
   5.00×10^{-4} mol./L·s = (0.016 L/mol.·s)(0.233 M)[H_2]
   [H_2] = 0.13 M

c. rate = k[ICl][H_2]
   3[ICl] = [H_2]
   substituting gives: rate = k[ICl](3[ICl])
   0.0934 mol./L·s = (0.016 L/mol.·s) [ICl](3[ICl])
   5.8 = [ICl](3[ICl]) = 3[ICl]^2
   1.9 = [ICl]^2
   [ICl] = 1.4 M

---

20. *Refer to Section 11.2 and Example 11.1.*

a. rate$_1$ = k[A]^m

   rate$_2$ = k[A]^m

$$\frac{\text{rate}_1}{\text{rate}_2} = \left(\frac{[A]_1}{[A]_2}\right)^m \quad \Rightarrow \quad \frac{0.288\,\text{mol./L}\cdot\text{min}}{0.245\,\text{mol./L}\cdot\text{min}} = \left(\frac{0.200\,\text{mol./L}}{0.170\,\text{mol./L}}\right)^m$$

$$1.18 = (1.18)^m \quad \Rightarrow \quad m = 1$$

b. rate = $k[Y]$

c. 0.288 mol./L·min = k(0.200 mol./L)

   k = 1.44/min

---

22. Refer to Section 11.1 and Example 11.2.

---

a. To determine the reaction order in [BF_3], select two experiments in which [NH_3] is constant (such as experiments 1 and 4, as used below).

$$\frac{\text{rate}_1}{\text{rate}_4} = \frac{k[BF_3]_1^m[NH_3]_1^n}{k[BF_3]_4^m[NH_3]_4^n} \quad \Rightarrow \quad \frac{\text{rate}_1}{\text{rate}_4} = \left(\frac{[BF_3]_1}{[BF_3]_4}\right)^m \left(\frac{[NH_3]_1}{[NH_3]_4}\right)^n$$

$$\frac{0.0341\,\text{mol./L}\cdot\text{s}}{0.102\,\text{mol./L}\cdot\text{s}} = \left(\frac{0.100\,\text{mol./L}}{0.300\,\text{mol./L}}\right)^m \left(\frac{0.100\,\text{mol./L}}{0.100\,\text{mol./L}}\right)^n$$

$0.334 = (0.333)^m(1)^n$
$0.334 = (0.333)^m$
$m = 1$ (The reaction is 1st order in BF_3.)

To determine the reaction order in [NH_3], select two experiments in which [BF_3] is constant (such as experiments 2 and 3) and repeat the process above.

$$\frac{0.159\,\text{mol./L}\cdot\text{s}}{0.0512\,\text{mol./L}\cdot\text{s}} = \left(\frac{0.200\,\text{mol./L}}{0.200\,\text{mol./L}}\right)^m \left(\frac{0.233\,\text{mol./L}}{0.750\,\text{mol./L}}\right)^n$$

$0.311 = (1)^m(0.311)^n$
$0.311 = (0.311)^n$
$n = 1$ (The reaction is 1st order in NH_3.)

Overall reaction order = $m + n = 1 + 1 = 2$

b. Rate = $k[BF_3][NH_3]$

c. Substitute values from any of the experiments into the rate equation and solve for k.
   0.0341 M/s = k(0.100 M)(0.100 M)
   0.0341 M/s = k(0.01 M^2)
   k = 3.41 s^{-1}·M^1 = 3.41 L/mol.·s

d. Rate = $k[BF_3][NH_3]$
   Rate = (3.41 L/mol.·s)(0.553 M)(0.300 M)
   Rate = 0.566 M/s

a. To determine the reaction order in $[I_2]$, select two experiments in which $[(C_2H_5)_2(NH)_2]$ is constant (such as experiments 1 and 2, as used below).

$$\frac{rate_1}{rate_2} = \frac{k[(C_2H_5)_2(NH)_2]_1^m[I_2]_1^n}{k[(C_2H_5)_2(NH)_2]_2^m[I_2]_2^n} \Rightarrow \frac{rate_1}{rate_2} = \left(\frac{[(C_2H_5)_2(NH)_2]_1}{[(C_2H_5)_2(NH)_2]_2}\right)^m \left(\frac{[I_2]_1}{[I_2]_2}\right)^n$$

$$\frac{1.08 \times 10^{-4} \text{ mol./L·h}}{1.56 \times 10^{-4} \text{ mol./L·h}} = \left(\frac{0.150 \text{ mol./L}}{0.150 \text{ mol./L}}\right)^m \left(\frac{0.250 \text{ mol./L}}{0.3620 \text{ mol./L}}\right)^n$$

$0.692 = (1)^m(0.691)^n$
$0.692 = (0.691)^n$
$n = 1$ (The reaction is 1st order in I_2.)

To determine the reaction order in $(C_2H_5)_2(NH)_2$, select two experiments in which $[I_2]$ is constant (such as experiments 3 and 4, as used below) and repeat the process above.

$$\frac{2.30 \times 10^{-4} \text{ mol./L·h}}{3.44 \times 10^{-4} \text{ mol./L·h}} = \left(\frac{0.200 \text{ mol./L}}{0.300 \text{ mol./L}}\right)^m \left(\frac{0.400 \text{ mol./L}}{0.400 \text{ mol./L}}\right)^n$$

$0.667 = (0.667)^m(1)^n$
$0.667 = (0.667)^m$
$m = 1$ (The reaction is 1st order in $(C_2H_5)_2(NH)_2$.)

Overall reaction order $= m + n = 1 + 1 = 2$

b. Rate $= k[(C_2H_5)_2(NH)_2][I_2]$

c. Substitute values from any of the experiments into the rate equation and solve for k.
1.08×10^{-4} M/h $= k(0.150$ $M)(0.250$ $M)$
1.08×10^{-4} M/h $= k(0.0375$ $M^2)$
$k = 2.88 \times 10^{-3}$ $h^{-1}·M^{-1} = 2.88 \times 10^{-3}$ L/mol.·h

d. Rate $= k[(C_2H_5)_2(NH)_2][I_2]$
5.00×10^{-4} M/h $= (2.88 \times 10^{-3}$ L/mol.·h$)([(C_2H_5)_2(NH)_2])(0.500$ $M)$
5.00×10^{-4} M/h $= (1.44 \times 10^{-3}$ /h$)([(C_2H_5)_2(NH)_2])$
$[(C_2H_5)_2(NH)_2] = 0.347$ mol./L

26. Refer to Section 11.2 and Example 11.2.

a. To determine the reaction order in $[CH_3COCH_3]$, select two experiments in which $[I_2]$ and $[H^+]$ are constant (such as experiments 1 and 2, as used below).

$$\frac{\text{rate}_1}{\text{rate}_2} = \frac{k[CH_3COCH_3]_1^m[I_2]_1^n[H^+]_1^p}{k[CH_3COCH_3]_2^m[I_2]_2^n[H^+]_2^p} \Rightarrow \frac{\text{rate}_1}{\text{rate}_2} = \left(\frac{[CH_3COCH_3]}{[CH_3COCH_3]}\right)^m\left(\frac{[I_2]}{[I_2]}\right)^n\left(\frac{[H^+]}{[H^+]}\right)^p$$

$$\frac{4.2\times10^{-6}\ \text{mol./L}\cdot\text{s}}{8.2\times10^{-6}\ \text{mol./L}\cdot\text{s}} = \left(\frac{0.80\ \text{mol./L}}{1.6\ \text{mol./L}}\right)^m\left(\frac{0.001\ \text{mol./L}}{0.001\ \text{mol./L}}\right)^n\left(\frac{0.20\ \text{mol./L}}{0.20\ \text{mol./L}}\right)^p$$

$0.51 = (0.50)^m(1)^n(1)^p$
$0.51 = (0.50)^m$
$m = 1$ (The reaction is 1st order in CH_3COCH_3.)

To determine the reaction order in $[H^+]$, select two experiments in which $[CH_3COCH_3]$ and $[I_2]$ are constant (such as experiments 1 and 3, as used below) and repeat the process above.

$$\frac{\text{rate}_1}{\text{rate}_3} = \frac{k[CH_3COCH_3]_1^m[I_2]_1^n[H^+]_1^p}{k[CH_3COCH_3]_3^m[I_2]_3^n[H^+]_3^p} \Rightarrow \frac{\text{rate}_1}{\text{rate}_3} = \left(\frac{[CH_3COCH_3]}{[CH_3COCH_3]}\right)^m\left(\frac{[I_2]}{[I_2]}\right)^n\left(\frac{[H^+]}{[H^+]}\right)^p$$

$$\frac{4.2\times10^{-6}\ \text{mol./L}\cdot\text{s}}{8.7\times10^{-6}\ \text{mol./L}\cdot\text{s}} = \left(\frac{0.80\ \text{mol./L}}{0.80\ \text{mol./L}}\right)^m\left(\frac{0.001\ \text{mol./L}}{0.001\ \text{mol./L}}\right)^n\left(\frac{0.20\ \text{mol./L}}{0.40\ \text{mol./L}}\right)^p$$

$0.48 = (1)^m(1)^n(0.50)^p$
$0.48 = (0.50)^p$
$p = 1$ (The reaction is 1st order in H^+.)

To determine the reaction order in $[I_2]$, select two experiments in which $[CH_3COCH_3]$ and $[H^+]$ are constant (such as experiments 1 and 4, as used below) and repeat the process above.

$$\frac{\text{rate}_1}{\text{rate}_4} = \frac{k[CH_3COCH_3]_1^m[I_2]_1^n[H^+]_1^p}{k[CH_3COCH_3]_4^m[I_2]_4^n[H^+]_4^p} \Rightarrow \frac{\text{rate}_1}{\text{rate}_3} = \left(\frac{[CH_3COCH_3]}{[CH_3COCH_3]}\right)^m\left(\frac{[I_2]}{[I_2]}\right)^n\left(\frac{[H^+]}{[H^+]}\right)^p$$

$$\frac{4.2\times10^{-6}\ \text{mol./L}\cdot\text{s}}{4.3\times10^{-6}\ \text{mol./L}\cdot\text{s}} = \left(\frac{0.80\ \text{mol./L}}{0.80\ \text{mol./L}}\right)^m\left(\frac{0.001\ \text{mol./L}}{0.0005\ \text{mol./L}}\right)^n\left(\frac{0.20\ \text{mol./L}}{0.20\ \text{mol./L}}\right)^p$$

$0.98 = (1)^m(2)^n(1)^p$
$0.98 = (2)^n$
$n = 0$ (The reaction is 0 order in I_2.)

b. Rate $= k[CH_3COCH_3][H^+][I_2]^0$
    Rate $= k[CH_3COCH_3][H^+]$

c. Substitute values from any of the experiments into the rate equation and solve for k.
    $4.2\times10^{-6}\ M/s = k(0.80\ M)(0.20\ M)$
    $4.2\times10^{-6}\ M/s = k(0.16\ M^2)$
    $k = 2.6\times10^{-5}\ s^{-1}\cdot M^{-1} = 2.6\times10^{-5}\ \text{L/mol.}\cdot\text{s}$

d. $[CH_3COCH_3] = 3(0.933\ M)$
    $[CH_3COCH_3] = 2.80\ M$
    rate $= (2.6\times10^{-5}\ \text{L/mol.}\cdot\text{s})(2.80M)(0.933M)$
    rate $= 6.8\times10^{-5}\ M/s$

a. To determine the reaction order in [SCN⁻], select two experiments in which $[Cr(H_2O)_6^{3+}]$ is constant (such as experiments 1 and 2, as used below).

$$\frac{\text{rate}_1}{\text{rate}_2} = \frac{k[Cr(H_2O)_6^{3+}]_1^m[SCN^-]_1^n}{k[Cr(H_2O)_6^{3+}]_2^m[SCN^-]_2^n} \quad \Rightarrow \quad \frac{\text{rate}_1}{\text{rate}_2} = \left(\frac{[Cr(H_2O)_6^{3+}]}{[Cr(H_2O)_6^{3+}]}\right)^m \left(\frac{[SCN^-]}{[SCN^-]}\right)^n$$

$$\frac{8.4 \times 10^{-4} \text{ mol./L} \cdot \text{min}}{6.5 \times 10^{-4} \text{ mol./L} \cdot \text{min}} = \left(\frac{0.025 \text{ mol./L}}{0.025 \text{ mol./L}}\right)^m \left(\frac{0.077 \text{ mol./L}}{0.060 \text{ mol./L}}\right)^n$$

$1.29 = (1)^m(1.28)^n$

$1.29 = (1.28)^n$

$n = 1$ (The reaction is 1st order in SCN⁻.)

To determine the reaction order in $[Cr(H_2O)_6^{3+}]$, select two experiments in which [SCN⁻] is constant (such as experiments 2 and 3, as used below) and repeat the process above.

$$\frac{8.4 \times 10^{-4} \text{ mol./L} \cdot \text{min}}{1.4 \times 10^{-2} \text{ mol./L} \cdot \text{min}} = \left(\frac{0.025 \text{ mol./L}}{0.042 \text{ mol./L}}\right)^m \left(\frac{0.077 \text{ mol./L}}{0.077 \text{ mol./L}}\right)^n$$

$0.60 = (0.60)^m(1)^n$

$0.60 = (0.60)^m$

$m = 1$ (The reaction is 1st order in $Cr(H_2O)_6^{3+}$.)

Overall reaction order $= m + n = 1 + 1 = 2$

Rate $= k[Cr(H_2O)_6^{3+}][SCN^-]$

b. Substitute values from any of the experiments into the rate equation and solve for k.
   6.5×10^{-4} $M/s = k(0.025 \ M)(0.060 \ M)$
   $k = 0.43$ min⁻¹·$M = 0.43$ L/mol.·min

c. Calculate $[Cr(H_2O)_6^{3+}]$ and [SCN⁻]. Then use the rate law and k determined above to calculate the rate.

$$15 \text{ mg KSCN} \times \frac{1 \text{ g KSCN}}{1000 \text{ mg KSCN}} \times \frac{1 \text{ mol. KSCN}}{97.19 \text{ g KSCN}} = 1.5 \times 10^{-4} \text{ mol. KSCN}$$

$$1.5 \times 10^{-4} \text{ mol. KSCN} \times \frac{1 \text{ mol. SCN}^-}{1 \text{ mol. KSCN}} = 1.5 \times 10^{-4} \text{ mol. SCN}^-$$

$$[SCN^-] = \frac{1.5 \times 10^{-4} \text{ mol. SCN}^-}{1.50 \text{ L}} = 1.0 \times 10^{-4} \ M \text{ SCN}^-$$

Rate $= k[Cr(H_2O)_6^{3+}][SCN^-]$
Rate $= (0.43$ L/mol.·min$)(0.0500 \ M)(1.0 \times 10^{-4} \ M)$
Rate $= 2.2 \times 10^{-6}$ M/min $= 2.2 \times 10^{-6}$ mol./L·min

Generate 2 graphs, ln [NOBr] vs. time and 1/[NOBr] vs. time. The one that yields a straight line is the one that indicates the order of the reaction (see Table 11.2).

time (min)	[NOBr]	ln[NOBr]	1/[NOBr]
0	0.0286	-3.554	35.0
6	0.0253	-3.677	39.5
12	0.0229	-3.777	43.7
18	0.0208	-3.873	48.1
24	0.0190	-3.963	52.6

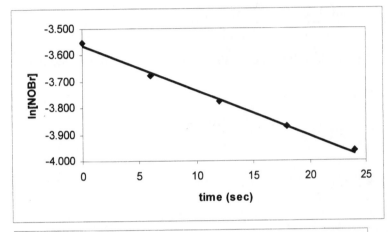

Note how the data points curve around the straight line plotted through the points. This graph is not linear, so the reaction is not 1st order.

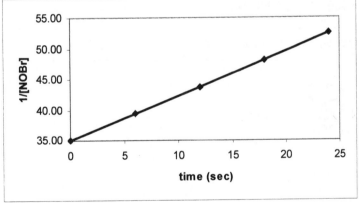

These data points are well represented by the straight line drawn through them. This graph is linear and the reaction is 2nd order.

rate = k[NOBr]^2

a. A 1st order reaction will yield a straight line when time is plotted against ln[HOF]. Note that the plot yields a straight line.

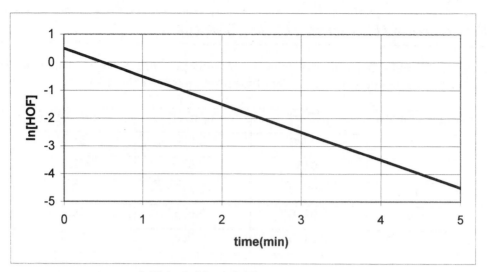

b. $k = -\text{slope} = -\dfrac{\Delta y}{\Delta x} = -\dfrac{0.500 - 2.00}{0 - 2.50} \dfrac{-2.50}{2.50\,\text{min}} = 1.00\,\text{min}^{-1}$

c. Use the graph to determine the $[\text{HOF}]_o$, the antiln of the point where the line crosses the y axis.

$[\text{HOF}]_o = e^{0.50} = 1.65\,M$

$k \cdot t = \ln\dfrac{[\text{HOF}]_o}{[\text{HOF}]} \Rightarrow (1.00\,\text{min}^{-1})(t) = \ln\dfrac{1.65\,M}{0.100\,M}$

$t = \dfrac{2.80}{1.00\,\text{min}^{-1}} = 2.80\,\text{min}$

d. rate = $k[\text{HOF}]$
   rate = $(1.00\,\text{min}^{-1})(0.0500\,M) = 0.0500\,M/\text{min}$

---

34. Refer to Section 11.3 and Examples 11.3 and 11.4.

a. $t = 400\,\text{min} - 200\,\text{min} = 200\,\text{min}$

$\ln\dfrac{[A]_o}{[A]} = kt \Rightarrow \ln\dfrac{0.0300\,M}{0.0200\,M} = k(200\,\text{min})$

$0.4055 = k(200\,\text{min})$
$k = 0.00203/\text{min}$

b. $t_{1/2} = \dfrac{0.693}{k} = \dfrac{0.693}{0.00203/\text{min}} = 341\,\text{min}$

c. $\ln\dfrac{[A]_o}{(0.0200\,M)} = (0.00203/\text{min})(400\,\text{min})$

$\ln[A]_0 - \ln(0.0200\,M) = (0.00203/\text{min})(400\,\text{min})$
$\ln[A]_0 + 3.912 = (0.00203/\text{min})(400\,\text{min})$
$\ln[A]_0 = 0.812 - 3.912$

$\ln[A]_0 = -3.100 \Rightarrow [A]_0 = e^{-3.100} = 0.0450\,M$

204

a. Since the volume of solution and molar mass of the drug are constant, we can substitute mass of the drug for concentration (volume and molar mass cancel).

$$\ln\frac{[A]_o}{[A]} = kt \implies \ln\frac{10.0\,g}{m_{drug}} = (0.215\ \text{mon.}^{-1})(12\ \text{mon.})$$

$$\ln(10.0) - \ln(m_{drug}) = 2.58$$

$$-\ln(m_{drug}) = 0.28$$

$$m_{drug} = e^{-0.28} = 0.758\ g$$

b. $t_{1/2} = \dfrac{0.693}{k} = \dfrac{0.693}{0.215\ \text{mon.}^{-1}} = 3.22\ \text{mon.}$

c. 65% of 10.0 g is 6.5 g. Thus, if 65% decomposed, 3.5 g (35%) are left.

$$k \cdot t = \ln\frac{[A]_o}{[A]} \implies (0.215\ \text{mon.}^{-1})(t) = \ln\frac{10.0\,g}{3.5\,g}$$

$$t = 4.9\ \text{mon.}$$

a. $\ln\dfrac{[C_2H_6]_o}{[C_2H_6]} = kt \implies \ln\dfrac{0.00839\,M}{0.00768\,M} = k(212\,s)$

$$\ln(1.09) = k(212\ s)$$
$$k = 4.17 \times 10^{-4}\,s^{-1}$$

b. Rate is requested in units of mol/L·h. Change the units of k to h^{-1}.

$$\frac{4.17 \times 10^{-4}}{s} \times \frac{60\,s}{1\,\min} \times \frac{60\,\min}{1\,h} = 1.50/h$$

$$\text{rate} = k[C_2H_6] = (1.50/h)(0.00422\,M) = 6.33 \times 10^{-3}\ \text{mol./L·h}$$

c. 27% of 0.00839 M is $(0.27)(0.00839\,M) = 0.00227\,M$

$$\ln\frac{[C_2H_6]_o}{[C_2H_6]} = kt \implies \ln\frac{0.00839\,M}{0.00227\,M} = 4.17 \times 10^{-4}/s(t)$$

$$1.31 = (4.17 \times 10^{-4}\,s^{-1})(t)$$
$$t = 3.14 \times 10^3\ s$$

$$3.14 \times 10^3\ s \times \frac{1\,\min}{60\,s} = 52\ \min$$

d. Let x = the fraction of C_2H_6 that remains; 1 - x = the fraction that decomposed.

$$t = 22\ \min \times \frac{60\,s}{1\,\min} = 1320\ s$$

$$\ln\frac{[C_2H_6]_o}{x[C_2H_6]_o} = kt \Rightarrow \ln\frac{1}{x} = (4.17\times10^{-4}/s)(1320\,s)$$

$\ln(1/x) = 0.55$

$1/x = e^{0.55}$

$1/x = 1.73$

$x = 0.58$

$1-x = 0.42$

42% is decomposed

---

40. Refer to Section 11.3 and Example 11.4.

The half-life is the time it takes for the initial sample to decrease by one-half of its initial amount. After 8.1 days, the amount of I-131 is 50% of the initial amount. After another 8.1 days, the amount is 25% of the initial amount. Thus, it takes $8.1 + 8.1 = 16.2$ days to reach 25% of the initial amount.

---

42. Refer to Section 11.3.

$$\ln\frac{[Na\text{-}24]_o}{[Na\text{-}24]} = kt \Rightarrow \ln\frac{0.050\,mg}{0.016\,mg} = (k)(24.9\,hr)$$

$\ln(3.1) = k(24.9\,hr)$

$k = 0.045\,hr^{-1}$

$$t_{1/2} = \frac{0.693}{k} = \frac{0.693}{0.045\,hr^{-1}} = 15\,hr$$

---

44. Refer to Section 11.3.

a. For a zero order reaction, the rate is independent of the concentration.

Rate $= k = 2.08 \times 10^{-4}$ mol./L·s

$$t_{1/2} = \frac{[A]_o}{2k} = \frac{0.250\,M}{2(2.08\times10^{-4}\,mol./L\cdot s)} = 6.01\times10^2\,s$$

$$6.01\times10^2\,s \times \frac{1\,min}{60\,s} = 10.0\,min$$

b. $[A]_o - [A] = kt$

$1.25\,M - 0.388\,M = (2.08 \times 10^{-4}\,mol./L\cdot s)(t)$

$t = 4.14 \times 10^3\,s$

$$4.14\times10^3\,s \times \frac{1\,min}{60\,s} \times \frac{1\,hr}{60\,min} = 1.2\,hr$$

46. Refer to Section 11.3.

a. $\dfrac{1}{[A]} - \dfrac{1}{[A]_o} = kt$

$\dfrac{1}{0.0841\,M} - \dfrac{1}{0.250\,M} = (k)(11.6\,h)$

$7.9\,M^{-1} = (k)(11.6\,h)$

$k = 0.68\,\text{L/mol·h}$

b. $t_{1/2} = \dfrac{1}{k[A]_o} = \dfrac{1}{(0.68\,\text{L/mol·h})(0.100\,\text{mol./L})} = 15\,h$

c. 39% is rearranged; that means (100-39) 61% is unreacted
$[A] = 0.61[A]_0 = 0.61(0.450\,M) = 0.275\,M$

$\dfrac{1}{0.275\,M} - \dfrac{1}{0.450\,M} = (0.68\,\text{L/mol·h})(t)$

$1.4\,\text{L/mol.} = (0.68\,\text{L/mol·h})t$
$t = 2.1\,h$

d. rate $= k[NH_4NCO]^2$
rate $= (0.68\,\text{L/mol·h})(0.839\,\text{mol./L})^2$
rate $= 0.48\,\text{mol./L·h}$

48. Refer to Section 11.3.

If 15% has decomposed, then (100-15) 85% is left.
85% of 0.300 is $(0.85)(0.300\,M) = 0.255\,M$.

$\dfrac{1}{[NOCl]} - \dfrac{1}{[NOCl]_o} = kt$

$\dfrac{1}{0.255\,M} - \dfrac{1}{0.300\,M} = (k)(0.20\,\text{min})$

$0.588\,M^{-1} = (k)(0.20\,\text{min})$

$k = 2.9\,M^{-1}\,\text{min}^{-1} = 2.9\,\text{L/mol·min}$

50. Refer to Section 11.5 and Figure 11.10.

Convert temperatures to kelvins, calculate $\ln(k)$, plot the data and calculate E_a from the slope.

k (L/mol·s)	0.048	2.3	49	590
$\ln(k)$	-3.04	0.833	3.89	6.38
T (K)	773	873	973	1073
$1/T$	1.29×10^{-3}	1.15×10^{-3}	1.03×10^{-3}	9.32×10^{-4}

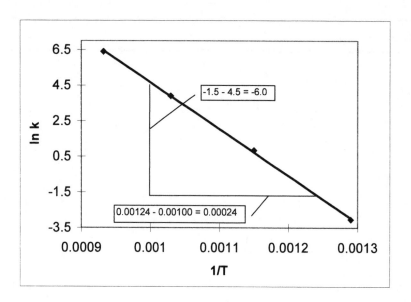

$$\text{slope} = \frac{-6.0}{0.00024\ \text{K}} = -2.5 \times 10^4\ \text{K}^{-1}$$

$$\text{slope} = -\frac{E_a}{R}$$

$E_a = -(R)(\text{slope}) = -(8.31\ \text{J/mol·K})(-2.5 \times 10^4\ \text{K}^{-1}) = 2.1 \times 10^5\ \text{J/mol.}$

$E_a = 2.1 \times 10^2\ \text{kJ/mol.}$

52. Refer to Section 11.5 and Example 11.7.

$$\ln\frac{k_2}{k_1} = \frac{E_a}{R}\left(\frac{1}{T_1} - \frac{1}{T_2}\right)$$

$35°C = 308\ \text{K}$

$22°C = 295\ \text{K}$

let x = fraction of reaction rate at temperature 2

rate is proportional to rate constant

then, $xk_1 = k_2$

$$\ln\frac{xk_1}{k_1} = \frac{6.5 \times 10^4\ \text{J/mol.}}{8.31\ \text{J/mol·K}}\left(\frac{1}{308} - \frac{1}{295}\right)$$

$\ln(x) = -1.12$

$x = 0.33 \rightarrow 33\%$

$100 - 33\% = 67\%$ decrease

a. $X_{25°C} = 7.2(25) - 32 = 148$ chirps/min
$X_{35°C} = 7.2(35) - 32 = 220$ chirps/min

b. Use the two point form of the Arrhenius equation and solve for E_a.

$$\ln\frac{k_{25°C}}{k_{35°C}} = \ln\frac{X_{35°C}}{X_{25°C}} = \frac{E_a}{R}\left(\frac{1}{T_1} - \frac{1}{T_2}\right)$$

$$\ln\frac{220 \text{ chirps/min}}{148 \text{ chirps/min}} = \frac{E_a}{8.31 \text{ J/mol.·K}}\left(\frac{1}{298 \text{ K}} - \frac{1}{308 \text{ K}}\right)$$

$$0.396 = E_a(1.3 \times 10^{-5} \text{ J/mol.})$$

$$E_a = 3.0 \times 10^4 \text{ J/mol.} = 30 \text{ kJ/mol.}$$

c. Percent increase for a 10°C rise in temperature is:
$148 + 148x = 220$
$x = 0.49 = 49\%$ increase

a. $\ln\dfrac{k_2}{k_1} = \dfrac{E_a}{R}\left(\dfrac{1}{T_1} - \dfrac{1}{T_2}\right)$

$$\ln\frac{k_2}{0.0174 \text{ L/mol.·h}} = \frac{1.82 \times 10^5 \text{ J/mol.}}{8.31 \text{ J/mol.·K}}\left(\frac{1}{1123 \text{ K}} - \frac{1}{973 \text{ K}}\right)$$

$\ln(k_2) - (-4.05) = -3.01$

$\ln(k_2) = -7.06$

$k_2 = e^{-7.05} = 8.59 \times 10^{-4}$ L/mol.·h

b. Use the two point form of the Arrhenius equation, setting $k_2 = \frac{1}{4}k_1$ (where k_1 corresponds to $T_1 = 850°C = 1123$ K).

$$\ln\frac{k_2}{k_1} = \ln\frac{\frac{1}{4}k_1}{k_1} = \frac{E_a}{R}\left(\frac{1}{T_1} - \frac{1}{T_2}\right)$$

$$\ln\frac{1}{4} = \frac{1.82 \times 10^5 \text{ J/mol.}}{8.31 \text{ J/mol.·K}}\left(\frac{1}{1123 \text{ K}} - \frac{1}{T_2}\right)$$

$$-1.39 = 2.19 \times 10^4 \text{ K}\left(\frac{1}{1123 \text{ K}} - \frac{1}{T_2}\right)$$

$$-6.33 \times 10^{-5} \text{ K}^{-1} = 8.90 \times 10^{-4} \text{ K}^{-1} - \frac{1}{T_2}$$

$$-9.53 \times 10^{-4} \text{ K}^{-1} = -\frac{1}{T_2}$$

$$T_2 = 1.05 \times 10^3 \text{ K} = 776°C$$

58. *Refer to Sections 11.3 and 11.5.*

$$t_{1/2} = \frac{0.693}{k} \implies k = \frac{0.693}{t_{1/2}}$$

at 25°C $\quad k = \dfrac{0.693}{2.81 \text{ s}} = 0.247 / \text{s}$

at 45°C $\quad k = \dfrac{0.693}{0.313 \text{ s}} = 2.21 / \text{s}$

$$\ln\frac{k_2}{k_1} = \frac{E_a}{R}\left(\frac{1}{T_1} - \frac{1}{T_2}\right)$$

$$\ln\frac{2.21}{0.247} = \frac{E_a}{8.31 \text{ J/mol.}\cdot\text{K}}\left(\frac{1}{298 \text{ K}} - \frac{1}{318 \text{ K}}\right)$$

$$2.19 = \frac{E_a}{8.31 \text{ J/mol.}\cdot\text{K}}\left(0.000211 \text{ K}^{-1}\right)$$

$$2.19 = \left(0.0000254 \text{ mol./J}\right)E_a$$

$$E_a = 8.62 \times 10^4 \text{ J/mol.} = 86.2 \text{ kJ/mol.}$$

60. *Refer to Section 11.7.*

a. rate = $k[NO][O_3]$

b. rate = $k[NO_2]^2$

c. rate = $k[K][HCl]$

62. *Refer to Section 11.7 and Example 11.7.*

The slow step is the rate determining step (RDS) and determines the rate equation. Thus:

$$\text{rate}_2 = k_2[N_2O_2][H_2]$$

However, N_2O_2 is an intermediate, and rate equations should not have intermediates. Use the fast equilibrium in the first step to develop an equation for $[N_2O_2]$. Bear in mind that the forward rate must equal the reverse rate for this to be an equilibrium. Consequently:

$$k_1[NO]^2 = \text{rate}_1 = \text{rate}_{-1} = k_{-1}[N_2O_2]$$

$$[N_2O_2] = k_1/k_{-1}[NO]^2$$

Substituting this into the first equation gives:

rate$_2$ = $k_2 k_1 / k_{-1}$[NO]^2[H$_2$]
rate$_2$ = k[NO]^2[H$_2$]   (*where k is the combined rate constant*)

Yes, the proposed mechanism is consistent with the rate expression.

---

64. *Refer to Section 11.7 and Example 11.7.*

Write a rate law for each of the proposed mechanisms as compare them to the experimental rate law.

Mechanism 1
The slow step is the rate determining step (RDS) and determines the rate equation. Thus:

rate = k[NO$_3$][NO]
However, NO$_3$ is an intermediate, and rate equations should not have intermediates. Use the fast equilibrium in the first step to develop an equation for [NO$_3$]. Bear in mind that the forward rate must equal the reverse rate for this to be an equilibrium. Consequently:
rate$_1$ = rate$_{-1}$

k_1[NO][O$_2$] = k_{-1}[NO$_3$]
k_1 / k_{-1}[NO][O$_2$] = [NO$_3$]

Substituting this into the first equation gives:

rate = $k k_1 / k_{-1}$[NO][O$_2$][NO]
rate = k[NO]^2[O$_2$]   (*where k is the combined rate constant*)

Mechanism 2
The slow step is the rate determining step (RDS) and determines the rate equation. Thus:

rate = k[N$_2$O$_2$][O$_2$]
However, N$_2$O$_2$ is an intermediate, and rate equations should not have intermediates. Use the fast equilibrium in the first step to develop an equation for [N$_2$O$_2$]. Bear in mind that the forward rate must equal the reverse rate for this to be an equilibrium. Consequently:
rate$_1$ = rate$_{-1}$

k_1[NO]^2 = k_{-1}[N$_2$O$_2$]
k_1 / k_{-1}[NO]^2 = [N$_2$O$_2$]

Substituting this into the first equation gives:

rate = $k k_1 / k_{-1}$[NO]^2[O$_2$]
rate = k[NO]^2[O$_2$]   (*where k is the combined rate constant*)

Thus proposed mechanisms 1 and 2 are both consistent with the observed rate law.

a. $t_{1/2} = \dfrac{0.693}{k} \Rightarrow k = \dfrac{0.693}{t_{1/2}}$

$k = \dfrac{0.693}{7.5 \times 10^2 \text{ min}} = 9.2 \times 10^{-4} / \text{min}$

$122.0 \text{ g} \times \dfrac{1 \text{ mol.}}{134.97 \text{ g}} = 0.9039 \text{ mol.}$

Assume the volume of the flask $= 1 \text{L}$

$[SO_2Cl_2]_0 = 0.9039 \ M$

$45.0 \text{ g} \times \dfrac{1 \text{ mol.}}{134.97 \text{ g}} = 0.333 \text{ mol.}$

$[SO_2Cl_2] = 0.333 \ M$

$\ln \dfrac{[SO_2Cl_2]_o}{[SO_2Cl_2]} = kt \Rightarrow \ln \dfrac{0.9039 \ M}{0.333 \ M} = 9.2 \times 10^{-4} / \text{min}(t)$

$\ln 2.71 = 9.2 \times 10^{-4} / \text{min}(t)$

$t = 1.08 \times 10^3 \text{ min}$

$1.08 \times 10^3 \text{ min} \times \dfrac{1 \text{ h}}{60 \text{ min}} = 18.0 \text{ h}$

b. $29.0 \text{ h} \times \dfrac{60 \text{ min}}{1 \text{ h}} = 1740 \text{ min}$

$\ln \dfrac{[SO_2Cl_2]_o}{[SO_2Cl_2]} = kt \Rightarrow \ln \dfrac{0.9039 \ M}{[SO_2Cl_2]} = 9.2 \times 10^{-4} / \text{min}(1740 \text{ min})$

$\ln(0.9039) - \ln[SO_2Cl_2] = 1.6$

$-\ln[SO_2Cl_2] = 1.7$

$[SO_2Cl_2] = e^{-1.7} = 0.183 \ M$

Since volume is 1L (see part (a) above), 0.183 mol remained.

Balanced equation shows that the amount of Cl_2 produced = amount of SO_2Cl_2 reacted.

0.9039 mol. – 0.183 mol. = 0.721 mol. Cl_2 produced

$PV = nRT$

(1.00 atm)(V) = (0.721 mol.)(0.0821 L•atm/mol. •K)(300 K)

$V = 17.8 \text{ L}$

$\ln \dfrac{k_2}{k_1} = \dfrac{E_a}{R} \left(\dfrac{1}{T_1} - \dfrac{1}{T_2} \right)$

$\ln \dfrac{k_2}{k_1} = \dfrac{8.5 \times 10^4 \text{ J/mol.}}{8.31 \text{ J/mol.} \cdot \text{K}} \left(\dfrac{1}{298} - \dfrac{1}{303} \right)$

$$\ln \frac{k_2}{k_1} = .57$$

$$\frac{k_2}{k_1} = 1.8$$

Thus, if $k_1 = 1$, $k_2 = 1.8$; that is an 80% increase.

70. Refer to Section 11.3.

$-\Delta[A] = k\Delta t$

Now expand the terms $\Delta[A]$ and Δt and substitute the expanded equations into the one above.

$\Delta[A] = [A] - [A]_o$

$\Delta t = t - t_o = t - 0 = t$ ($t_o = 0$ since the reaction starts at time zero)

$-([A] - [A]_o) = kt$

$-[A] + [A]_o = kt$

$[A]_o - [A] = kt$ (see Table 11.2)

   or

$[A] = [A]_o - kt$

72. Refer to Section 11.3 and Table 11.2.

In both the trials depicted, the amount of reactant is reduced to half, and thus the time is the half-life. In a 0 order reaction, half-life increases as initial concentration increases. In a 1st order reaction, half-life is independent of the initial concentration. In a 2nd order reaction, half-life decreases as initial concentration increases. The decomposition of B is 0 order because the half-life increases as the initial concentration increases.

0 order: half-life = $[A]_o/2k$

1st order: half-life = $0.693/k$

2nd order: half-life = $1/k[A]_o$

74. Refer to Section 11.1 and Figure 11.1.

$$\text{rate} = \frac{-\Delta[X]}{2\Delta t} = \frac{\Delta[Y]}{2\Delta t} = \frac{\Delta[Z]}{\Delta t}$$

Curve 1 represents a decrease in concentration. Thus, it must represent a change in [X], since X is the only reactant.

Curve 2 represents an increase in concentration approximately half that of the decrease in [X]. Thus, this is the change in [Z]

Curve 3 represents an increase in concentration equal in magnitude to the decrease in [X]. Thus, the species involved must have the same coefficient in the balanced chemical reaction as [X]. This curve represents the change in [Y].

---

76. *Refer to Section 11.5.*

---

a. The fastest reaction will be the one with the smallest activation energy barrier: **A**.

b. The reaction with the largest half-life will be the slowest reaction, which will be the one with the largest activation energy barrier: **C**.

c. The reaction with the largest rate will be the fastest reaction, which will be the one with the smallest activation energy barrier: **A**.

---

78. *Refer to Section 11.3.*

After 2 min, 4A have reacted to form 4B and 4C. There are 6A unreacted. Thus,
$[A]_0 = 10$
$[A] = 6$

$$\ln\frac{10}{6} = k(2\,\text{min}) \Rightarrow k = 0.26/\text{min}$$

Use k to determine [A] when $t = 3$ min

$$\ln\frac{10}{[A]} = (0.26/\text{min})(3\,\text{min})$$

$\ln(10) - \ln[A] = (0.26/\text{min})(3\,\text{min})$
$\ln[A] = 1.52$
$[A] = 4.6 \approx 5$
This is ½ initial concentration
$t_{1/2} = 3$ min

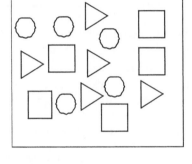

When $t = 4$ min:

$$\ln\frac{10}{[A]} = (0.26/\text{min})(4\,\text{min})$$

$\ln(10) - \ln[A] = (0.26/\text{min})(4\,\text{min})$
$[A] = 3.5 \approx 4$

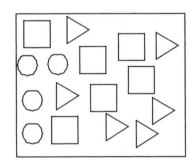

80. Refer to Sections 11.2 and 11.5.

a. (1) rate = $k(0.10\ M)(0.10\ M) = 0.010k$
(2) rate = $k(0.15\ M)(0.15\ M) = 0.023k$
(3) rate = $k(0.06\ M)(1.0\ M) = 0.06k$

Test tube (1) will have the smallest rate.

b. When the temperature increases, rate increases.
Since rate and k are directly proportional, k also increases.
E_a is constant for a given reaction, thus E_a remains the same. The increase in rate is due to an increase in the kinetic motion of the molecules.

82. Refer to Sections 11.2, 11.4, and 11.6, and Chapter 8.

a. $\Delta H^\circ = \Sigma\ \Delta H^\circ_{f(prod)} - \Sigma\ \Delta H^\circ_{f(react)}$

$\Delta H^\circ = (2\ \text{mol.})(26.48\ \text{kJ/mol.}) - [(1\ \text{mol.})(62.44\ \text{kJ/mol.}) + (1\ \text{mol.})(0.0\ \text{kJ/mol.})]$

$\Delta H^\circ = -9.48\ \text{kJ}$

$\Delta H^\circ = E_{a(forward)} - E_{a(reverse)}$

$E_{a(reverse)} = 165\ \text{kJ/mol.} - (-9.48\ \text{kJ/mol.}) = 174\ \text{kJ/mol.}$

b. $k_{forward} = Ae^{-Ea/RT}$

$138\ \text{L/mol.·s} = Ae^{(-165000\ \text{J})/(8.31\ \text{J/mol.·K})(973\ \text{K})}$

$138\ \text{L/mol.·s} = Ae^{-20.4}$

$138\ \text{L/mol.·s} = A(1.38 \times 10^{-9})$

$A = 1.00 \times 10^{11}$

$k_{reverse} = Ae^{-Ea/RT}$

$k_{reverse} = (1.00 \times 10^{11})\ e^{(-174000\ \text{J})/(8.31\ \text{J/mol.·K})(973\ \text{K})}$

$k_{reverse} = 46\ \text{L/mol.·s}$

c. $\text{rate}_{reverse} = k_{reverse}\ [\text{HI}]^2$
$\text{rate}_{reverse} = (46\ \text{L/mol.·s})(0.200\ M)^2$
$\text{rate}_{reverse} = 1.8\ \text{mol./L·s}$

83. Refer to Section 11.3.

For first order reactions $\text{rate} = \dfrac{-d[\text{A}]}{\text{a} \cdot dt} = k[\text{A}]$

$-d[\text{A}] = \text{a}k[\text{A}]dt \quad \Rightarrow \quad \int \dfrac{-d[\text{A}]}{[\text{A}]} = \int \text{a}kdt$

$$\ln \frac{[A]_0}{[A]} = a\,kt$$

Since both a and k are constants, they can be replaced with a new constant k.

$$\ln \frac{[A]_0}{[A]} = kt$$

---

84. *Refer to Section 11.2 and Example 11.2.*

To determine the order with respect to A, use two trials in which [A] changes and [B] and [C] are constant (such as experiments 1 and 4, as used below), and solve for m.

$$\frac{rate_1}{rate_2} = \frac{k[A]_1^m[B]_1^n[C]_1^p}{k[A]_2^m[B]_2^n[C]_2^p}$$

$$\frac{4x}{x} = \left(\frac{0.40}{0.20}\right)^m \left(\frac{0.40}{0.40}\right)^n \left(\frac{0.10}{0.10}\right)^p$$

$4 = (2)^m$

$m = 2$ Thus the reaction is 2nd order in A

Determining the order with respect to B involves a small twist of the usual procedure. Use two trials in which [B] changes and [C] is constant (such as experiments 2 and 3, as used below), and substitute $m = 2$ into the equation.

$$\frac{8x}{x} = \left(\frac{0.40}{0.20}\right)^2 \left(\frac{0.40}{0.20}\right)^n \left(\frac{0.20 \text{ mol./L}}{0.20 \text{ mol./L}}\right)^p$$

$2 = (2)^n$

$n = 1$ Thus the reaction is 1st order in B

To determine the order with respect to C, use two trials in which [C] changes (such as experiments 1 and 2, as used below) and substitute $m = 2$ and $n = 1$ into the equation.

$$\frac{8x}{x} = \left(\frac{0.40}{0.20}\right)^2 \left(\frac{0.40}{0.40}\right)^1 \left(\frac{0.20}{0.10}\right)^p$$

$2 = (2)^p$

$p = 1$ Thus the reaction is 1st order in C

rate = $k[A]^2[B][C]$

a. $\text{rate} = \dfrac{-d[A]}{a \cdot dt} = k[A]^2$

$\displaystyle\int \dfrac{-d[A]}{[A]^2} = \int a k\, dt$

$\dfrac{1}{[A]} - \dfrac{1}{[A]_o} = a\, kt$

Since both a and k are constants, they can be replaced with a new constant k.

$\dfrac{1}{[A]} - \dfrac{1}{[A]_o} = kt$

b. $\text{rate} = \dfrac{-d[A]}{a \cdot dt} = k[A]^3$

$\displaystyle\int \dfrac{-d[A]}{[A]^3} = \int a k\, dt$

$\dfrac{1}{2[A]^2} - \dfrac{1}{2[A]_o{}^2} = a\, kt$

Since both a and k are constants, they can be replaced with a new constant k.

$\dfrac{1}{[A]^2} - \dfrac{1}{[A]_o{}^2} = 2\, kt$

---

86. Refer to Section 11.3.

At 3 times per day, $\quad \Delta t = \dfrac{24 \text{ hr}}{3} = 8 \text{ hr}$

$t_{1/2} = 2.0 \text{ days} = 48 \text{ hr}$

$\text{saturation value} = \dfrac{X}{1 - 10^{-0.30\frac{\Delta t}{t_{1/2}}}} = \dfrac{0.100 \text{ g}}{1 - 10^{-0.30\frac{8}{48}}}$

$\text{saturation value} = \dfrac{0.100 \text{ g}}{1 - 10^{-0.05}} = \dfrac{0.100 \text{ g}}{1 - 0.89} = 0.91 \text{ g}$

In the second part of the problem, the goal is to find a mass (X) of drug that gives a saturation value of 0.500g

$$\text{saturation value} = 0.500\,\text{g} = \frac{X}{1-0.89} = \frac{X}{0.11}$$

$X = 0.055\,\text{g} = 55\,\text{mg}$

Thus a dose of 55 mg, three times a day would produce an accumulation of 0.500 g drug in the body, and thus the patient would experience side effects. To prevent the side effects, a pharmacist would assign a dosage of 54 mg, three times a day.

Chapter 12: Gaseous Chemical Equilibrium

---

2. *Refer to Section 12.1.*

---

a. 75 s. After 75 s, P_A and P_B remain constant.

b. After 45 s, [A] is still decreasing, so the rate of the forward reaction must be **greater** than the rate of the reverse reaction.
After 90 s, the [A] is no longer changing, so the rates of the forward and reverse reactions must be **equal**.

---

4. *Refer to Section 12.1.*

Time (min)	0	1	2	3	4	5	6
P_A (atm)	1.000	0.778	**0.580**	**0.415**	**0.355**	0.325	**0.325**
P_B (atm)	0.400	**0.326**	0.260	**0.205**	0.185	**0.175**	0.175
P_C (atm)	0.000	**0.148**	**0.280**	0.390	**0.430**	**0.450**	**0.450**

1 min: 1.000 atm A – 0.788 atm A = 0.222 atm A

$$0.222 \text{ atm A} \times \frac{1\text{B}}{3\text{A}} = 0.0740 \text{ atm B}$$

0.400 atm B – 0.0740 atm B = 0.326 atm B

$$0.222 \text{ atm A} \times \frac{2\text{C}}{3\text{A}} = 0.148 \text{ atm C}$$

2 min: 0.400 atm B – 0.260 atm B = 0.140 atm B

$$0.140 \text{ atm B} \times \frac{3\text{A}}{1\text{B}} = 0.420 \text{ atm A}$$

1.000 atm A – 0.420 atm A = 0.580 atm A

$$0.140 \text{ atm B} \times \frac{2\text{C}}{1\text{B}} = 0.280 \text{ atm C}$$

3 min: $0.390 \text{ atm C} \times \dfrac{1\text{B}}{2\text{C}} = 0.195 \text{ atm B}$

0.400 atm B – 0.195 atm B = 0.205 atm B

$$0.390 \text{ atm C} \times \frac{3A}{2C} = 0.585 \text{ atm A}$$

$$1.000 \text{ atm A} - 0.585 \text{ atm A} = 0.415 \text{ atm A}$$

4 min: $0.400 \text{ atm B} - 0.185 \text{ atm B} = 0.215 \text{ atm B}$

$$0.215 \text{ atm B} \times \frac{3A}{1B} = 0.645 \text{ atm A}$$

$$1.000 \text{ atm A} - 0.645 \text{ atm A} = 0.355 \text{ atm A}$$

$$0.215 \text{ atm B} \times \frac{2C}{1B} = 0.430 \text{ atm C}$$

5 min: $1.000 \text{ atm A} - 0.325 \text{ atm A} = 0.675 \text{ atm A}$

$$0.675 \text{ atm A} \times \frac{1B}{3A} = 0.225 \text{ atm B}$$

$$0.400 \text{ atm B} - 0.225 \text{ atm B} = 0.175 \text{ atm B}$$

$$0.675 \text{ atm A} \times \frac{2C}{3A} = 0.450 \text{ atm C}$$

6 min: Since P_B has not changed from 5 min, P_A and P_C will also equal the values from 5 min.

---

6. *Refer to Section 12.2 and Examples 12.1 and 12.2.*

Recall that liquids and solids do not show up in equilibrium expressions.

a. $K = \dfrac{(P_{H_2})^3 (P_{CO})}{(P_{CH_4})}$

b. $K = \dfrac{(P_{NO})^4 (P_{H_2O})^6}{(P_{NH_3})^4 (P_{O_2})^5}$

c. $K = P_{CO_2}$

d. $K = \dfrac{1}{(P_{NH_3})(P_{HCl})}$

---

8. *Refer to Section 12.2 and Example 12.2.*

Remember, only gases and dissolved species enter the equilibrium constant expression.

a. $K = \dfrac{P_{Ni(CO)_4}}{(P_{CO})^4}$

b. $K = \dfrac{[H^+][F^-]}{[HF]}$

c. $K = \dfrac{[Cl^-]^2}{(P_{Cl_2})[Br^-]^2}$

---

10. Refer to Section 12.2 and Example 12.2.

a. $C_3H_6O(l) \rightleftharpoons C_3H_6O(g)$ $\qquad\qquad\qquad K = P_{C_3H_6O}$

Note that the reactant is a liquid and therefore does not appear in the equilibrium constant expression.

b. $7H_2(g) + 2NO_2(g) \rightleftharpoons 2NH_3(g) + 4H_2O(g)$ $\qquad K = \dfrac{(P_{NH_3})^2(P_{H_2O})^4}{(P_{H_2})^7(P_{NO})^2}$

c. $H_2S(g) + Pb^{2+}(aq) \rightleftharpoons PbS(s) + 2H^+(aq)$ $\qquad K = \dfrac{[H^+]^2}{(P_{H_2S})[Pb^{2+}]}$

---

12. Refer to Section 12.2.

Those species not in the K_{eq} expression, but needed to balance the reactions, must be present in the reaction as either a solid or a liquid.

a. $2SO_2(g) + 2H_2O(g) \rightleftharpoons 2H_2S(g) + 3O_2(g)$

b. $IF(g) \rightleftharpoons {}^1\!/_2 F_2(g) + {}^1\!/_2 I_2(g)$

c. $Cl_2(g) + 2Br^-(aq) \rightleftharpoons Br_2(l) + 2Cl^-(aq)$ (see problem 8c.)

d. $2NO_3^-(aq) + 8H^+(aq) \rightleftharpoons 2NO(g) + 4H_2O(g) + 3Cu^{2+}(aq)$

   $3Cu(s) + 2NO_3^-(aq) + 8H^+(aq) \rightleftharpoons 2NO(g) + 4H_2O(g) + 3Cu^{2+}(aq)$

---

14. Refer to Section 12.2 and Example 12.1.

Write the balanced reaction for the equilibrium and relate that to the one given.

a. $SO_2(g) + {}^1\!/_2 O_2(g) \rightleftharpoons SO_3(g)$

This equation is ${}^1\!/_2$ times the given equation, so the given K must be raised to the ${}^1\!/_2$ power.
$K = (0.76)^{1/2} = 0.87$

b. $2SO_3(g) \rightleftharpoons 2SO_2(g) + O_2(g)$

This equation is the reverse of the equation in part a, so take the reciprocal of that K.

$K = 1/(0.76) = 1.3$

221

As with thermochemical equations in Chapter 8, rearrange the given equations as necessary so that they sum the goal equation. Remember that multiplying through by two means K must be squared. To calculate K for the overall equation, multiply the K values of the rearranged equations.

$$C(s) + CO_2(g) \rightleftharpoons 2CO(g) \qquad\qquad K_1 = 2.4 \times 10^{-9}$$
$$2Cl_2(g) + 2CO(g) \rightleftharpoons 2COCl_2(g) \qquad K_2 = 1/(8.8 \times 10^{-13})^2$$

$$\overline{C(s) + CO_2(g) + 2Cl_2(g) + 2CO(g) \rightleftharpoons 2CO(g) + 2COCl_2(g)}$$

$$C(s) + CO_2(g) + 2Cl_2(g) \rightleftharpoons 2COCl_2(g)$$

$$K = K_1 K_2$$
$$= (2.4 \times 10^{-9})(1.3 \times 10^{24})$$
$$= 3.1 \times 10^{15}$$

Reversing the first equation and adding it to the second gives the equation for the formation of 2 moles NOBr. Divide that equation by two and take the square root of K. Remember that reversing an equation means taking the reciprocal of the equilibrium constant.

$$N_2(g) + O_2(g) \rightleftharpoons 2NO(g) \qquad\qquad K_1 = (1 \times 10^{-30})^{-1} = 1 \times 10^{30}$$
$$2NO(g) + Br_2(g) \rightleftharpoons 2NOBr(g) \qquad K_2 = 8 \times 10^1$$

$$N_2(g) + O_2(g) + 2NO(g) + Br_2(g) \rightleftharpoons 2NOBr(g) + 2NO(g) \qquad K = K_1 K_2 = 8 \times 10^{31}$$
$$\tfrac{1}{2}N_2(g) + \tfrac{1}{2}O_2(g) + \tfrac{1}{2}Br_2(g) \rightleftharpoons NOBr(g) \qquad K = (8 \times 10^{31})^{\frac{1}{2}} = 9 \times 10^{15}$$

$$K = \frac{P_{CH_3OH}}{(P_{CO})(P_{H_2})^2} = \frac{0.0512 \text{ atm}}{(0.814 \text{ atm})(0.274 \text{ atm})^2} = 0.838$$

Note that K is usually expressed without units.

a. $CH_4(g) + 2H_2S(g) \rightleftharpoons CS_2(g) + 4H_2(g)$

b. Use the ideal gas law equation to calculate the pressure of each gas, then substitute those values into the equation for the equilibrium constant.

$$P_{CH_4} = \frac{n}{V}RT = \frac{0.00142\,mol.}{1\,L}(0.0821\,L \cdot atm/mol. \cdot K)(1123\,K) = 0.131\,atm$$

$$P_{H_2S} = \frac{n}{V}RT = \frac{6.14 \times 10^{-4}\,mol.}{1\,L}(0.0821\,L \cdot atm/mol. \cdot K)(1123\,K) = 0.0566\,atm$$

$$P_{CS_2} = \frac{n}{V}RT = \frac{0.00266\,mol.}{1\,L}(0.0821\,L \cdot atm/mol. \cdot K)(1123\,K) = 0.245\,atm$$

$$P_{H_2} = \frac{n}{V}RT = \frac{0.00943\,mol.}{1\,L}(0.0821\,L \cdot atm/mol. \cdot K)(1123\,K) = 0.869\,atm$$

$$K = \frac{(P_{H_2})^4(P_{CS_2})}{(P_{CH_4})(P_{H_2S})^2} = \frac{(0.869\,atm)^4(0.245\,atm)}{(0.131\,atm)(0.0566\,atm)^2} = 333$$

24. Refer to Section 12.3 and Example 12.4.

$P_{eq\,(O_2)} = 0.216\,atm$, thus $\Delta P_{(O_2)} = 0.216\,atm$
$\Delta P_{(SO_2)} = 2\Delta P_{(O_2)} = 0.432\,atm$
$\Delta P_{(SO_3)} = -0.432\,atm$

	$2SO_3(g)$	$\rightleftharpoons$	$2SO_2(g)$	$+$	$O_2(g)$
P_o	0.541 atm		0 atm		0 atm
ΔP	-0.432 atm		0.432 atm		0.216 atm
P_{eq}	0.109 atm		0.432 atm		0.216 atm

$$K = \frac{(P_{SO_2})^2(P_{O_2})}{(P_{SO_3})^2} = \frac{(0.432\,atm)^2(0.216\,atm)}{(0.109\,atm)^2} = 3.39$$

26. Refer to Section 12.4 and Example 12.5.

a. $\quad K = \frac{(P_{PCl_3})(P_{Cl_2})}{(P_{PCl_5})} = 26$

$$Q = \frac{(P_{PCl_3})(P_{Cl_2})}{(P_{PCl_5})} = \frac{(0.90)(0.45)}{0.012} = 34$$

Since $Q \neq K$, the system is **not** at equilibrium.

b. $\quad Q > K$, so the reaction will proceed to the **left** (more reactants).

28. Refer to Section 12.4 and Example 12.5.

a. $K = \dfrac{(P_{NO})^2 (P_{O_2})}{(P_{NO_2})^2}$

$Q = \dfrac{(P_{NO})^2 (P_{O_2})}{(P_{NO_2})^2} = \dfrac{(0.10)^2 (0.10)}{(0.10)^2} = 0.10$

$Q < K$, therefore the reaction proceeds to the **right** (more products)

b. $Q = \dfrac{(P_{NO})^2 (P_{O_2})}{(P_{NO_2})^2} = \dfrac{(0.0116)^2 (0)}{(0.0848)^2} = 0$

$Q < K$, therefore the reaction proceeds to the **right** (more products). If the partial pressure of one of the products is zero, then the reaction can only proceed to the right, and calculations were actually unnecessary.

c. $Q = \dfrac{(P_{NO})^2 (P_{O_2})}{(P_{NO_2})^2} = \dfrac{(0.040)^2 (0.010)}{(0.20)^2} = 4.0 \times 10^{-4}$

$Q < K$, therefore the reaction proceeds to the **right** (more products)

30. Refer to Section 12.4.

$K = \dfrac{(P_{NH_3})^2}{(P_{N_2})(P_{H_2})^3} \Rightarrow 1.5 \times 10^{-5} = \dfrac{(0.015)^2}{(1.2)(P_{H_2})^3}$

$(P_{H_2})^3 = 12.5 \text{ atm}$

$(P_{H_2}) = 2.3 \text{ atm}$

32. Refer to Section 12.4.

$2BrCl(g) \rightleftharpoons Br_2(g) + Cl_2(g)$

$K = \dfrac{(P_{Br_2})(P_{Cl_2})}{(P_{BrCl})^2} \Rightarrow 0.040 = \dfrac{(0.0493)(0.0493)}{(P_{BrCl})^2}$

$(P_{BrCl})^2 = 0.061$

$P_{BrCl} = 0.25 \text{ atm}$

$$P_{total} = P_{NO} + P_{NO_2} + P_{O_2} \Rightarrow 1.25 = P_{NO} + P_{NO_2} + 0.515$$

$$0.735 \text{ atm} = P_{NO} + P_{NO_2}$$

let $x = P_{NO}$; then $0.735 - x = P_{NO2}$

$$K = \frac{(P_{NO})^2 (P_{O_2})}{(P_{NO_2})^2} \Rightarrow 0.87 = \frac{(x)^2 (0.515)}{(0.735 - x)^2}$$

$$\sqrt{1.69} = \sqrt{\frac{(x)^2}{(0.735 - x)^2}}$$

$$1.3 = \frac{x}{0.735 - x} \Rightarrow x = 0.42$$

$$P_{NO_2} = 0.32 \text{ atm}$$

$$P_{NO} = 0.42 \text{ atm}$$

---

36. *Refer to Section 12.4 and Example 12.6.*

Set ΔP to x, calculate P_{eq} and then apply the resulting expressions to the equilibrium equation. Then solve for x.

	$N_2(g)$	+	$O_2(g)$	$\rightleftharpoons$	$2NO(g)$
P_o	0.300 atm		0.300 atm		0.300 atm
ΔP	-x atm		-x atm		+2x atm
P_{eq}	0.300 - x atm		0.300 - x atm		0.300 + 2x atm

$$K = \frac{(P_{NO})^2}{(P_{N_2})(P_{O_2})} \Rightarrow 0.0255 = \frac{(0.300 + 2x)^2}{(0.300 - x)(0.300 - x)}$$

$$\sqrt{0.0255} = \sqrt{\frac{(0.300 + 2x)^2}{(0.300 - x)^2}} \Rightarrow 0.160 = \frac{0.300 + 2x}{0.300 - x}$$

$$0.0480 - 0.300 = 2x + 0.160x$$

$$x = -0.117$$

$$P_{eq}(N_2) = P_{eq}(O_2) = 0.300 - (-0.117) = 0.417 \text{ atm}$$
$$P_{eq}(NO) = 0.300 + 2(-0.117) = 0.066 \text{ atm}$$

	$SO_2(g)$	$+$	$NO_2(g)$	$\rightleftharpoons$	$NO(g)$	$+$	$SO_3(g)$
P_o	1.25 atm		1.25 atm		1.25 atm		1.25 atm
ΔP	$-x$ atm		$-x$ atm		$+x$ atm		$+x$ atm
P_{eq}	$1.25 - x$ atm		$1.25 - x$ atm		$1.25 + x$ atm		$1.25 + x$ atm

a. $K = \dfrac{(P_{NO})(P_{SO_3})}{(P_{SO_2})(P_{NO_2})} \Rightarrow 84.7 = \dfrac{(1.25 + x)(1.25 + x)}{(1.25 - x)(1.25 - x)}$

$\sqrt{84.7} = \sqrt{\dfrac{(1.25 + x)^2}{(1.25 - x)^2}} \Rightarrow 9.20 = \dfrac{1.25 + x}{1.25 - x}$

$11.5 - 9.20x = 1.25 + x$

$x = 1.00$

$P_{eq}(SO_2) = P_{eq}(NO_2) = 1.25 - 1.00 = 0.25$ atm
$P_{eq}(NO) = P_{eq}(SO_3) = 1.25 + 1.00 = 2.25$ atm

b. $P_{init} = \Sigma P_o = 1.25 + 1.25 + 1.25 + 1.25 = 5.00$ atm
$P_{final} = \Sigma P_{eq} = 0.25 + 0.25 + 2.25 + 2.25 = 5.00$ atm
$P_{init} = P_{final}$; This is always true when $\Delta n_g = 0$.
No, the relation is not true of all gaseous reactions. The relation depends on the stoichiometry of the reaction.

a. Write the equilibrium expression, bearing in mind that NH_4I is a solid, and therefore does not take part in the equation.

$K = \dfrac{(P_{NH_3})(P_{HI})}{1} = (P_{NH_3})(P_{HI})$

Since equimolar amounts of NH_3 and HI are produced the pressure of each will be equal:
$K = (P_{NH_3})(P_{HI}) = x^2 \Rightarrow 0.215 = x^2$

$P_{NH_3} = P_{HI} = x = 0.464$ atm

$P_{total} = P_{NH_3} + P_{HI} = 0.464 + 0.464 = 0.928$ atm

b. Use the ideal gas equation to calculate the amount of NH_3 (or HI) that was produced, then calculate the amount of NH_4I this represents.

$$PV = nRT$$
$$(0.464 \text{ atm})(5.0 \text{ L}) = (n)(0.0821 \text{ L·atm/mol·K})(673 \text{ K})$$
$$n = 0.042 \text{ mol.}$$

$$0.042 \text{ mol. NH}_3 \times \frac{1 \text{ mol. NH}_4\text{I}}{1 \text{ mol. NH}_3} \times \frac{145 \text{ g NH}_4\text{I}}{1 \text{ mol. NH}_4\text{I}} = 6.1 \text{ g NH}_4\text{I}$$

---

42. Refer to Section 12.4.

	$2HI(g)$	$\rightleftharpoons$	$I_2(g)$	+	$H_2(g)$
P_o	0.200 atm		0 atm		0 atm
ΔP	-2x atm		+x atm		+x atm
P_{eq}	0.200 - 2x atm		x atm		x atm

$$K = \frac{(P_{I_2})(P_{H_2})}{(P_{HI})^2} \Rightarrow 0.0169 = \frac{(x)(x)}{(0.200 - 2x)^2}$$

$$\sqrt{0.0169} = \sqrt{\frac{x^2}{(0.200 - 2x)^2}}$$

$$0.130 = \frac{x}{0.200 - 2x}$$

$$x = 0.021$$
$$P_{eq}(I_2) = P_{eq}(H_2) = 0.021 \text{ atm}$$

---

44. Refer to Section 12.5 and Example 12.8.

a. (1) Removing O_2 means removing a reactant. The equilibrium will shift towards the reactant side, causing an **increase** in the amount of ammonia at equilibrium.

   (2) Adding N_2 means adding a product. The equilibrium will shift towards the reactant side, causing an **increase** in the amount of ammonia at equilibrium.

   (3) Adding a liquid, such as water, will not change the partial pressures of the gases involved in the reaction, thus adding water has **no effect** in the amount of ammonia at equilibrium.

   (4) Expanding the container at constant pressure means increasing the volume. The equilibrium will shift towards the side with the greater moles of gas. There are 7 moles of gaseous reactants and 2 moles of gaseous products. The equilibrium will shift towards the reactant side, causing an **increase** in the amount of ammonia at equilibrium.

(5) Increasing the temperature in an exothermic reaction will cause the reverse reaction to occur. The equilibrium will shift towards the reactant side, causing an **increase** in the amount of ammonia at equilibrium.

b. None of the factors listed above will increase K. The only factor above that changes K is **(5)** which causes a decrease in K.

---

46. Refer to Section 12.5 and Example 12.8.

---

a. $4 \text{ mol.} (g) \rightleftharpoons 1 \text{ mol.} (g)$
A pressure decrease would shift the equilibrium to more moles of gas, to the **left**.

b. $1 \text{ mol.} (g) \rightleftharpoons 2 \text{ mol.} (g)$
A pressure decrease would shift the equilibrium to more moles of gas, to the **right**.

c. $1 \text{ mol.} (g) \rightleftharpoons 1 \text{ mol.} (g)$
Since there is an equal number of moles, **no change** would occur.

---

48. Refer to Section 12.5 and Examples 12.6 and 12.7.

---

a. $K = \dfrac{(P_{COBr_2})}{(P_{CO})(P_{Br_2})} \Rightarrow \dfrac{(0.12)}{(1.00)(0.65)} = 0.18$

b. Change P_o for Br_2 to 0.50 atm and then solve for x.

	CO(g)	+	Br$_2$(g)	$\rightleftharpoons$	COBr$_2$(g)
P_o	1.00 atm		0.50 atm		0.12 atm
ΔP	-x atm		-x atm		+x atm
P_{eq}	1.00 - x atm		0.50 - x atm		0.12 + x atm

$K = \dfrac{(P_{COBr_2})}{(P_{CO})(P_{Br_2})} \Rightarrow 0.18 = \dfrac{(0.12 + x)}{(1.00 - x)(0.50 - x)}$

$0.090 - 0.27x + 0.18\,x^2 = 0.12 + x \quad \Rightarrow \quad 0.18x^2 - 1.27x - 0.03 = 0$

$x = \dfrac{-b \pm \sqrt{b^2 - 4ac}}{2a} = \dfrac{-(1.27) \pm \sqrt{(-1.27)^2 - (4)(0.18)(-0.03)}}{2(0.18)}$

$x = -0.01/0.36$ or $2.55/0.36$

$x = -0.03$ or 7.08

If $x = 7.08$, the pressure of COBr$_2$ becomes negative. A negative pressure is impossible, so this value is rejected.

228

P_{eq} (COBr$_2$) = 0.12 + (-0.03) = 0.09 atm
P_{eq} (CO) = 1.00 – (-0.03) = 1.03 atm
P_{eq} (Br$_2$) = 0.50 – (-0.03) = 0.53 atm

50. Refer to Sections 8.5 and 12.5 and Appendix 1.

First calculate ΔH^0_{rxn} as in Chapter 8.

$\Delta H_{reaction} = \Sigma \, \Delta H_f^o \text{ (products)} - \Sigma \, \Delta H_f^o \text{ (reactants)}$

$\Delta H_{reaction} = [(1 \text{ mol.})(\Delta H_f^o \text{ O}_2(g)) + (2 \text{ mol.})(\Delta H_f^o \text{ SO}_2(g))] - [(2 \text{ mol.})(\Delta H_f^o \text{ SO}_3(g))]$

$\Delta H_{reaction} = [(1 \text{ mol.})(0.0 \text{ kJ/mol.}) + (2 \text{ mol.})(-296.8 \text{ kJ/mol.})] - [(2 \text{ mol.})(-395.7 \text{ kJ/mol.})]$

$\Delta H_{reaction} = 197.8 \text{ kJ} = 197800 \text{ J}$

Use the van't Hoff equation, and solve for K_2.

$$\ln \frac{K_2}{K_1} = \frac{\Delta H^\circ}{R}\left[\frac{1}{T_1} - \frac{1}{T_2}\right] \Rightarrow \ln \frac{K_2}{1.32} = \frac{197800 \text{ J/mol.}}{8.31 \text{ J/mol.} \cdot \text{K}}\left[\frac{1}{900 \text{ K}} - \frac{1}{828 \text{ K}}\right]$$

$\ln(K_2) - \ln(1.32) = -2.30$

$\ln(K_2) = -2.30 + 0.28 = -2.02$

$K_2 = e^{-2.02} = 0.132$

52. Refer to Section 12.5.

Set $K_1 = 0.40K_2$ and substitute into the van't Hoff equation.

$$\ln \frac{K_2}{K_1} = \frac{\Delta H^\circ}{R}\left[\frac{1}{T_1} - \frac{1}{T_2}\right] \Rightarrow \ln \frac{K_2}{0.40 K_2} = \frac{\Delta H^\circ}{8.31 \text{ J/mol.} \cdot \text{K}}\left[\frac{1}{323 \text{ K}} - \frac{1}{310 \text{ K}}\right]$$

$0.92 = \Delta H^\circ(-1.56 \times 10^{-5} \text{ mol./J})$

$\Delta H^\circ = -5.9 \times 10^4 \text{ J/mol.}$

$\Delta H^\circ = -59 \text{ kJ/mol.}$

54. Refer to Section 12.2 and Chapter 5.

$PV = nRT$

$$P_{O_2} = \frac{(1.00 \text{ mol.})(0.0821 \text{ L} \cdot \text{atm/mol.} \cdot \text{K})(1800 \text{ K})}{5.0 \text{ L}} = 30 \text{ atm}$$

$$K = \frac{(P_O)^2}{P_{O_2}}$$

$$1.7 \times 10^{-8} = \frac{(P_O)^2}{30 \text{ atm}}$$

$(P_O)^2 = 5.1 \times 10^{-7}$

$P_O = 7.1 \times 10^{-4}$ atm

$$\% \text{ dissociated} = \frac{\Delta P_{O_2}}{P_{eq(O_2)}} \times 100$$

This calculation can be made using pressure since pressure and amount are directly related via the ideal gas law.

P_o (O) = 0 atm

P_{eq} (O) = 7.1 x 10^{-4} atm

ΔP (O) = 7.1 x 10^{-4} = 2x

$x = (7.1 \times 10^{-4})/2 = 3.6 \times 10^{-4}$ atm $= \Delta P$ (O$_2$)

$$\% \text{ dissociated} = \frac{3.6 \times 10^{-4} \text{ atm}}{30 \text{ atm}} \times 100 = 1.2 \times 10^{-3} \%$$

The number of oxygen atoms can be calculated using the ideal gas law.

$PV = nRT$

$(7.1 \times 10^{-4} \text{ atm})(5.0 \text{ L}) = (n)(0.0821 \text{ L} \cdot \text{atm/mol} \cdot \text{K})(1800 \text{ K})$

$n = 2.4 \times 10^{-5}$ mol.

$$2.4 \times 10^{-5} \text{ mol.} \times \frac{6.022 \times 10^{23} \text{ atoms}}{1 \text{ mol.}} = 1.4 \times 10^{19} \text{ atoms}$$

---

56. *Refer to Section 12.5.*

Calculate $K_{700°C}$ using the ideal gas law equation to determine the partial pressures. Then repeat the calculation for $K_{600°C}$.

$$K_{700°C} = \frac{(P_{CO})^2}{P_{CO_2}} = \frac{\left(\dfrac{(0.10 \text{ mol.})(0.0821 \text{ L} \cdot \text{atm/mol} \cdot \text{K})(973 \text{ K})}{2.0 \text{ L}}\right)^2}{\dfrac{(0.20 \text{ mol.})(0.0821 \text{ L} \cdot \text{atm/mol} \cdot \text{K})(973 \text{ K})}{2.0 \text{ L}}} = 2.0$$

If 0.040 additional mol. of C(s) are formed, then an additional 0.040 mol. CO$_2$(g) are also formed (giving 0.24 mol.) and the amount of CO(g) is decreased by 2(0.040) to 0.020 mol.

$$K_{600°C} = \frac{(P_{CO})^2}{P_{CO_2}} = \frac{\left(\dfrac{(0.020\,\text{mol.})(0.0821\,\text{L}\cdot\text{atm/mol.}\cdot\text{K})(873\,\text{K})}{2.0\,\text{L}}\right)^2}{\dfrac{(0.24\,\text{mol.})(0.0821\,\text{L}\cdot\text{atm/mol.}\cdot\text{K})(873\,\text{K})}{2.0\,\text{L}}} = 0.06$$

58. Refer to Section 12.1 and Figure 12.2.

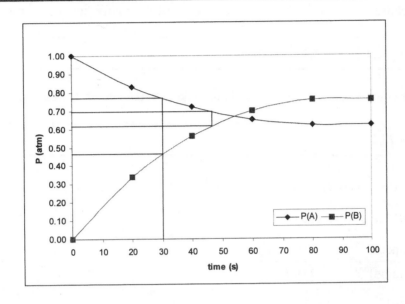

a. $P_A = 0.78$, $P_B = 0.47$

b. 0.62 atm; equilibrium was established after 80 seconds.

c. 0.61 atm; see the graph above.

60. Refer to Chapter 12.

a. **EQ**   Once the system is at equilibrium, pressures do not change.
b. **GT**   Pressures of reactants will decrease until equilibrium is established.
c. **MI**   A K value for either the forward or the reverse reaction is needed.
d. **EQ**   By definition.
e. **EQ** K is not changed when products or reactants are added or removed.
f. **EQ**   A catalyst changes the rate, but not the equilibrium constant.
g. **MI** K for 1 mol. is equal to the square root of K for the system given. In order to compare the two K values, one of them needs to be given; the other one can be calculated from the given value.
h. **MI**   The sign of ΔH is required.
i. **EQ**   C is a solid. Addition of C will not shift equilibrium.

If the pressure of Q is increased to 1.5 atm, the system will respond by shifting the equilibrium to the right, producing more R and consequently reacting Q. The pressure of Q will decrease. The equilibrium pressure of Q will be between the initial pressure of 0.44 atm and the increased pressure of 1.5 atm.

(**b**) and (**c**) are between 1.5 and 0.44 atm, so these could be the equilibrium pressures of Q.

64. *Refer to Sections 12.3 and 12.5.*

The value of the equilibrium constant has nothing to do with the rate of the reaction.

66. *Refer to Section 12.2.*

The symbol $\prod$ is the product symbol, where : $\prod x_i = x_1 \times x_2 \times x_3 \times ... x_n$

$[\text{products}] = n_p/V$ and $[\text{reactants}] = n_r/V$

$$K_c = \frac{\prod [\text{products}]^{\text{coef}}}{\prod [\text{reactants}]^{\text{coef}}} = \frac{\prod (n_p/V)^{n_p}}{\prod (n_r/V)^{n_r}}$$

$n_p = P_p V/RT$ and $n_r = P_r V/RT$

$$K_c = \frac{\prod (P_p/RT)^{n_p}}{\prod (P_r/RT)^{n_r}} = \frac{\left(\dfrac{1}{RT}\right)^{n_p} \prod (P_p)^{n_p}}{\left(\dfrac{1}{RT}\right)^{n_r} \prod (P_r)^{n_r}}$$

Since, $K = \dfrac{\prod (P_p)^{n_p}}{\prod (P_r)^{n_r}}$, substituting gives

$$K_c = K \times \frac{\left(\dfrac{1}{RT}\right)^{n_p}}{\left(\dfrac{1}{RT}\right)^{n_r}} = K \times \left(\frac{1}{RT}\right)^{n_p - n_r}$$ but: $n_p - n_r = \Delta n_g$

$$K_c = K \times \left(\frac{1}{RT}\right)^{\Delta n_g} \quad \Rightarrow \quad K = K_c \times (RT)^{\Delta n_g}$$

Also see the solution given in the answers in Appendix 6.

67. Refer to Section 12.4.

	$2NH_3(g)$	$\rightleftharpoons$	$N_2(g)$	$+$	$3H_2(g)$
P_o	1.00 atm		0 atm		0 atm
ΔP	-2x atm		x atm		3x atm
P_{eq}	1.00 - 2x atm		x atm		3x atm

$$K = 2.5 = \frac{(P_{N_2})(P_{H_2})^3}{(P_{NH_3})^2} = \frac{(x)(3x)^3}{(1-2x)^2} = \frac{27x^4}{1.00 - 4x + 4x^2}$$

$$2.5 - 10x + 10x^2 = 27x^4 \quad \Rightarrow \quad 2.5 = 10x - 10x^2 + 27x^4$$

This is best solved by successive approximations. See p 363. As a first approximation, $x \sim 2.5/10$.
Successive approximations gives: $x = 0.33$.

$$P_{NH_3} = 1 - 2(0.33) = 0.34 \text{ atm}$$

$$P_{H_2} = 3(0.33) = 0.99 \text{ atm}$$

$$P_{N_2} = 0.33 \text{ atm}$$

68. Refer to Section 12.4.

Use the titration data to calculate the amount of I_2 that was present at equilibrium. Then calculate the moles of each species and calculate K using moles instead of pressures (this is acceptable since the R, T and V terms would cancel in the equilibrium constant expression).

$$0.0370 \text{ L} \times \frac{0.200 \text{ mol. } S_2O_3^{2-}}{1 \text{ L}} \times \frac{1 \text{ mol. } I_2}{2 \text{ mol. } S_2O_3^{2-}} = 3.70 \times 10^{-3} \text{ mol. } I_2$$

$$3.20 \text{ g HI} \times \frac{1 \text{ mol. HI}}{127.9 \text{ g HI}} = 2.50 \times 10^{-2} \text{ mol. HI}$$

At equilibrium: $HI = 2.50 \times 10^{-2} - 2(3.70 \times 10^{-3}) = 0.0176$ mol.
$I_2 = 3.70 \times 10^{-3}$ mol.
$H_2 = 3.70 \times 10^{-3}$ mol.

$$K = \frac{(P_{I_2})(P_{H_2})}{(P_{HI})^2} = \frac{(3.7 \times 10^{-3})(3.7 \times 10^{-3})}{(0.0176)^2} = 0.0442$$

	$SO_3(g)$	$\rightleftharpoons$	$SO_2(g)$	$+$	$\frac{1}{2}O_2(g)$
P_0	1.00 atm		0 atm		0 atm
ΔP	-2x atm		2x atm		x atm
P_{eq}	1.00 - 2x atm		2x atm		x atm

$$K = 0.45 = \frac{(P_{SO_2})(P_{O_2})^{\frac{1}{2}}}{(P_{SO_3})} = \frac{(2x)(x)^{\frac{1}{2}}}{1.00 - 2x} = \frac{2x^{\frac{3}{2}}}{1.00 - 2x}$$

$$0.45 - 0.90x = 2x^{\frac{3}{2}}$$

This is best solved by successive approximations.

$x = 0.24$

$$P_{SO_3} = 1.00 - 2(0.24) = 0.52$$

	$I_2(g)$	$\rightleftharpoons$	$2I(g)$
P_0	1.00 atm		0 atm
ΔP	-x atm		2x atm
P_{eq}	1.00 - x atm		2x atm

Total pressure at equilibrium is 40% greater than initial pressure, so total pressure equals 1.4 and $1 - x + 2x = 1.4$, and $x = 0.4$

$$K = \frac{(P_I)^2}{P_{I_2}} = \frac{(0.8)^2}{0.6} = 1.1$$

A 50% yield indicates that half the reactants were converted to product, and the partial pressure changes accordingly.

$$K = \frac{(P_{XeF_4})}{(P_{Xe})(P_{F_2})^2} = \frac{(0.10)}{(0.10)(0.20)^2} = 25$$

A 75% yield would mean that the pressure of Xe would decrease by 75% (to 0.05 atm) and the amount of XeF_4 would increase by that amount (to 0.15 atm). Let x = initial pressure of F_2. The F_2 would decrease by twice the amount of Xe (to $x - 2(0.15)$).

$$K = 25 = \frac{(P_{XeF_4})}{(P_{Xe})(P_{F_2})^2} = \frac{(0.15)}{(0.050)(x - 0.30)^2} \quad \Rightarrow \quad 0 = x^2 - 0.60x - 0.03$$

$$x = \frac{-b \pm \sqrt{b^2 - 4ac}}{2a} = \frac{0.60 \pm \sqrt{(0.60)^2 - 4(1)(-0.03)}}{2(1)} = 0.65$$

x = initial pressure of F_2 = 0.65 atm

---

72. Refer to Section 12.4.

a. Use the ideal gas law to determine the pressure of the benzyl alcohol, then use an equilibrium table and the equilibrium expression to solve for x.

$$1.50 \text{ g} \times \frac{1 \text{ mol.}}{108.1 \text{ g}} = 0.0139 \text{ mol.}$$

$$P = \frac{nRT}{V} = \frac{(0.0139 \text{ mol.})(0.0821 \text{ L} \cdot \text{atm/mol.} \cdot \text{K})(523 \text{ K})}{2.0 \text{ L}} = 0.30 \text{ atm}$$

	$C_6H_5CH_2OH(g)$	$\rightleftharpoons$	$C_6H_5CHO(g)$	+	$H_2(g)$
P_o	0.30 atm		0 atm		0 atm
ΔP	$-x$ atm		x atm		x atm
P_{eq}	0.30 - x atm		x atm		x atm

$$K = 0.56 = \frac{x^2}{0.30 - x} \quad \Rightarrow \quad x^2 + 0.56x - 0.168 = 0$$

$$x = \frac{-b \pm \sqrt{b^2 - 4ac}}{2a} = \frac{-0.56 \pm \sqrt{(0.56)^2 - 4(1)(-0.168)}}{2(1)} = 0.22 \text{ atm}$$

b. $$n = \frac{PV}{RT} = \frac{(0.30 - 0.22 \text{ atm})(2.0 \text{ L})}{(0.0821 \text{ L} \cdot \text{atm/mol.} \cdot \text{K})(523 \text{ K})} = 4 \times 10^{-3} \text{ mol.}$$

$$4 \times 10^{-3} \text{ mol.} \times \frac{108.1 \text{ g}}{1 \text{ mol.}} = 0.4 \text{ g}$$

Chapter 13: Acids and Bases

---

2. *Refer to Section 13.1 and Example 13.1.*

The acid is the reactant that donates an H^+, while the base is the reactant that accepts the H^+. The conjugate acid/base is the product that received/donated an H^+.

	Acid	Base	acid/conj. base pair	conj. acid/base pair
a.	H_2O	CN^-	H_2O/OH^-	HCN/CN^-
b.	H_3O^+	HCO_3^-	H_3O^+/H_2O	H_2CO_3/HCO_3^-
c.	$HC_2H_3O_2$	HS^-	$HC_2H_3O_2/C_2H_3O_2^-$	H_2S/HS^-

---

4. *Refer to Section 13.1.*

a. CHO_2^- **Base.** As an anion, it is likely to act as an H^+ acceptor.

b. NH_4^+ **Acid.** Since the molecule has a positive charge, it is unlikely to accept an H^+ with another positive charge.

c. HSO_3^- **Acid or base** (amphoteric). As an anion, it can act as an H^+ acceptor. With an H^+ attached to an oxygen, it can also act as an H^+ donor.

---

6. *Refer to Section 13.1.*

Write the equation for the dissociation of the species (HB) into H^+ and B^-. The resulting B^- will be the conjugate base of the acid HB.

a. $HCO_3^- \rightarrow H^+ + \mathbf{CO_3^{2-}}$

b. $[Cu(H_2O)(OH)_3]^- \rightarrow H^+ + \mathbf{[Cu(OH)_4]^{2-}}$

c. $HNO_2 \rightarrow H^+ + \mathbf{NO_2^-}$

d. $(CH_3)_2NH_2^+ \rightarrow H^+ + \mathbf{(CH_3)_2NH}$

e. $H_2SO_3 \rightarrow H^+ + \mathbf{HSO_3^-}$

8. *Refer to Section 13.3 and Example 13.2.*

See page 614 for rules for significant figures in Logarithms and Inverse Logarithms.

a. $pH = -\log[H^+] = -\log(1.0) = 0$ pH < 7, thus the solution is acidic.
     $pOH = 14.00 - pH = 14.00 - 0 = 14.00$

b. $pH = -\log[H^+] = -\log(1.7 \times 10^{-4}) = 3.78$ pH < 7, thus the solution is acidic.
     $pOH = 14.00 - pH = 14.00 - 3.78 = 10.22$

c. $pH = -\log[H^+] = -\log(6.8 \times 10^{-8}) = 7.17$ pH > 7, thus the solution is basic.
     $pOH = 14.00 - pH = 14.00 - 7.17 = 6.83$

d. $pH = -\log[H^+] = -\log(9.3 \times 10^{-11}) = 10.03$ pH > 7, thus the solution is basic.
     $pOH = 14.00 - pH = 14.00 - 10.03 = 3.97$

10. *Refer to Section 13.3 and Example 13.3.*

See page 614 for rules for significant figures in Logarithms and Inverse Logarithms.

a. $[H^+] = 10^{-pH} = 10^{-9.0} = 1 \times 10^{-9}\ M$

     $[OH^-] = \dfrac{K_w}{[H^+]} = \dfrac{1.0 \times 10^{-14}}{1 \times 10^{-9}} = 1 \times 10^{-5}\ M$

b. $[H^+] = 10^{-pH} = 10^{-3.2} = 6 \times 10^{-4}\ M$

     $[OH^-] = \dfrac{K_w}{[H^+]} = \dfrac{1.0 \times 10^{-14}}{6 \times 10^{-4}} = 2 \times 10^{-11}\ M$

c. $[H^+] = 10^{-pH} = 10^{+1.05} = 11\ M$

     $[OH^-] = \dfrac{K_w}{[H^+]} = \dfrac{1.0 \times 10^{-14}}{11} = 9.1 \times 10^{-16}\ M$

d. $[H^+] = 10^{-pH} = 10^{-7.46} = 3.5 \times 10^{-8}\ M$

     $[OH^-] = \dfrac{K_w}{[H^+]} = \dfrac{1.0 \times 10^{-14}}{3.5 \times 10^{-8}} = 2.9 \times 10^{-7}\ M$

12. *Refer to Section 13.3 and Example 13.2.*

See page 614 for rules for significant figures in Logarithms and Inverse Logarithms.

Calculate the pH and pOH of both solutions and compare.

Sol. X: $pH = 11.7$
$pOH = 14.00 - pH = 14.00 - 11.7 = 2.3$

Sol. Y: $pOH = -\log[OH^-] = -\log(4.5 \times 10^{-2}) = 1.35$
$pH = 14.00 - pOH = 14.00 - 1.35 = 12.65$
Solution Y is more basic (higher pH and thus lower $[H^+]$)
Solution X has a higher pOH.

14. *Refer to Section 13.3 and Example 13.2.*

See page 614 for rules for significant figures in Logarithms and Inverse Logarithms.
Calculate $[H^+]$ for solutions A, B and C. Then use the results to answer parts a, b and c.

a. $[H^+]_A = 10^{-pH} = 10^{-12.32} = 4.8 \times 10^{-13} \ M$

$[H^+]_B = 3(4.8 \times 10^{-13} \ M) = 1.4 \times 10^{-12} \ M$

$pH_C = pH_A/2 \ = 12.32/2 = 6.16$

$[H^+]_C = 10^{-pH} = 10^{-6.16} = 6.9 \times 10^{-7} \ M$

b. $pH_B = -\log[H^+] = -\log(1.4 \times 10^{-12}) = 11.84$
$pH_C = 6.16$ (see calculation above)

c. $pH_A = 12.32$ pH is greater than 7, therefore solution A is basic
$pH_B = 11.84$ pH is greater than 7, therefore solution B is basic
$pH_C = 6.16$ pH is less than 7, therefore solution C is acidic

16. *Refer to Section 13.3 and Example 13.2.*

a. $[H^+]_{\text{milk of magnesia}} = 10^{-pH} = 10^{-10.5} = 3 \times 10^{-11} \ M$

b. $[H^+]_{\text{gastric juice}} = 10^{-pH} = 10^{-1.5} = 3 \times 10^{-2} \ M$

$$\frac{[H^+]_{\text{gastric juice}}}{[H^+]_{\text{milk of magnesia}}} = \frac{3 \times 10^{-2}}{3 \times 10^{-11}} = 1 \times 10^{9}$$

See page 614 for rules for significant figures in Logarithms and Inverse Logarithms.
Calculate the concentration of the hydrochloric acid. [H^+] will equal [HCl] since this is a
strong monoprotic acid.

a. $\dfrac{37.5 \text{ g HCl}}{100 \text{ g solution}} \times \dfrac{1.00 \text{ g}}{1.00 \text{ mL}} \times \dfrac{1000 \text{ mL}}{1 \text{ L}} \times \dfrac{1 \text{ mol. HCl}}{36.46 \text{ g HCl}} = 10.3 \ M$

pH = -log[H^+] = -log(10.3) = -1.013

1.75 L would have the same concentration as 0.175 L, and thus the same pH.

b. $\dfrac{22 \text{ g HBr}}{0.479 \text{ L}} \times \dfrac{1 \text{ mol. HBr}}{80.91 \text{ g HBr}} = 0.57 \ M$

pH = -log[H^+] = -log(0.57) = 0.24

$\dfrac{22 \text{ g HBr}}{0.0479 \text{ L}} \times \dfrac{1 \text{ mol. HBr}}{80.91 \text{ g HBr}} = 5.7 \ M$

pH = -log[H^+] = -log(5.7) = -0.76

Calculate the moles of H^+ in each solution. Add those values together and divide by the total
volume (L) to get concentration. Remember that both acids are strong acids and ionize
completely. Consequently, [H^+] = [HX]

HCl: $0.145 \text{ L HCl} \times \dfrac{0.575 \text{ mol. HCl}}{1 \text{ L}} \times \dfrac{1 \text{ mol. H}^+}{1 \text{ mol. HCl}} = 0.0834 \text{ mol. H}^+$

HNO$_3$:   [H^+] = $10^{-1.39}$ = 0.041 M

$0.493 \text{ L HNO}_3 \times \dfrac{0.041 \text{ mol. HNO}_3}{1 \text{ L}} \times \dfrac{1 \text{ mol. H}^+}{1 \text{ mol. HNO}_3} = 0.020 \text{ mol. H}^+$

moles H^+ = 0.0834 mol. + 0.020 mol. = 0.103 mol.

$[H^+] = \dfrac{0.103 \text{ mol.}}{0.143 \text{ L} + 0.493 \text{ L}} = 0.162 \ M$

pH = -log[H^+] = -log(0.162) = 0.790

22. *Refer to Section 13.3.*

Calculate the concentration of OH^- in the final solution, and from that the pOH, then pH and finally $[H^+]$.

a. $$0.0450\,\text{L} \times \frac{0.0921\,\text{mol. Ba(OH)}_2}{1\,\text{L}} \times \frac{2\,\text{mol. OH}^-}{1\,\text{mol. Ba(OH)}_2} = 8.29 \times 10^{-3}\,\text{mol. OH}^-$$

$$[\text{OH}^-] = \frac{8.29 \times 10^{-3}\,\text{mol. OH}^-}{0.3500\,\text{L}} = 0.0237\,M$$

$\text{pOH} = -\log(0.0237) = 1.625$

$\text{pH} = 14.00 - \text{pOH} = 14.00 - 1.625 = 12.38$

$[H^+] = 10^{-12.38} = 4.2 \times 10^{-13}\,M$

b. $$4.68\,\text{g NaOH} \times \frac{1\,\text{mol. NaOH}}{40.0\,\text{g NaOH}} \times \frac{1\,\text{mol. OH}^-}{1\,\text{mol. NaOH}} = 0.117\,\text{mol. OH}^-$$

$$[\text{OH}^-] = \frac{0.117\,\text{mol. OH}^-}{0.635\,\text{L}} = 0.184\,M$$

$\text{pOH} = -\log(0.184) = 0.735$

$\text{pH} = 14.00 - \text{pOH} = 14.00 - 0.735 = 13.26$

$[H^+] = 10^{-13.26} = 5.5 \times 10^{-14}\,M$

24. *Refer to Section 13.3.*

Convert pH to pOH. Convert pOH to moles of OH^-. Calculate the moles of OH^- in each solution. Then calculate $[OH^-]_{final}$ by dividing the total moles of OH^- by the total volume of solution. Then calculate pOH and convert to pH.

KOH: $\text{pOH} = 14.00 - \text{pH} = 14.00 - 12.66 = 1.34$

$$[\text{OH}^-] = 10^{-\text{pOH}} = 10^{-1.34} = 4.6 \times 10^{-2}\,M$$

$$0.139\,\text{L} \times \frac{0.046\,\text{mol. KOH}}{1\,\text{L}} = 0.0064\,\text{mol. OH}^-$$

$Ba(OH)_2$: $\text{pOH} = 14.00 - \text{pH} = 14.00 - 11.79 = 2.21$

$$[\text{OH}^-] = 10^{-\text{pOH}} = 10^{-2.21} = 6.2 \times 10^{-3}\,M$$

$$0.293\,\text{L} \times \frac{6.2 \times 10^{-3}\,\text{mol. OH}^-}{1\,\text{L}} = 0.0018\,\text{mol. OH}^-$$

$$[OH^-]_{final} = \frac{0.0064 \text{ mol. OH}^- + 0.0018 \text{ mol. OH}^-}{0.139 \text{ L} + 0.293 \text{ L}} = 0.019 \, M$$

pOH = -log[OH⁻] = -log(0.019) = 1.72

pH = 14.00 - pOH = 14.00 - 1.72 = 12.28

---

26. Refer to Section 13.4 and Example 13.4.

a. One of the water molecules attached to the metal becomes a hydroxide and the hydrogen ion is transferred to the water to form the hydronium ion.

$[Zn(H_2O)_3OH]^+(aq) + H_2O(l) \rightleftharpoons [Zn(H_2O)_2(OH)_2](aq) + H_3O^+(aq)$

b. $HSO_4^-(aq) + H_2O(l) \rightleftharpoons SO_4^{2-}(aq) + H_3O^+(aq)$

c. $HNO_2(aq) + H_2O(l) \rightleftharpoons NO_2^-(aq) + H_3O^+(aq)$

d. $[Fe(H_2O)_6]^{2+}(aq) + H_2O(l) \rightleftharpoons [Fe(H_2O)_5OH]^+(aq) + H_3O^+(aq)$

e. $HC_2H_3O_2(aq) + H_2O(l) \rightleftharpoons C_2H_3O_2^-(aq) + H_3O^+(aq)$

f. $H_2PO_4^-(aq) + H_2O(l) \rightleftharpoons HPO_4^{2-}(aq) + H_3O^+(aq)$

---

28. Refer to Section 13.4 and Example 13.4.

Since each of the species is identified as an acid, the ionization equation will follow the general format: $HB(aq) \rightleftharpoons H^+(aq) + B^-(aq)$, and $K = K_a$.

a. $HSO_3^-(aq) \rightleftharpoons H^+(aq) + SO_3^{2-}(aq)$

$$K_a = \frac{[H^+][SO_3^{2-}]}{[HSO_3^-]}$$

b. $HPO_4^{2-}(aq) \rightleftharpoons H^+(aq) + PO_4^{3-}(aq)$

$$K_a = \frac{[H^+][PO_4^{3-}]}{[HPO_4^{2-}]}$$

c. $HNO_2(aq) \rightleftharpoons H^+(aq) + NO_2^-(aq)$

$$K_a = \frac{[H^+][NO_2^-]}{[HNO_2]}$$

a. $pK_a = -\log K_a = -\log(1.8 \times 10^{-4}) = 3.74$

b. $pK_a = -\log K_a = -\log(6.8 \times 10^{-8}) = 7.17$

c. $pK_a = -\log K_a = -\log(4.9 \times 10^{-11}) = 10.31$

32. *Refer to Section 13.4.*

a. The stronger the acid, the greater the degree of ionization and thus the larger the K_a.

$$C < D < B < A$$

b. The largest K_a corresponds to the smallest pK_a.

A has the smallest pK_a.

34. *Refer to Section 13.4 and Example 13.5.*

Calculate $[HC_7H_6NO_2]_0$, and then the equilibrium concentrations for all three species. Use those values to calculate K_a.

$$[HC_7H_6NO_2]_0 = \frac{0.263 \text{ mol.}}{0.7500 \text{ L}} = 0.351 \, M$$

	$HC_7H_6NO_2(aq)$	$\rightleftharpoons$	$C_7H_6NO_2^-(aq)$	$+$	$H^+(aq)$
[ ]$_0$	0.351		0		0
Δ[ ]	-2.6×10^{-3}		$+2.6 \times 10^{-3}$		$+2.6 \times 10^{-3}$
[ ]$_{eq}$	0.348		2.6×10^{-3}		2.6×10^{-3}

$$K_a = \frac{[H^+][C_7H_6NO_2^-]}{[HC_7H_6NO_2]} = \frac{(2.60 \times 10^{-3})(2.6 \times 10^{-3})}{0.348} = 1.9 \times 10^{-5}$$

36. *Refer to Section 13.4 and Example 13.5.*

Calculate $[HC_4H_3N_2O_3]_0$ from the mass and volume and $[H^+]_{eq}$ from the pH. Then calculate the equilibrium concentrations for all three species. Since there was no H^+ before the acid dissolved, $[H^+]_{eq}$ is the change in concentration. Use these values to calculate K_a.

$$9.00 \text{ g } HC_4H_3N_2O_3 \times \frac{1 \text{ mol.}}{128.09 \text{ g}} = 0.0703 \text{ mol.}$$

$$[HC_4H_3N_2O_3]_0 = \frac{0.0703 \text{ mol.}}{0.325 \text{ L}} = 0.216 \ M$$

$$[H^+]_{eq} = 10^{-pH} = 10^{-2.34} = 0.0046 \ M$$

	$HC_4H_3N_2O_3 \ (aq)$	$\rightleftharpoons$	$C_4H_3N_2O_3^- (aq)$	$+$	$H^+(aq)$
$[\ ]_0$	0.216		0		0
$\Delta[\ ]$	-0.0046		+0.0046		+0.0046
$[\ ]_{eq}$	0.211		0.0046		0.0046

$$K_a = \frac{[H^+][C_4H_3N_2O_2^-]}{[HC_4H_3N_2O_2]} = \frac{(0.0046)(0.0046)}{0.211} = 1.0 \times 10^{-4}$$

---

38. Refer to Section 13.4 and Example 13.7.

Consider the 5% rule: if $\dfrac{x}{a} \leq 0.05$, then $x \leq 0.05a$. Substituting this relationship into the

equation for K_a gives: $K_a = \dfrac{x^2}{a} = \dfrac{(0.05a)^2}{a} = (2.5 \times 10^{-3})a$ and finally: $K_a/a < 2.5 \times 10^{-3}$.

Thus one can apply the 5% rule when $K_a/a < 2.5 \times 10^{-3}$. This provides a method for determining *a priori* if the 5% rule applies.

a. Let Penicillin = HPen

$$[HPen]_0 = \frac{0.187 \text{ mol.}}{0.725 \text{ L}} = 0.258 \ M$$

	$HPen(aq)$	$\rightleftharpoons$	$Pen^-(aq)$	$+$	$H^+(aq)$
$[\ ]_0$	0.258		0		0
$\Delta[\ ]$	$-x$		$+x$		$+x$
$[\ ]_{eq}$	$0.258 - x$		x		x

$$K_a = \frac{[H^+][Pen^-]}{[HPen]} \Rightarrow 1.7 \times 10^{-3} = \frac{(x)(x)}{0.258 - x} = \frac{x^2}{0.258 - x}$$

$$K_a/a = 1.7 \times 10^{-3} \ / \ 0.258 = 6.7 \times 10^{-3}$$
$$K_a/a = 6.7 \times 10^{-3} > 2.5 \times 10^{-3} \qquad \text{(The 5\% rule does not apply.)}$$

$$4.4 \times 10^{-4} - 1.7 \times 10^{-3}(x) = x^2$$

$$x^2 + 1.7 \times 10^{-3}(x) - 4.4 \times 10^{-4} = 0$$

$$x = \frac{-b \pm \sqrt{b^2 - 4ac}}{2a} = \frac{-1.7 \times 10^{-3} \pm \sqrt{(1.7 \times 10^{-3})^2 - 4(1)(-4.4 \times 10^{-4})}}{2(1)} = 2.1 \times 10^{-2}$$

$[H^+] = x = 2.1 \times 10^{-2} \, M$
(the other possible answer gives a negative concentration which is not possible)

b. $[HPen]_0 = \dfrac{127\,g}{0.725\,L} \times \dfrac{1\,mol.}{356\,g} = 0.357 \, M$

	HPen(aq)	$\rightleftharpoons$	Pen$^-$(aq)	+	H$^+$(aq)
[ ]$_0$	0.357		0		0
Δ[ ]	-x		+x		+x
[ ]$_{eq}$	0.357 - x		x		x

$$K_a = \frac{[H^+][Pen^-]}{[HPen]} \Rightarrow 1.7 \times 10^{-3} = \frac{(x)(x)}{0.357 - x} = \frac{x^2}{0.357 - x}$$

$K_a/a = 1.7 \times 10^{-3} / 0.357 = 4.8 \times 10^{-3}$
$K_a/a = 4.8 \times 10^{-3} > 2.5 \times 10^{-3}$ $\qquad$ (thus the 5% rule does not apply)

$8.4 \times 10^{-4} - 1.7 \times 10^{-3}(x) = x^2$

$x^2 + 1.7 \times 10^{-3}(x) - 8.4 \times 10^{-4} = 0$

$$x = \frac{-b \pm \sqrt{b^2 - 4ac}}{2a} = \frac{-1.7 \times 10^{-3} \pm \sqrt{(1.7 \times 10^{-3})^2 - 4(1)(-8.4 \times 10^{-4})}}{2(1)} = 2.8 \times 10^{-2}$$

$[H^+] = x = 2.8 \times 10^{-2} \, M$
(the other possible answer gives a negative concentration which is not possible)

---

40. Refer to Section 13.4 and Problem 38 (above).

HBarb(aq) $\rightleftharpoons$ Barb$^-$(aq) + H$^+$(aq)

a. $K_a = \dfrac{[H^+][Barb^-]}{[HBarb]} \Rightarrow 1.1 \times 10^{-4} = \dfrac{x^2}{0.673 - x}$

$K_a/a = 1.1 \times 10^{-4} / 0.673 = 1.63 \times 10^{-4}$
$K_a/a = 1.63 \times 10^{-4} < 2.5 \times 10^{-3}$ $\qquad$ (The 5% rule applies. See the explanation
$\qquad\qquad\qquad\qquad\qquad\qquad\qquad\qquad\qquad\qquad$ accompanying problem 38.)

$K_a = 1.1 \times 10^{-4} = \dfrac{x^2}{0.673} \Rightarrow x = 8.6 \times 10^{-3}$

$$[H^+] = x = 8.6 \times 10^{-3} \, M$$

b. $[OH^-] = \dfrac{K_w}{[H^+]} = \dfrac{1.00 \times 10^{-14}}{8.6 \times 10^{-3}} = 1.2 \times 10^{-12} \, M$

c. $pH = -\log[H^+] = -\log(8.6 \times 10^{-3}) = 2.07$

d. % ionization $= \dfrac{[H^+]_{eq}}{[HBarb]_0} \times 100\% = \dfrac{8.6 \times 10^{-3}}{0.673} \times 100\% = 1.3\%$

---

42. Refer to Section 13.4, Table 13.2, and Example 13.7.

Consider the 5% rule: if $\dfrac{x}{a} \le 0.05$, then $x \le 0.05a$. Substituting this relationship into the

equation for K_a gives: $K_a = \dfrac{x^2}{a} = \dfrac{(0.05a)^2}{a} = (2.5 \times 10^{-3})a$ and finally: $K_a/a < 2.5 \times 10^{-3}$.

Thus one can apply the 5% rule when $K_a/a < 2.5 \times 10^{-3}$. This provides a method for determining *a priori* if the 5% rule applies.

a. $[HC_7H_5O_2]_0 = \dfrac{0.288 \text{ mol.}}{0.726 \text{ L}} = 0.397 \, M$

	$HC_7H_5O_2(aq)$	$\rightleftharpoons$	$C_7H_5O_2^-(aq)$	+	$H^+(aq)$
$[\ ]_0$	0.397		0		0
$\Delta[\ ]$	-x		+x		+x
$[\ ]_{eq}$	0.397 - x		x		x

$$K_a = \dfrac{[H^+][C_7H_5O_2^-]}{[HC_7H_5O_2]} \;\Rightarrow\; 6.6 \times 10^{-5} = \dfrac{(x)(x)}{0.397 - x} = \dfrac{x^2}{0.397 - x}$$

$K_a/a = 6.6 \times 10^{-5} / 0.397 = 1.66 \times 10^{-4}$
$K_a/a = 1.66 \times 10^{-4} < 2.5 \times 10^{-3}$ (The 5% rule applies.)

$$K_a = 6.6 \times 10^{-5} = \dfrac{x^2}{0.397} \;\Rightarrow\; x = 5.12 \times 10^{-3}$$

$[H^+] = x = 5.12 \times 10^{-3} \, M$
$pH = -\log[H^+] = 2.29$

% ionization $= \dfrac{[H^+]_{eq}}{[HB]_0} \times 100\% = \dfrac{5.12 \times 10^{-3}}{0.397} \times 100\% = 1.29\%$

$$NH_4^+(aq) \rightleftharpoons NH_3(aq) + H^+(aq)$$

$$K_a = \frac{[H^+][NH_3]}{[NH_4^+]} \Rightarrow 5.6 \times 10^{-10} = \frac{x^2}{0.39 - x}$$

$K_a/a = 5.6 \times 10^{-10} / 0.39 = 1.4 \times 10^{-9}$
$K_a/a = 1.4 \times 10^{-9} < 2.5 \times 10^{-3}$ (The 5% rule applies. See the explanation accompanying problem 38.)

$$K_a = 5.6 \times 10^{-10} = \frac{x^2}{0.39} \Rightarrow x = 1.5 \times 10^{-5}$$

$[H^+] = x = 1.5 \times 10^{-5} \, M$

pH $= -\log[H^+] = -\log(1.5 \times 10^{-5}) = 4.82$

$H_3PO_4(aq) \rightleftharpoons H^+(aq) + H_2PO_4^-(aq)$	$K_1 = 7.1 \times 10^{-3}$
$H_2PO_4^-(aq) \rightleftharpoons H^+(aq) + HPO_4^{2-}(aq)$	$K_2 = 6.2 \times 10^{-8}$
$HPO_4^{2-}(aq) \rightleftharpoons H^+(aq) + PO_4^{3-}(aq)$	$K_3 = 4.5 \times 10^{-13}$
$H_3PO_4(aq) \rightleftharpoons 3H^+(aq) + PO_4^{3-}(aq)$	$K = K_1 \times K_2 \times K_3 = 2.0 \times 10^{-22}$

Assume that all the H^+ comes from the dissociation of $H_2C_6H_6O_6$ and calculate $[H^+]$ and $[HC_6H_6O_6^-]$

$$H_2C_6H_6O_6 (aq) \rightleftharpoons H^+(aq) + HC_6H_6O_6^-(aq)$$

$$K_{a_1} = \frac{[H^+][HC_6H_6O_6^-]}{[H_2C_6H_6O_6]} \Rightarrow 7.9 \times 10^{-5} = \frac{x^2}{0.63 - x}$$

$K_a/a = 7.9 \times 10^{-5} / 0.63 = 1.3 \times 10^{-4}$
$K_a/a = 1.3 \times 10^{-4} < 2.5 \times 10^{-3}$ (The 5% rule applies. See the explanation accompanying problem 38.)

$$K_{a_1} = 7.9 \times 10^{-5} = \frac{x^2}{0.63} \Rightarrow x = 7.1 \times 10^{-3}$$

$[H^+] = [HC_6H_6O_6^-] = x = 7.1 \times 10^{-3} \, M$

pH $= -\log[H^+] = -\log(7.1 \times 10^{-3}) = 2.15$

Now use the second dissociation constant to calculate $[C_6H_6O_6^{2-}]$.

$$K_{a_2} = \frac{[H^+][C_6H_6O_6^{2-}]}{[HC_6H_6O_6^-]} = [C_6H_6O_6^{2-}]$$

Recall that $[H^+] = [HC_6H_6O_6^-]$, thus these terms cancel.

$[C_6H_6O_6^{2-}] = 1.6 \times 10^{-12} \, M$

---

50. Refer to Section 13.5 and Example 13.10.

a. $(CH_3)_3N(aq) + H_2O(l) \rightleftharpoons (CH_3)_3NH^+(aq) + OH^-(aq)$

b. $PO_4^{3-}(aq) + H_2O(l) \rightleftharpoons HPO_4^{2-}(aq) + OH^-(aq)$

c. $HPO_4^{2-}(aq) + H_2O(l) \rightleftharpoons H_2PO_4^-(aq) + OH^-(aq)$

d. $H_2PO_4^-(aq) + H_2O(l) \rightleftharpoons H_3PO_4(aq) + OH^-(aq)$

e. $HS^-(aq) + H_2O(l) \rightleftharpoons H_2S(aq) + OH^-(aq)$

f. $C_2H_5NH_2(aq) + H_2O(l) \rightleftharpoons C_2H_5NH_3^+(aq) + OH^-(aq)$

---

52. Refer to Section 13.5 and Table 13.4.

The stronger the base, the greater the K_b. Also, stronger bases have higher pH's.

a. KOH is a strong base, thus K_b will be large and pH will be high.

b. NaCN $K_b = 1.7 \times 10^{-5}$

c. HCO_3^- $K_b = 2.3 \times 10^{-8}$

d. $Ba(OH)_2$ is a strong base and will have a large K_b and high pH. Upon dissociating, however, this base produces $2OH^-$. Therefore, 0.1 M $Ba(OH)_2$ will have a higher pH than a 0.1 M NaOH solution.

    $Ba(OH)_2 >$ KOH $>$ NaCN $>$ NaHCO$_3$

---

54. Refer to Section 13.5.

At 25°C, $K_a \times K_b = K_w = 1.0 \times 10^{-14}$. Substitute the given value for K_b and solve for K_a.

a. $K_a = \dfrac{K_w}{K_b} = \dfrac{1.0 \times 10^{-14}}{1.5 \times 10^{-9}} = 6.7 \times 10^{-6}$

b. $K_a = \dfrac{K_w}{K_b} = \dfrac{1.0 \times 10^{-14}}{3.8 \times 10^{-10}} = 2.6 \times 10^{-5}$

---

56. Refer to Section 13.5.

a. Since acrylate is a weak base (like ammonia), it will react with water to produce OH^-.

$C_3H_3O_2^-(aq) + H_2O(l) \rightleftharpoons HC_3H_3O_2(aq) + OH^-(aq)$

b. $K_b = \dfrac{K_w}{K_a} = \dfrac{1.0 \times 10^{-14}}{5.5 \times 10^{-5}} = 1.8 \times 10^{-10}$

c. $1.61 \text{ g } NaC_3H_3O_2 \times \dfrac{1 \text{ mol.}}{94.04 \text{ g}} = 0.0171 \text{ mol.}$

$\dfrac{0.0171 \text{ mol.}}{0.835 \text{ L}} = 0.0205 M$

Use K_b to calculate $[OH^-]$. Then calculate pOH and convert that to pH.

$K_b = \dfrac{[HC_3H_3O_2][OH^-]}{[C_3H_3O_2^-]} \implies 1.8 \times 10^{-10} = \dfrac{x^2}{0.0205 - x}$

$K_b/b = 1.8 \times 10^{-10} / 0.0205 = 8.8 \times 10^{-9}$
$K_b/b = 8.8 \times 10^{-9} < 2.5 \times 10^{-3}$

(The 5% rule applies. See the explanation accompanying problem 38.)

$K_b = 1.8 \times 10^{-10} = \dfrac{x^2}{0.0205} \implies x = 1.9 \times 10^{-6}$

$[OH^-] = x = 1.9 \times 10^{-6} M$
pOH = -log$[OH^-]$ = -log(1.9×10^{-6}) = 5.72
pH = 14.00 - pOH = 14.00 - 5.72 = 8.28

---

58. Refer to Section 13.5.

Calculate pOH from pH, then calculate $[OH^-]$. [HCN] will equal $[OH^-]$. Use K_b to calculate [NaCN]. Finally, calculate the mass of NaCN necessary to give the calculated concentration.

$NaCN(aq) + H_2O(l) \rightleftharpoons HCN(aq) + NaOH(aq)$

pOH = 14.00 - pH = 14.00 - 12.10 = 1.90
$[OH^-] = 10^{-pOH} = 10^{-1.90} = 0.013 M$

$$K_b = \frac{[HCN][NaOH]}{[NaCN]} \Rightarrow 1.7 \times 10^{-5} = \frac{(0.013)^2}{x}$$

$$x = [NaCN] = 9.9 \ M$$

$$0.425 \ L \times \frac{9.9 \ mol. \ NaCN}{1 \ L} \times \frac{49.01 \ g \ NaCN}{1 \ mol. \ NaCN} = 2.1 \times 10^2 \ g$$

---

60. Refer to Sections 13.5 and 13.6, Example 13.12 and Table 13.5.

Consider the ions that form when the salt dissolves. Label the ions as acidic, basic or spectator, and add the effects.

	Salt	Cation	Anion	Solution
a.	$FeCl_3$	Fe^{3+} (acidic)	Cl^- (spectator)	acidic
b.	BaI_2	Ba^{2+} (spectator)	I^- (spectator)	neutral
c.	NH_4NO_2	NH_4^+ (acidic)	NO_2^- (basic)	acidic*
d.	Na_2HPO_4	Na^+ (spectator)	HPO_4^{2-} (acidic or basic)	basic*
e.	K_3PO_4	K^+ (spectator)	PO_4^{3-} (basic)	basic

* For these problems, one must compare K_a to K_b to determine if the acid or the base dominates. See problem 62 below.

---

62. Refer to Section 13.6.

Write the equations showing the reactions of the acidic and basic ions (from problem 60 above) with water. The spectator ions can be ignored.

a. $Fe(H_2O)_6^{3+}(aq) + H_2O(l) \rightleftharpoons [Fe(H_2O)_5(OH)]^{2+}(aq) + H_3O^+(aq)$

b. No ionic equation since both ions are spectator ions.

c. $NH_4^+(aq) + H_2O(l) \rightleftharpoons NH_3(aq) + H_3O^+(aq)$ $K_a = 5.6 \times 10^{-10}$
$NO_2^-(aq) + H_2O(l) \rightleftharpoons HNO_2(aq) + OH^-(aq)$ $K_b = 1.7 \times 10^{-11}$
       $K_a > K_b$, so the solution is acidic.

d. $HPO_4^{2-}(aq) + H_2O(l) \rightleftharpoons PO_4^{3-}(aq) + H_3O^+(aq)$ $K_a = 4.5 \times 10^{-13}$
$HPO_4^{2-}(aq) + H_2O(l) \rightleftharpoons H_2PO_4^-(aq) + OH^-(aq)$ $K_b = 1.6 \times 10^{-7}$
       $K_b > K_a$, so the solution is basic.

e. $PO_4^{3-}(aq) + H_2O(l) \rightleftharpoons HPO_4^{2-}(aq) + OH^-(aq)$

The first three salts contain the same spectator cation (K^+), thus one can consider the K_b's of the anions. The last three salts contain the same spectator anion (Cl^-), thus one can consider the K_a's of the cations.

OH^-	strong base	K_b = very large	pH $\approx$ 14
F^-	weak base	$K_b = 1.4 \times 10^{-11}$	pH = 8.1
Cl^-	spectator	neutral	pH = 7
K^+	spectator	neutral	pH = 7
Zn^{2+}	weak acid	$K_a = 3.3 \times 10^{-10}$	pH = 5.2
H^+	strong acid	K_a = very large	pH $\approx$ 0

$HCl < ZnCl_2 < KCl < KF < KOH$

There are many correct answers to this problem.

a. Since NH_4^+ is an acidic ion, salts with basic anions with $K_b > K_a$ will give basic salts.
       NH_4ClO, $(NH_4)_2CO_3$

b. Since CO_3^{2-} is a basic ion, salts with neutral cations will give basic salts.
       K_2CO_3, $MgCO_3$

c. Since Br^- is a spectator ion, salts with neutral cations will give neutral salts.
       $NaBr$, $MgBr_2$

d. Since ClO_4^- is a spectator ion, salts with acidic cations will give acidic salts.
       $Al(ClO_4)_3$, $Zn(ClO_4)_2$

a. NH_4^+ $K_a = 5.6 \times 10^{-10}$
    $C_2H_3O_2^-$ $K_b = 5.6 \times 10^{-10}$
    $K_a = K_b$ neutral

b. NH_4^+ $K_a = 5.6 \times 10^{-10}$
    $H_2PO_4^-$ $K_b = 1.4 \times 10^{-12}$
    $K_a > K_b$ acidic

c. Al^{3+} $K_a = 1.2 \times 10^{-5}$
    NO_2^- $K_b = 1.7 \times 10^{-11}$
    $K_a > K_b$ acidic

d. NH_4^+ $K_a = 5.6 \times 10^{-10}$
    F^- $K_b = 1.4 \times 10^{-11}$
    $K_a > K_b$ acidic

---

70. Refer to Section 13.6, Chapter 5 and Table 13.4.

Calculate $[NH_3]$ from the pH, then use the ideal gas equation to calculate the volume.

$$NH_3(aq) + H_2O(l) \rightleftharpoons NH_4^+(aq) + OH^-(aq)$$

$pOH = 14.00 - pH = 14.00 - 11.55 = 2.45$
$[OH^-] = [NH_4^+] = 10^{-pOH} = 10^{-2.45} = 3.5 \times 10^{-3} \, M$

$$K_b = \frac{[NH_4^+][OH^-]}{[NH_3]} \Rightarrow 1.8 \times 10^{-5} = \frac{(3.5 \times 10^{-3})^2}{x}$$

$x = [NH_3] = 0.68 \, M$

$$4.00 \, L \times \frac{0.68 \, mol. \, NH_3}{1 \, L} = 2.7 \, mol. \, NH_3$$

$$V = \frac{nRT}{P} = \frac{(2.7 \, mol.)(0.0821 \, L \cdot atm/mol. \cdot K)(298 \, K)}{1 \, atm} = 66 \, L$$

---

72. Refer to Section 13.4.

The products are a weak acid (HOCl) and a strong acid (HCl). The contribution to the pH from the weak acid will be negligible in comparison to the contribution from the strong acid. Thus, calculate the $[H^+]$ using the strong acid which completely dissociates. No consideration of equilibrium is necessary.

$[H^+] = 10^{-1.19} = 0.065 \, M$
$V = 1.00 \, L$, thus there are 0.065 mol. of H^+.

$$0.065 \, M \times \frac{1 \, mol. \, Cl_2}{1 \, mol. \, H^+} \times \frac{70.9 \, g}{1 \, mol. \, Cl_2} = 4.6 \, g$$

---

74. Refer to Chapter 13.

---

a. **False**. Weak bases do not completely react to form HB. Thus, $[HB] < 0.10 \, M$.

b. **True**. $B^- + H_2O \rightleftharpoons HB + OH^-$. The concentrations of the two products are approximately equal.

c. **True**. See (a) above. Thus, there are more reactants than products at equilibrium.

d. **False.** $[H^+] = \dfrac{1.0 \times 10^{-14}}{[OH^-]}$.

e. **False.** For weak bases, $[OH^-] < [B^-]$. Use K_b to calculate $[OH^-]$ in order to calculate pH.

---

76. *Refer to Chapter 13 and to Tables 13.2 and 13.4.*

Determine if the substance in each beaker is a strong or weak acid or base.

Beaker A has HI, a strong acid
Beaker B has HNO_2, a weak acid, $K_a = 6.0 \times 10^{-4}$
Beaker C has NaOH, a strong base
Beaker D has $Ba(OH)_2$, a strong base
Beaker E has NH_4Cl, a weak acid, $K_a = 5.6 \times 10^{-10}$
Beaker F has $C_2H_5NH_2$, a weak base

a. **LT**   Strong acids have greater $[H^+]$ than weak acids. High $[H^+]$ corresponds to low pH.

b. **LT**   pH of NaOH = $14.00 - \log(0.1)$
pH of $Ba(OH)_2 = 14.00 - \log(0.2)$; there are two moles of OH^- for each mole $Ba(OH)_2$.

c. **EQ**   Strong acids and bases completely dissociate. % ionization = 100%.

d. **LT** $K_a (HNO_2) > K_a (NH_4^+)$ Thus, $[H^+]$ from $HNO_2 > [H^+]$ from NH_4^+
Thus, pH of $HNO_2 <$ pH of NH_4^+.

e. **LT**   Acids have lower pH values than bases.

f. **GT**   Strong bases have higher pH values than weak bases.

---

78. *Refer to Sections 13.1 and 13.4.*

Box one shows $2H^+$, $2B^-$ and 5HB. This is an acid that is partially ionized in solution, and thus must be a **weak acid**.

Box two shows $5H^+$, $5B^-$ and no HB. This is an acid that is completely ionized in solution, and thus must be a **strong acid**.

---

80. *Refer to Sections 13.1 – 13.3.*

Dissolve the 0.10 mol. of the solid in one liter of water and measure the pH. For a strong acid, pH = $-\log(0.10) = 1.00$. For a strong base, pH = $14.00 - 1.00 = 13.00$. A weak acid would have $[H^+]_{eq} < 0.10$ and thus, pH > 1.00. A weak base would have $[H^+]_{eq} > 1.0 \times 10^{-13}$ and thus, pH < 13.00. If the pH > 7.00, the solid is a base; if pH < 13.00, it is a weak base. If pH < 7.00, solid is an acid; if pH > 1.00, it is a weak acid.

Write the net ionic equation for the reaction, then calculate $\Delta H°$ for the reaction. ($\Delta H° = \Sigma \Delta H_f°$ (products) $- \Sigma \Delta H_f°$ (reactants)). Calculate $\Delta H°$ for the 0.10 mol. given in the problem.

a. $Na^+(aq) + OH^-(aq) + H^+(aq) + Cl^-(aq) \rightleftharpoons Na^+(aq) + Cl^-(aq) + H_2O(l)$
$OH^-(aq) + H^+(aq) \rightleftharpoons H_2O(l)$

$\Delta H° = [(1 \text{ mol.})(-285.8 \text{ kJ/mol.})] - [(1 \text{ mol.})(-230.0 \text{ kJ/mol.}) + (1\text{mol.})(0.0 \text{ kJ/mol.})]$
$\Delta H° = -55.8 \text{ kJ}$ (This is for 1 mole, calculate for 0.10 moles.)

$\Delta H°_{(0.10 \text{ mol.})} = -55.8 \text{ kJ/mol.} \times 0.10 \text{ mol.} = -5.58 \text{ kJ}$

b. $HF(aq) + Na^+(aq) + OH^-(aq) \rightleftharpoons Na^+(aq) + F^-(aq) + H_2O(l)$
$HF(aq) + OH^-(aq) \rightleftharpoons F^-(aq) + H_2O(l)$

$\Delta H° = [(1 \text{ mol.})(-332.6 \text{ kJ/mol.}) + (1 \text{ mol.})(-285.5 \text{ kJ/mol.})]$
$\qquad\qquad - [(1 \text{ mol.})(-320.1 \text{ kJ/mol.}) + (1 \text{ mol.})(-230.0 \text{ kJ/mol.})]$
$\Delta H° = -68.3 \text{ kJ}$ (This is for 1 mole, calculate for 0.10 moles.)

$\Delta H°_{(0.10 \text{ mol.})} = -68.3 \text{ kJ/mol.} \times 0.10 \text{ mol.} = -6.83 \text{ kJ}$

Write K_a expressions for the concentrated acid $[HA]_c$ and the diluted acid $[HA]_d$, assuming that $[HA]_0 = [HA]_{eq}$ (which is true for a weak acid). Recalling that K_a is constant for a given acid, equate the two expressions and solve for $[H^+]$ and substitute the results into the percent ionization expression. Note the we are using the equation:
$$HA(aq) \rightleftharpoons H^+(aq) + A^-(aq)$$

$$K_a = \frac{[H^+]_c[A^-]_c}{[HA]_c} = \frac{x^2}{[HA]_c}, \text{ where } [H^+]_c = [A^-]_c = x$$

$$K_a = \frac{[H^+]_d[A^-]_d}{[HA]_d} = \frac{y^2}{[HA]_d}, \text{ where } [H^+]_d = [A^-]_d = y$$

If $[HA]_d$ is one tenth the concentration of $[HA]_c$, then $10[HA]_d = [HA]_c$ and

$$K_a = \frac{x^2}{[HA]_c} = \frac{y^2}{[HA]_d} \Rightarrow \frac{x^2}{10[HA]_d} = \frac{y^2}{[HA]_d} \Rightarrow x^2 = 10y^2$$

$$x = y\sqrt{10}$$

Substituting into the percent ionization expression:

$$\% \text{ ion.}_{(conc.)} = \frac{[H^+]_c}{[HA]_c} \times 100\% = \frac{100 \cdot x}{[HA]_c} = \frac{100(y\sqrt{10})}{10[HA]_d} = \frac{y \cdot 10\sqrt{10}}{[HA]_d}$$

$$\% \text{ ion.}_{(dil.)} = \frac{[H^+]_d}{[HA]_d} \times 100\% = \frac{y}{[HA]_d} \times 100\% = \frac{100 \cdot y}{[HA]_d}$$

Obviously the two terms differ by a factor of $\sqrt{10}$,

$$\frac{y \cdot 100}{[HA]_d} = \frac{y \cdot 10\sqrt{10}}{[HA]_d} \cdot \sqrt{10} \qquad \text{thus: } \% \text{ ion.}_{(dil.)} = \left(\% \text{ ion.}_{(conc.)}\right) \times \sqrt{10}$$

---

83. *Refer to Section 13.4 and Chapter 10.*

Recall that freezing point lowering is a colligative property and thus depends on the total moles of solute (ions and molecules) present in solution. Determine the moles of H^+, B^- and HB in solution, calculate the molality of the solution and the freezing point depression. Assume one liter of solution.

$$1\,L \times \frac{1000\,mL}{1\,L} \times \frac{1.006\,g}{1\,mL} = 1006\,g \text{ solution}$$

$$1006\,g \text{ solution} \times \frac{5.00\,g\,HC_2H_3O_2}{100\,g \text{ solution}} = 50.3\,g\,HC_2H_3O_2$$

$$50.3\,g\,HC_2H_3O_2 \times \frac{1\,mol.}{60.05\,g} = 0.838\,mol.\,HC_2H_3O_2$$

	$HC_2H_3O_2(aq)$	$\rightleftharpoons$	$C_2H_3O_2^-(aq)$	+	$H^+(aq)$
$[\ ]_0$	0.838		0		0
$\Delta[\ ]$	-x		+x		+x
$[\ ]_{eq}$	0.838 - x		x		x

$$K_a = 1.8 \times 10^{-5} = \frac{[H^+][C_2H_3O_2^-]}{[HC_2H_3O_2]} = \frac{x^2}{0.838 - x}$$

$$K_a/a = 1.8 \times 10^{-5} / 0.838 = 2.1 \times 10^{-5}$$
$$K_a/a = 2.1 \times 10^{-5} < 2.5 \times 10^{-3}$$

(The 5% rule applies. See the explanation accompanying problem 38.)

$$1.8 \times 10^{-5} = \frac{x^2}{0.838} \quad \Rightarrow \quad 1.5 \times 10^{-5} = x^2$$

$$x = 3.9 \times 10^{-3} \, M$$

moles solute = $[HC_2H_3O_2] + [C_2H_3O_2^-] + [H^+] = (0.838 - x) + x + x = 0.838 + x$
moles solute = $0.838 + 0.0039 = 0.842$

$$\text{molality} = \frac{\text{mol. solute}}{\text{kg solvent}} = \frac{0.842 \text{ mol.}}{1.006 \text{ kg solution} - 0.0503 \text{ kg } HC_2H_3O_2} = 0.881 \, m$$

$\Delta T_f = k_f (m) = 1.86 \, °C/m \, (0.877 \, m) = 1.64°C$

$T_f = 0°C - 1.64°C = -1.64°C$

---

84. Refer to Section 13.4 and Chapter 10.

Assume 1 L of solution and complete dissociation of $Ca(OH)_2$ in the solution. Then calculate $[OH^-]$ for a saturated $Ca(OH)_2$ solution. Then calculate pOH and pH.

$$\frac{0.153 \text{ g } Ca(OH)_2}{100 \text{ mL}} \times \frac{1000 \text{ mL}}{1 \text{ L}} \times \frac{1 \text{ mol. } Ca(OH)_2}{74.096 \text{ g}} \times \frac{2 \text{ mol. } OH^-}{1 \text{ mol. } Ca(OH)_2} = 0.0413 \, M \, OH^-$$

pOH = $-\log[OH^-] = -\log(0.0413) = 1.384$
pH = $14.00 - pOH = 14.00 - 1.384 = 12.62$

Chapter 14: Equilibria in Acid-Base Solutions

---

2. *Refer to Sections 4.2, and 13.4, and Table 13.2.*

Write the reactions, eliminating the spectator ions. Bear in mind that soluble salts ionize, thus they exist in solution as ions. Weak acids and bases (Table 13.2) do not ionize significantly, thus they exist in solution as the undissociated acid or base.

a. $NaC_2H_3O_2(aq) + HNO_3(aq) \rightleftharpoons HC_2H_3O_2(aq) + NaNO_3(aq)$
$NaC_2H_3O_2$ and $NaNO_3$ are salts and HNO_3 is a strong acid, thus all three dissociate in solution, while $HC_2H_3O_2$ is a weak acid and thus remains undissociated.
$Na^+(aq) + C_2H_3O_2^-(aq) + H^+(aq) + NO_3^-(aq) \rightleftharpoons HC_2H_3O_2(aq) + Na^+(aq) + NO_3^-(aq)$
$C_2H_3O_2^-(aq) + H^+(aq) \rightleftharpoons HC_2H_3O_2(aq)$

b. $2HBr(aq) + Sr(OH)_2(aq) \rightleftharpoons SrBr_2(aq) + 2H_2O(l)$
HBr and $Sr(OH)_2$ are strong acids and bases (respectively) and thus dissociate completely in solution.
$2H^+(aq) + 2Br^-(aq) + Sr^{+2}(aq) + 2OH^-(aq) \rightleftharpoons Sr^{+2}(aq) + 2Br^-(aq) + 2H_2O(l)$
$2H^+(aq) + 2OH^-(aq) \rightleftharpoons 2H_2O(l)$
$H^+(aq) + OH^-(aq) \rightleftharpoons H_2O(l)$

c. $HClO(aq) + NaCN(aq) \rightleftharpoons NaClO(aq) + HCN(aq)$
$HClO$ and HCN are weak acids and do not dissociate; $NaCN$ and $NaClO$ are salts and do dissociate in solution.
$HClO(aq) + Na^+(aq) + CN^-(aq) \rightleftharpoons Na^+(aq) + ClO^-(aq) + HCN(aq)$
$HClO(aq) + CN^-(aq) \rightleftharpoons ClO^-(aq) + HCN(aq)$

d. $HNO_2(aq) + NaOH(aq) \rightleftharpoons NaNO_2(aq) + H_2O(l)$
HNO_2 is a weak acid and does not dissociate; $NaOH$ is a strong base and $NaNO_2$ is a salt, thus both dissociate in solution.
$HNO_2(aq) + Na^+(aq) + OH^-(aq) \rightleftharpoons Na^+(aq) + NO_2^-(aq) + H_2O(l)$
$HNO_2(aq) + OH^-(aq) \rightleftharpoons NO_2^-(aq) + H_2O(l)$

---

4. *Refer to Chapters 4 and 13 and Problem 2 above.*

Write the reactions, eliminating the spectator ions. Bear in mind that weak acids and bases (Table 13.2) do not ionize significantly, thus they exist in solution as the undissociated acid or base.

a. $NH_4NO_3(aq) + OH^-(aq) \rightleftharpoons NH_3(aq) + H_2O(l) + NO_3^-(aq)$
$NH_4^+(aq) + NO_3^-(aq) + OH^-(aq) \rightleftharpoons NH_3(aq) + H_2O(l) + NO_3^-(aq)$
$NH_4^+(aq) + OH^-(aq) \rightleftharpoons NH_3(aq) + H_2O(l)$

b. $NaH_2PO_4(aq) + OH^-(aq) \rightleftharpoons Na^+(aq) + HPO_4^{2-}(aq) + H_2O(l)$
     $Na^+(aq) + H_2PO_4^-(aq) + OH^-(aq) \rightleftharpoons Na^+(aq) + HPO_4^{2-}(aq) + H_2O(l)$
     $H_2PO_4^-(aq) + OH^-(aq) \rightleftharpoons HPO_4^{2-}(aq) + H_2O(l)$

c. $[Al(H_2O)_6]^{3+}(aq) + OH^-(aq) \rightleftharpoons [Al(H_2O)_5OH]^{2+}(aq) + H_2O(l)$

---

6.   Refer to Chapter 13.

Recall that K for the reverse reaction equals the inverse of K for the forward reaction, and that when adding reactions, the K's are multiplied.

a. $C_2H_3O_2^-(aq) + H^+(aq) \rightleftharpoons HC_2H_3O_2(aq)$ $K = 1/K_a = 5.6 \times 10^4$

b. $H^+(aq) + OH^-(aq) \rightleftharpoons H_2O(l)$ $K = 1/K_w = 1.0 \times 10^{14}$

c. $HClO(aq) \rightleftharpoons H^+(aq) + ClO^-(aq)$ $K_a = 2.8 \times 10^{-8}$
     $H^+(aq) + CN^-(aq) \rightleftharpoons HCN(aq)$ $K = 1/K_a = 1.7 \times 10^9$
     $HClO(aq) + CN^-(aq) \rightleftharpoons ClO^-(aq) + HCN(aq)$ $K = K_a \times 1/K_a = 48$

d. $HNO_2(aq) \rightleftharpoons H^+(aq) + NO_2^-(aq)$ $K_a = 6.0 \times 10^{-4}$
     $H^+(aq) + OH^-(aq) \rightleftharpoons H_2O(l)$ $K = 1/K_w = 1.0 \times 10^{14}$
     $HNO_2(aq) + OH^-(aq) \rightleftharpoons NO_2^-(aq) + H_2O(l)$ $K = K_a \times 1/K_w = 6.0 \times 10^{10}$
     $K = 1/K_w = 1.0 \times 10^{14}$

---

8.   Refer to Chapter 13 and Problem 4 above.

---

a. $NH_4^+(aq) + OH^-(aq) \rightleftharpoons NH_3(aq) + H_2O(l)$ $K = 1/K_b = 5.6 \times 10^4$

b. $H_2PO_4^-(aq) \rightleftharpoons HPO_4^{2-}(aq) + H^+(aq)$ $K_a = 6.2 \times 10^{-8}$
     $H^+(aq) + OH^-(aq) \rightleftharpoons H_2O(l)$ $K = 1/K_w = 1.0 \times 10^{14}$
     $H_2PO_4^-(aq) + OH^-(aq) \rightleftharpoons HPO_4^{2-}(aq) + H_2O(l)$ $K = K_a \times 1/K_w = 6.2 \times 10^6$

c. $[Al(H_2O)_6]^{3+}(aq) + OH^-(aq) \rightleftharpoons [Al(H_2O)_5OH]^{2+}(aq) + H_2O(l)$ $K = 1/K_b = 1.2 \times 10^9$

---

10. Refer to Section 14.1.

Calculate [H$^+$], and from that, pH and [OH$^-$].

a. $[H^+] = K_a \times \dfrac{[HB]}{[B^-]} = 6.2 \times 10^{-8} \times \dfrac{0.335}{0.335} = 6.2 \times 10^{-8}\ M$

     $pH = -\log[H^+] = -\log(6.2 \times 10^{-8}) = 7.21$

$$[OH^-] = \frac{K_w}{[H^+]} = \frac{1.0 \times 10^{-14}}{6.2 \times 10^{-8}} = 1.6 \times 10^{-7} \; M$$

b. $\quad [H^+] = K_a \times \frac{[HB]}{[B^-]} = 6.2 \times 10^{-8} \times \frac{0.335}{0.100} = 2.1 \times 10^{-7} \; M$

$\quad$ pH = -log[H$^+$] = -log(2.1 x 10^{-7}) = 6.68

$$[OH^-] = \frac{K_w}{[H^+]} = \frac{1.0 \times 10^{-14}}{2.1 \times 10^{-7}} = 4.8 \times 10^{-8} \; M$$

c. $\quad [H^+] = K_a \times \frac{[HB]}{[B^-]} = 6.2 \times 10^{-8} \times \frac{0.335}{0.0750} = 2.8 \times 10^{-7} \; M$

$\quad$ pH = -log[H$^+$] = -log(2.8 x 10^{-7}) = 6.55

$$[OH^-] = \frac{K_w}{[H^+]} = \frac{1.0 \times 10^{-14}}{2.8 \times 10^{-7}} = 3.6 \times 10^{-8} \; M$$

d. $\quad [H^+] = K_a \times \frac{[HB]}{[B^-]} = 6.2 \times 10^{-8} \times \frac{0.335}{0.0300} = 6.9 \times 10^{-7} \; M$

$\quad$ pH = -log[H$^+$] = -log(6.9 x 10^{-7}) = 6.16

$$[OH^-] = \frac{K_w}{[H^+]} = \frac{1.0 \times 10^{-14}}{6.9 \times 10^{-7}} = 1.4 \times 10^{-8} \; M$$

---

12. *Refer to Section 14.1.*

A buffer should work until all the acid ($H_2PO_4^-$) or all the base (HPO_4^{2-}) is consumed. The buffer capacity will be equal to the moles of the acid or base present in the buffer. If we assume one liter of buffer, then the capacity will equal the concentration of the acid or base. The concentration of acid is the same in each case (0.335 *M*), thus one liter of solution will be able to buffer 0.335 mol. of base.

a. $\quad [HPO_4^{2-}]$ = 0.335 *M*, thus 1 L will be able to buffer 0.335 mol. acid.

b. $\quad [HPO_4^{2-}]$ = 0.100 *M*, thus 1 L will be able to buffer 0.100 mol. acid.

c. $\quad [HPO_4^{2-}]$ = 0.0750 *M*, thus 1 L will be able to buffer 0.0750 mol. acid.

d. $\quad [HPO_4^{2-}]$ = 0.0300 *M*, thus 1 L will be able to buffer 0.0300 mol. acid.

14. Refer to Section 14.1.

$[HF] = 0.0237\ M$

$$[KF] = \frac{0.037\ \text{mol.}}{0.135\ \text{L}} = 0.27\ M$$

$$[H^+] = K_a \times \frac{[HB]}{[B^-]} = 6.9 \times 10^{-4} \times \frac{0.0237}{0.27} = 6.0 \times 10^{-5}\ M$$

$$pH = -\log[H^+] = -\log(6.0 \times 10^{-5}) = 4.22$$

16. Refer to Section 14.1, Example 14.2 and Tables 13.2 and 14.1.

For the ideal buffer, $pH \approx pK_a$. Look for an acid with a pK_a approximately equal to the desired pH.

a. pK_a of $HC_2H_3O_2$ is 4.74, thus use $HC_2H_3O_2/\ C_2H_3O_2^-$

b. pK_a of NH_4^+ is 9.25, thus use NH_4^+/NH_3

c. pK_a of HCO_3^- is 10.32, thus use $HCO_3^-/\ HCO_3^{2-}$

18. Refer to Section 14.1, Example 14.2 and Table 13.2.

Calculate the moles of the acid and base, then calculate the $[H^+]$ and pH.

a. Calculate $[H^+]$ from pH, then calculate the acid/base ratio of the buffer.

$$[H^+] = 10^{-pH} = 10^{-3.0} = 1 \times 10^{-3}\ M$$

$$[H^+] = K_a \times \frac{[HB]}{[B^-]} \Rightarrow 1 \times 10^{-3} = 1.9 \times 10^{-4} \times \frac{[HCHO_2]}{[CHO_2^-]}$$

$$\frac{[HCHO_2]}{[CHO_2^-]} = 5$$

b. $NaCHO_2$ will dissociate to Na^+ and CHO_2^-. Use this fact and the equation from part (a) to calculate $[HCHO_2]$.

$$\frac{[HCHO_2]}{[CHO_2^-]} = 5 \Rightarrow \frac{[HCHO_2]}{0.139} = 5$$

$[HCHO_2] = 0.7\ M$, thus 0.7 mol. would need to be added to the liter of solution.

c. $\dfrac{[HCHO_2]}{[CHO_2^-]} = 5 \Rightarrow \dfrac{0.159}{[CHO_2^-]} = 5$

$[CHO_2^-] = 0.03\ M$

$0.3500\,L \times \dfrac{0.03\ mol.\ CHO_2^-}{1\,L} \times \dfrac{1\,mol.\ NaCHO_2}{1\,mol.\ CHO_2^-} \times \dfrac{68.01\ g\ NaCHO_2}{1\,mol.\ NaCHO_2} = 0.7\ g\ NaCHO_2$

d. $\dfrac{[HCHO_2]}{[CHO_2^-]} = 5 \Rightarrow \dfrac{[HCHO_2]}{0.500} = 5$

$[HCHO_2] = 2\ M$, since the volume of the formate solution is one liter, 2 mol of $HCHO_2$ is needed.

$2\,mol.\ HCHO_2 \times \dfrac{1\,L}{0.236\,mol.\ HCHO_2} = 1 \times 10^1\ L$

---

20. Refer to Section 14.1.

a. $5.50\ g \times \dfrac{1\,mol.\ NH_4Cl}{53.49\ g\ NH_4Cl} = 0.103\ mol.\ NH_4Cl$

$[H^+] = K_a \times \dfrac{n_{NH_4^+}}{n_{NH_3}} = 5.6 \times 10^{-10} \times \dfrac{0.103\,mol.\ NH_4^+}{0.0188\,mol.\ NH_3} = 3.1 \times 10^{-9}\ M$

$pH = -\log[H^+] = -\log(3.1 \times 10^{-9}) = 8.51$

b. Note that volume was not used in the calculations above. Thus we would expect the pH to remain the same, independent of the volume of solution.

---

22. Refer to Section 14.1.

Use the equilibrium expression to calculate the moles of CH_3NH_2, then use $PV = nRT$ to calculate the volume.

$K_b = \dfrac{[OH^-][CH_3NH_3^+]}{[CH_3NH_2]} \Rightarrow [OH^-] = K_b \times \dfrac{[CH_3NH_2]}{[CH_3NH_3^+]}$

$[OH^-] = K_b \times \dfrac{n_{CH_3NH_2}}{n_{CH_3NH_3^+}}$

$0.750\,L \times \dfrac{0.588\,mol.\ CH_3NH_3^+}{1\,L} = 0.441\ mol.$

$pOH = 14.00 - 9.80 = 4.20$

$[OH^-] = 10^{-4.20} = 6.3 \times 10^{-5}\ M$

$$6.3 \times 10^{-5} = 4.2 \times 10^{-4} \times \frac{n_{CH_3NH_2}}{0.441 \, \text{mol.}} \Rightarrow n_{CH_3NH_2} = 6.6 \times 10^{-2} \, \text{mol.}$$

$PV = nRT$

$(1.2 \, \text{atm})(V) = (6.6 \times 10^{-2} \, \text{mol.})(0.0821 \, \text{L·atm/mol.·K})(300\text{K})$

$V = 1.4 \, \text{L}$

---

24. Refer to Section 14.1 and Example 14.4.

The equation for the reaction is: $HX \rightleftharpoons X^- + H^+$. Calculate $[H^+]$ and K_a. Then calculate $[H^+]$ and pH for the solution after addition of HX.

$$[H^+] = 10^{-3.92} = 1.2 \times 10^{-4} \, M$$

$$K_a = \frac{[H^+][X^-]}{[HX]} = \frac{(1.2 \times 10^{-4})^2}{0.043} = 3.3 \times 10^{-7}$$

$$[H^+] = K_a \times \frac{[HX]}{[X^-]} = 3.3 \times 10^{-7} \times \frac{0.043}{0.021} = 6.8 \times 10^{-7} \, M$$

$$pH = -\log(6.8 \times 10^{-7}) = 6.17$$

---

26. Refer to Section 14.1 and Example 14.3.

---

a. Calculate the moles of HCO_3^- and CO_3^{2-} and then calculate $[H^+]$ and pH.

$$0.355 \, \text{L} \times \frac{0.200 \, \text{mol.}}{1 \, \text{L}} = 0.0710 \, \text{mol.} \, HCO_3^-$$

$$0.355 \, \text{L} \times \frac{0.134 \, \text{mol.}}{1 \, \text{L}} = 0.0476 \, \text{mol.} \, CO_3^{2-}$$

$$[H^+] = K_a \times \frac{n_{HCO_3^-}}{n_{CO_3^{2-}}} = 4.7 \times 10^{-11} \times \frac{0.0710}{0.0476} = 7.0 \times 10^{-11} \, M$$

$$pH = -\log(7.0 \times 10^{-11}) = 10.15$$

b. The HCl will convert CO_3^{2-} to HCO_3^-. Calculate the moles of each that are present after the HCl addition. Then calculate $[H^+]$ and pH.

$$n_{HCO_3^-} = 0.0710 \, \text{mol.} + 0.0300 \, \text{mol.} = 0.101 \, \text{mol.}$$

$$n_{CO_3^{2-}} = 0.0476 \, \text{mol.} - 0.0300 \, \text{mol.} = 0.0176 \, \text{mol.}$$

$$[H^+] = K_a \times \frac{n_{HCO_3^-}}{n_{CO_3^{2-}}} = 4.7 \times 10^{-11} \times \frac{0.101}{0.0176} = 2.7 \times 10^{-10} \, M$$

$$pH = -\log(2.7 \times 10^{-10}) = 9.57$$

c. The NaOH will convert HCO_3^- to CO_3^{2-}. Calculate the moles of each that are present after the NaOH addition. Then calculate $[H^+]$ and pH.

$$n_{HCO_3^-} = 0.0710 \text{ mol.} - 0.0300 \text{ mol.} = 0.0410 \text{ mol.}$$

$$n_{CO_3^{2-}} = 0.0476 \text{ mol.} + 0.0300 \text{ mol.} = 0.0776 \text{ mol.}$$

$$[H^+] = K_a \times \frac{n_{HCO_3^-}}{n_{CO_3^{2-}}} = 4.7 \times 10^{-11} \times \frac{0.0410}{0.0776} = 2.5 \times 10^{-11} \, M$$

$$pH = -\log(2.5 \times 10^{-11}) = 10.60$$

---

28. *Refer to Section 14.1.*

a. $$[H^+] = K_a \times \frac{n_{HCO_3^-}}{n_{CO_3^{2-}}} = 4.7 \times 10^{-11} \times \frac{0.0710}{0.0476} = 7.0 \times 10^{-11} \, M$$

$$pH = -\log(7.0 \times 10^{-11}) = 10.15$$

Note that the pH hasn't changed. This should be expected since the *moles* of HCO_3^- and CO_3^{2-} did not change upon dilution.

b. Begin by calculating the moles of HCO_3^- and CO_3^{2-} in 0.710 L of the diluted buffer.

$$0.710 \, L \times \frac{0.0710 \text{ mol.}}{10.0 \, L} = 0.00504 \text{ mol.} HCO_3^-$$

$$0.710 \, L \times \frac{0.0476 \text{ mol.}}{10.0 \, L} = 0.00338 \text{ mol.} CO_3^{2-}$$

The HCl will convert CO_3^{2-} to HCO_3^-. However, the moles of HCl are greater than the moles of CO_3^{2-}, thus all the CO_3^{2-} will be converted to HCO_3^- and excess HCl will remain. Calculate the moles of HCO_3^- and HCl that are present.

$$n_{HCl \,(remaining)} = n_{HCl} - n_{CO_3^{2-}} = 0.0300 \text{ mol.} - 0.00338 \text{ mol.} = 0.0266 \text{ mol.} HCl$$

$$n_{HCO_3^- \,(final)} = n_{HCO_3^- \,(initial)} + n_{CO_3^{2-}} = 0.00504 \text{ mol.} + 0.00338 \text{ mol.} = 0.00842 \text{ mol.} HCO_3^-$$

The excess HCl will then react with the HCO_3^-, converting it to H_2CO_3. But again, an excess of acid will remain. Given the low concentration of the H_2CO_3 (relative to the HCl) and the fact that H_2CO_3 is a weak acid, it can be ignored, and the pH calculations can be based entirely on the remaining HCl.

$$n_{HCl \,(remaining)} = n_{HCl} - n_{HCO_3^-} = 0.0266 \text{ mol.} - 0.00842 \text{ mol.} = 0.0182 \text{ mol.} HCl$$

$$[H^+] = \frac{n_{HCl}}{L \text{ solution}} = \frac{0.0182 \text{ mol.}}{0.710 \text{ L}} = 0.0256 \, M$$

$$pH = -\log(0.0256) = 1.591$$

Note: If one neglects $HCO_3^- + H^+ \rightarrow H_2CO_3$, then one would get:

$$[H^+] = \frac{n_{HCl}}{L \text{ solution}} = \frac{0.0266 \text{ mol.}}{0.710 \text{ L}} = 0.0375 \, M$$

$$pH = -\log(0.0375) = 1.426$$

c. The addition of 0.0030 mol. NaOH will convert HCO_3^- to CO_3^{2-}. Calculate the moles of each that are present after the NaOH addition. Then calculate $[H^+]$ and pH.

$$n_{HCO_3^-} = 0.00504 \text{ mol.} - 0.0030 \text{ mol.} = 0.0020 \text{ mol.}$$

$$n_{CO_3^{2-}} = 0.00338 \text{ mol.} + 0.0030 \text{ mol.} = 0.0064 \text{ mol.}$$

$$[H^+] = K_a \times \frac{n_{HCO_3^-}}{n_{CO_3^{2-}}} = 4.7 \times 10^{-11} \times \frac{0.0020}{0.0064} = 1.5 \times 10^{-11} \, M$$

$$pH = -\log(1.5 \times 10^{-11}) = 10.82$$

d. The pH of the diluted buffer is unchanged, but the pH of 28b is dramatically different from that of 26b, due to the decreased capacity of the diluted buffer. The pH's after base addition are little different since the buffer capacity was not exceeded.

e. Diluting the buffer does not affect the pH, but a diluted buffer has less capacity than an equal volume of a more concentrated buffer.

---

30. Refer to Section 14.1 and Table 13.2.

---

a. $$[H^+] = K_a \times \frac{[HC_3H_5O_2]}{[C_3H_5O_2^-]} = 1.4 \times 10^{-5} \times 4.50 = 6.3 \times 10^{-5} \, M$$

$$pH = -\log(6.3 \times 10^{-5}) = 4.20$$

b. Set $[HC_3H_5O_2]_i = 4.50$ and $[C_3H_5O_2^-]_i = 1.00$. If 27% of the $HC_3H_5O_2$ is converted to $C_3H_5O_2^-$, we get:

$$[HC_3H_5O_2]_f = [HC_3H_5O_2]_i - 0.27 \, [HC_3H_5O_2]_i = 4.50 - 0.27(4.50) = 3.28 \, M$$

$$[C_3H_5O_2^-]_f = [C_3H_5O_2^-]_i + 0.27[C_3H_5O_2^-]_i = 1.00 + 0.27(4.50) = 2.22 \, M$$

$$[H^+] = K_a \times \frac{[HC_3H_5O_2]}{[C_3H_5O_2^-]} = 1.4 \times 10^{-5} \times \frac{3.28}{2.22} = 2.1 \times 10^{-5} \, M$$

$$pH = -\log(2.1 \times 10^{-5}) = 4.68$$

c. New pH = 4.20 + 1.00 = 5.20

$$[H^+] = 10^{-pH} = 10^{-5.20} = 6.3 \times 10^{-6} \, M$$

$$[H^+] = K_a \times \frac{[HC_3H_5O_2]}{[C_3H_5O_2^-]} \Rightarrow 6.3 \times 10^{-5} = 1.4 \times 10^{-5} \times \frac{[HC_3H_5O_2]}{[C_3H_5O_2^-]} \Rightarrow \frac{[HC_3H_5O_2]}{[C_3H_5O_2^-]} = 0.45$$

---

32. Refer to Section 14.1.

A buffer is made of a weak acid and its conjugate base, or a weak base and its conjugate acid. Consider the reaction between the reagents. Those reactions that result in roughly equal amounts of F⁻ and HF will act as buffers. In all cases, the initial amount of $Sr(OH)_2$ is:

$$0.6500 \, L \times \frac{0.40 \, mol. \, Sr(OH)_2}{1 \, L} = 0.26 \, mol. \, Sr(OH)_2$$

The reaction to consider is: $2 \, HF(aq) + Sr(OH)_2(aq) \rightarrow SrF_2(aq) + 2 \, H_2O(l)$

$$0.26 \, mol. \, Sr(OH)_2 \times \frac{1 \, mol. \, SrF_2}{1 \, mol. \, Sr(OH)_2} = 0.26 \, mol. \, SrF_2$$

$$0.26 \, mol. \, Sr(OH)_2 \times \frac{2 \, mol. \, HF}{1 \, mol. \, Sr(OH)_2} = 0.52 \, mol. \, HF$$

a. In this case, 0.52 mol.of HF will react, giving 0.26 mol. SrF_2, leaving 0.48 mol. HF. Thus, both the acid (HF) and the conjugate base (F⁻) will be present in solution, and this mixture will act as a buffer.

b. In this case, 0.52 mol. of HF will react, giving 0.26 mol. SrF_2, leaving 0.23 mol. HF. Thus, both the acid (HF) and the conjugate base (F⁻) will be present in solution, and this mixture will act as a buffer.

c. In this case, 0.30 mol. of HF will react, giving 0.15 mol. SrF_2, leaving 0 mol. HF. Thus, only the conjugate base (F⁻) will be present in solution, and this mixture will not act as a buffer.

d. The reaction to consider is: $2NaF(aq) + Sr(OH)_2(aq) \rightarrow 2 \, H_2O(l) + SrF_2(aq)$
There is no weak acid in this equation, thus, no buffer can be formed.

e. The reaction to consider is: $2HCl(aq) + Sr(OH)_2(aq) \rightarrow 2\ H_2O(l) + SrCl_2(aq)$
There is no weak acid in this equation, thus, no buffer can be formed.

Thus a and b will act as buffers.

---

34. _Refer to Section 14.1 and Example 14.3._

Consider the products of the reaction of pyridine and hydrochloric acid. Determine the amounts of each, then calculate the pH.

$$C_5H_5N(aq) + HCl(aq) \rightarrow C_5H_5NH^+(aq) + Cl^-(aq)$$

$$25.0\ mL \times \frac{0.978\ g\ C_5H_5N}{1\ mL} \times \frac{1\ mol.\ C_5H_5N}{79.1\ g} = 0.309\ mol.\ C_5H_5N$$

$$0.0650\ L\ HCl \times \frac{2.75\ mol.\ HCl}{1\ L} = 0.180\ mol.\ HCl$$

Since C_5H_5N and HCl react in a 1:1 ratio, HCl is the limiting reactant and is thus completely consumed. The amount of $C_5H_5NH^+$ after the reaction is 0.180 mol.; the amount of C_5H_5N is:

$$0.309\ mol. - 0.180\ mol. = 0.129\ mol.\ C_5H_5N$$

$$K_a = \frac{K_w}{K_b} = \frac{1 \times 10^{-14}}{1.5 \times 10^{-9}} = 6.7 \times 10^{-9}$$

$$[H^+] = K_a \times \frac{n_{C_5H_5NH^+}}{n_{C_5H_5N}} = 6.7 \times 10^{-6} \times \frac{0.180}{0.129} = 9.3 \times 10^{-6}\ M$$

$$pH = -\log(9.3 \times 10^{-6}) = 5.03$$

---

36. _Refer to Section 14.1._

Consider the products of the reaction of ethanolamine and hydrochloric acid. Determine the amounts of each, then calculate the pH.

$$C_2H_5ONH_2(aq) + HCl(aq) \rightarrow C_2H_5ONH_3^+(aq) + Cl^-(aq)$$

$$0.0500\ L\ HCl \times \frac{1.0\ mol.\ HCl}{1\ L} = 0.050\ mol.\ HCl$$

$$0.1000\ L\ C_2H_5ONH_2 \times \frac{1.20\ mol.\ C_2H_5ONH_2}{1\ L} = 0.120\ mol.\ C_2H_5ONH_2$$

Since $C_2H_5ONH_2$ and HCl react in a 1:1 ratio, HCl is the limiting reactant and is thus completely consumed. The amount of $C_2H_5ONH_3^+$ after the reaction is 0.0500 mol.; the amount of $C_2H_5ONH_2$ is:

0.120 mol. - 0.050 mol. = 0.070 mol. $C_2H_5ONH_2$

$$[H^+] = K_a \times \frac{n_{C_2H_5ONH_3^+}}{n_{C_2H_5ONH_2}} = 3.6 \times 10^{-10} \times \frac{0.050}{0.070} = 2.6 \times 10^{-10} \ M$$

$$pH = -\log(2.6 \times 10^{-10}) = 9.59$$

---

38. *Refer to Section 14.1.*

a. $[H^+] = 10^{-pH} = 10^{-7.40} = 4.0 \times 10^{-8} \ M$

$$[H^+] = K_a \times \frac{[H_2PO_4^-]}{[HPO_4^{2-}]} \Rightarrow 4.0 \times 10^{-8} = 6.2 \times 10^{-8} \times \frac{[H_2PO_4^-]}{[HPO_4^{2-}]}$$

$$\frac{[H_2PO_4^-]}{[HPO_4^{2-}]} = 0.65$$

b. $[H^+] = 10^{-pH} = 10^{-6.80} = 1.6 \times 10^{-7} \ M$

$$[H^+] = K_a \times \frac{n_{H_2PO_4^-}}{n_{HPO_4^{2-}}} \Rightarrow 1.6 \times 10^{-7} = 6.2 \times 10^{-8} \times \frac{n_{H_2PO_4^-}}{n_{HPO_4^{2-}}}$$

$$\frac{n_{H_2PO_4^-}}{n_{HPO_4^{2-}}} = 2.6$$

To calculate the amount of HPO_4^{2-} ions that were converted, set $n_{HPO_4^{2-}} = 1$ in part (a) above. Then $n_{H_2PO_4^-} = 0.65$. Then set x = moles HPO_4^{2-} converted. Substituting gives:

$$2.6 = \frac{0.65 + x}{1.0 - x}$$

$x = 0.54$, thus 54% is converted.

c. $[H^+] = 10^{-pH} = 10^{-7.80} = 1.6 \times 10^{-8} \ M$

$$[H^+] = K_a \times \frac{n_{H_2PO_4^-}}{n_{HPO_4^{2-}}} \Rightarrow 1.6 \times 10^{-8} = 6.2 \times 10^{-8} \times \frac{n_{H_2PO_4^-}}{n_{HPO_4^{2-}}}$$

$$\frac{n_{H_2PO_4^-}}{n_{HPO_4^{2-}}} = 0.26$$

To calculate the amount of $H_2PO_4^{2-}$ ions that were converted, set $n_{H_2PO_4^-} = 1$ in part (a) above, then $n_{HPO_4^{2-}} = 1.5$. Then set x = moles HPO_4^{2-} converted. Substituting gives:

$$2.6 = \frac{1-x}{1.5+x}$$

$x = 0.48$, thus 48% is converted.

---

40. Refer to Sections 14.2 and 14.3, Example 14.8 and Figures 14.5 - 14.7.

For an indicator to be suitable for a given titration, the color change (end point) must occur at a pH corresponding to the equivalence point of the titration. At that point, small additions of the titrant result in dramatic changes in pH, and thus a sudden (definitive) color change. This occurs along the steep portion of the titration curves shown in Figures 14.5 -14.7.

a. This is the titration of a weak acid with a strong base (see Figure 14.6). An indicator that changes color between pH 8 to pH 10 would be ideal.
    Phenolphthalein changes color in this range.

b. This is the titration of a weak acid with a strong base (see Figure 14.6). An indicator that changes color between pH 8 to pH 10 would be ideal.
    Phenolphthalein changes color in this range.

c. This is the titration of a strong acid with a strong base (see Figure 14.5). An indicator that changes color between pH 4 to pH 10 would be ideal.
    Any of the three indicators change color in this range.

d. This is the titration of a weak base with a strong acid (see Figure 14.7). An indicator that changes color between pH 2 to pH 6 would be ideal.
    Methyl orange changes color in this range.

---

42. Refer to Section 14.2, Equation 14.3, and Example 14.5.

a. $pK_a = -\log K_a$
    $K_a = 10^{-pK_a}$
    $K_a = 10^{-7.4} = 4 \times 10^{-8}$

b. A color change occurs when $pH \approx pK_a$
    The range will be from $\dfrac{[HIn]}{[In^-]} \geq 10$ to $\dfrac{[HIn]}{[In^-]} \leq 0.1$

    Using equation 14.3:
    $$\frac{[HIn]}{[In^-]} = \frac{[H^+]}{K_a}$$

This rearranges to form:

$$[H^+] = K_a \times \frac{[HIn]}{[In^-]} \quad \Rightarrow \quad -\log[H^+] = -\log K_a - \log\frac{[HIn]}{[In^-]}$$

$$pH = pK_a - \log\frac{[HIn]}{[In^-]} = 7.4 - \log 10 = 6.4$$

$$pH = pK_a - \log\frac{[HIn]}{[In^-]} = 7.4 - \log 0.1 = 8.4$$

Thus the range is from pH 6.4 to 8.4.

c. When the pH = pK_a = 7.4 one would be half way between red and yellow, thus the color would be **orange**.

---

44. Refer to Section 14.3 and Example 14.5.

---

a. $HBr(aq) + KOH(aq) \rightarrow KBr(aq) + H_2O(l)$
    $H^+(aq) + Br^-(aq) + K^+(aq) + OH^-(aq) \rightarrow K^+(aq) + Br^-(aq) + H_2O(l)$
    $H^+(aq) + OH^-(aq) \rightarrow H_2O(l)$

b. At the equivalence point, all the base has reacted with the acid. Thus, the species present are K^+, Br^-, and H_2O.

c. $0.03500\,L \times \dfrac{0.257\,mol.\,HBr}{1\,L} \times \dfrac{1\,mol.\,KOH}{1\,mol.\,HBr} \times \dfrac{1\,L}{0.1375\,mol.\,KOH} = 0.0654\,L$

d. Before KOH is added, pH can be calculated from the HBr. Since HBr is a strong acid,

    $[H^+] = [HBr]$

    $pH = -\log(0.257) = 0.590$

e. At half-way to the equivalence point, half of the HBr is neutralized. That means half as much base as needed to react the equivalence point will have been added. The total volume will be 0.03500 L + ½(0.0654 L) = 0.0677 L.

    $0.03500\,L \times \dfrac{0.257\,mol.\,HBr}{1\,L} \times \dfrac{1}{2} = 0.00450\,mol.\,HBr = 0.00450\,mol.\,H^+$

    $[H^+] = 0.00450\,mol.\,/\,0.0677\,L = 0.0665\,M$
    $pH = -\log[H^+] = -\log(0.0665) = 1.177$

f. At the equivalence point, the only species present are the spectator ions and water, thus the pH will be 7.

a. $C_{17}H_{19}O_3N(aq) + HCl(aq) \rightleftharpoons C_{17}H_{19}O_3NH^+(aq) + Cl^-(aq)$

$C_{17}H_{19}O_3N(aq) + H^+(aq) + Cl^-(aq) \rightleftharpoons C_{17}H_{19}O_3NH^+(aq) + Cl^-(aq)$

$C_{17}H_{19}O_3N(aq) + H^+(aq) \rightleftharpoons C_{17}H_{19}O_3NH^+(aq)$

b. At the equivalence point, the $C_{17}H_{19}O_3N$ is converted to $C_{17}H_{19}O_3NH^+$. As a weak acid, however, a small amount will dissociate. Thus all the species in the ionic equation ($H^+(aq), Cl^-(aq), C_{17}H_{19}O_3N(aq)$ and $C_{17}H_{19}O_3NH^+(aq)$) are present.

c. $0.0500 \, L \times \dfrac{0.1500 \, mol. \, C_{17}H_{19}O_3N}{1 \, L} \times \dfrac{1 \, mol. \, HCl}{1 \, mol. \, C_{17}H_{19}O_3N} = 0.0075 \, mol. \, HCl$

$0.0075 \, mol. \, HCl \times \dfrac{1 \, L}{0.1045 \, mol. \, HCl} = 0.0718 \, L \, HCl = 71.8 \, mL \, HCl$

d. At the beginning of the titration:

	$C_{17}H_{19}O_3N$ *(aq)*	+	$H_2O(l)$	$\rightleftharpoons$	$C_{17}H_{19}O_3NH^+(aq)$	+	$OH^-(aq)$
$[\ ]_o$	0.1500		---		0		0
$\Delta[\ ]$	$-x$		---		$+x$		$+x$
$[\ ]_{eq}$	$0.1500 - x$		---		x		x

$K_b = 7.4 \times 10^{-7} = \dfrac{[C_{17}H_{19}O_3NH^+][OH^-]}{[C_{17}H_{19}O_3N]} = \dfrac{x^2}{0.1500 - x}$

$K_b/b = 7.4 \times 10^{-7} / 0.1500 = 4.9 \times 10^{-6}$

$K_b/b = 4.9 \times 10^{-6} < 2.5 \times 10^{-3}$ (The 5% rule applies. See the explanation accompanying Chapter 13, Problem 38.)

$K_b = 7.4 \times 10^{-7} = \dfrac{x^2}{0.1500}$

$x^2 = 1.1 \times 10^{-7}$

$x = [OH^-] = 3.3 \times 10^{-4} \, M$

$pOH = -\log (3.3 \times 10^{-4}) = 3.48$

$pH = 14.00 - pOH = 14.00 - 3.48 = 10.52$

e. At half-way to the equivalence point, half of the base will remain. The other half will have been converted to $C_{17}H_{19}O_3NH^+$. The concentrations of $C_{17}H_{19}O_3N$ and $C_{17}H_{19}O_3NH^+$ are equal. Consequently, $[OH^-] = K_b$.

$K_b = \dfrac{[OH^-][C_{17}H_{19}O_3NH^+]}{[C_{17}H_{19}O_3N]} \Rightarrow K_b = [OH^-] = 7.4 \times 10^{-7} \, M$

$pOH = -\log(7.4 \times 10^{-7}) = 6.13$

$pH = 14.00 - 6.13 = 7.87$

f. At the equivalence point, all the $C_{17}H_{19}O_3N$ has been converted to $C_{17}H_{19}O_3NH^+$, giving 0.0075 mol.

$$[C_{17}H_{19}O_3NH^+] = \frac{0.0075 \text{ mol. } C_{17}H_{19}O_3NH^+}{0.0500 \text{ L} + 0.00718 \text{ L}} = 0.062 \ M$$

	$C_{17}H_{19}O_3NH^+$ (aq)	$\rightleftharpoons$	H^+(aq)	+	$C_{17}H_{19}O_3N$ (aq)
[ ]$_o$	0.062		0		0
Δ[ ]	-x		+x		+x
[ ]$_{eq}$	0.062 - x		x		x

$$K_a = \frac{K_w}{K_b} = \frac{1.0 \times 10^{-14}}{7.4 \times 10^{-7}} = 1.4 \times 10^{-8}$$

$$K_a = \frac{[H^+][C_{17}H_{19}O_3N]}{[C_{17}H_{19}O_3NH^+]} \implies 1.4 \times 10^{-8} = \frac{x^2}{0.062 - x}$$

$K_a/a = 1.4 \times 10^{-8} / 0.062 = 2.2 \times 10^{-7}$
$K_a/a = 2.2 \times 10^{-7} < 2.5 \times 10^{-3}$ (The 5% rule applies. See the explanation accompanying Chapter 13, Problem 38.)

$$K_a = \frac{[H^+][C_{17}H_{19}O_3N]}{[C_{17}H_{19}O_3NH^+]} \implies 1.4 \times 10^{-8} = \frac{x^2}{0.062}$$

$x^2 = 8.7 \times 10^{-10}$

$x = [H^+] = 2.9 \times 10^{-5} \ M$

pH = -log (2.9 x 10^{-5}) = 4.54

---

48. Refer to Section 14.3 and Problem 46 (above).

$PV = nRT$
(1.00 atm)(1.00L) = n(0.0821 L·atm/mol.·K)(298K)
n = 4.09 x 10^{-2} mol.
[NH$_3$] = 4.09 x 10^{-2} mol./0.725 L = 5.64 x 10^{-2} M

a. Before the titration, consider the reaction of ammonia with water.

	NH$_3$(aq)	+	H$_2$O(l)	$\rightleftharpoons$	H^+(aq)	+	NH$_4^+$(aq)
[ ]$_o$	0.0564		---		0		0
Δ[ ]	-x		---		+x		+x
[ ]$_{eq}$	0.0564 - x		---		x		x

$$K_b = \frac{[NH_4^+][OH^-]}{[NH_3]} \Rightarrow 1.8 \times 10^{-5} = \frac{x^2}{0.0564 - x}$$

$K_b/b = 1.8 \times 10^{-5} / 0.0564 = 3.2 \times 10^{-4}$

$K_b/b = 3.2 \times 10^{-4} < 2.5 \times 10^{-3}$ (The 5% rule applies. See the explanation accompanying Chapter 13, Problem 38.)

$$K_b = \frac{[NH_4^+][OH^-]}{[NH_3]} \Rightarrow 1.8 \times 10^{-5} = \frac{x^2}{0.0564}$$

$x^2 = 1.0 \times 10^{-6}$

$x = [OH^-] = 1.0 \times 10^{-3} M$

$pOH = -\log (1.0 \times 10^{-3}) = 3.00$

$pH = 14.00 - 3.00 = 11.00$

b. At half-way to the equivalence point, $[OH^-] = K_b$ (see Problem 46 (above)).
$[OH^-] = 1.8 \times 10^{-5} M$
$pOH = -\log(1.8 \times 10^{-5}) = 4.74$
$pH = 14.00 - 4.74 = 9.26$

c. At the equivalence point, all the ammonia has been converted to ammonium ion. The volume of the solution has increased, thus calculate the volume of base added and recalculate $[NH_4^+]$. Then calculate $[H^+]$ using K_a, and from that, pH.

$$0.0500 \, L \times \frac{5.64 \times 10^{-2} \, mol.}{1 \, L} = 2.82 \times 10^{-3} \, mol. \, NH_4^+$$

$$2.82 \times 10^{-3} \, mol. \, NH_4^+ \times \frac{1 \, mol. \, HNO_3}{1 \, mol. \, NH_4^+} \times \frac{1 \, L \, solution}{0.2193 \, mol. \, HNO_3} = 0.0129 \, L$$

$$[NH_4^+] = \frac{2.82 \times 10^{-3} \, mol.}{0.0500 \, L + 0.0129 \, L} = 0.0448 \, M$$

	$NH_4^+(aq)$	$\rightleftharpoons$	$H^+(aq)$	$+$	$NH_3(aq)$
$[\,]_0$	0.0448		0		0
$\Delta[\,]$	-x		+x		+x
$[\,]_{eq}$	0.0448 - x		x		x

$$K_a = \frac{[NH_3][H^+]}{[NH_4^+]} \Rightarrow 5.6 \times 10^{-10} = \frac{x^2}{0.0448 - x}$$

$K_b/b = 5.6 \times 10^{-10} / 0.0448 = 1.2 \times 10^{-8}$

$K_b/b = 1.2 \times 10^{-8} < 2.5 \times 10^{-3}$ (The 5% rule applies.)

$$K_a = \frac{[NH_3][H^+]}{[NH_4^+]} \Rightarrow 5.6 \times 10^{-10} = \frac{x^2}{0.0448}$$

$x^2 = 2.5 \times 10^{-11}$

$x = [H^+] = 5.0 \times 10^{-6} M$

$pH = -\log(5.0 \times 10^{-6}) = 5.30$

---

50. *Refer to Section 14.3 and Problem 46 (above).*

$RNH(aq) + HCl(aq) \rightarrow RNH_2^+(aq) + Cl^-(aq)$
$RNH(aq) + H^+(aq) \rightarrow RNH_2^+(aq)$

a. $0.05990 \, L \times \dfrac{0.925 \, mol. \, HCl}{1 \, L} \times \dfrac{1 \, mol. \, RNH}{1 \, mol. \, HCl} = 0.0554 \, mol. \, RNH$

$\dfrac{2.500 \, g \, RNH}{0.0554 \, mol. \, RNH} = 45.1 \, g/mol. \, RNH$

b. As seen in Problem (46e) above, at halfway to the equivalence point, $[OH^-] = K_b$.
   $pOH = 14.00 - 10.77 = 3.23$
   $[OH^-] = 10^{-3.23} = 5.9 \times 10^{-4}$
   $K_b = 5.9 \times 10^{-4}$

c. $K_a = \dfrac{K_w}{K_b} = \dfrac{1.0 \times 10^{-14}}{5.9 \times 10^{-4}} = 1.7 \times 10^{-11}$

---

52. *Refer to Chapters 8 and 12.*

a. $\Delta H_{reaction} = \Sigma \, \Delta H_f^\circ \, _{(products)} - \Sigma \, \Delta H_f^\circ \, _{(reactants)}$
   $\Delta H_{reaction} = [\Delta H_f^\circ \, H^+(aq) + \Delta H_f^\circ \, F^-(aq)] - [\Delta H_f^\circ \, HF(aq)]$
   $\Delta H_{reaction} = [(1 \, mol.)(0.0 \, kJ/mol.) + (1 \, mol.)(-332.6 \, kJ/mol.)] - [(1 \, mol.)(-320.1 \, kJ/mol.)]$
   $\Delta H_{reaction} = -12.5 \, kJ = -12500 \, J$

$$\ln\frac{K_2}{K_1} = \frac{\Delta H^\circ}{R}\left[\frac{1}{T_1} - \frac{1}{T_2}\right] \Rightarrow \ln\frac{K_2}{6.9 \times 10^{-4}} = \frac{12500 \, J/mol.}{8.31 \, J/mol. \cdot K}\left[\frac{1}{298 \, K} - \frac{1}{373 \, K}\right]$$

$\ln K_2 - (-7.28) = -1.5 \times 10^{-3} K(6.75 \times 10^{-4} K^{-1})$
$\ln K_2 = -1.01 - 7.28 = -8.29$
$K_2 = e^{-8.28} = 2.5 \times 10^{-4}$

$$K_a = \frac{[H^+][F^-]}{[HF]} \Rightarrow 2.5 \times 10^{-4} = \frac{x^2}{0.100}$$

$x = [H^+] = 5.0 \times 10^{-3} M$

$pH = -\log(5.0 \times 10^{-3}) = 2.30$

b. $K_a = \dfrac{[H^+][F^-]}{[HF]} \Rightarrow 6.9 \times 10^{-4} = \dfrac{x^2}{0.100}$

$x = [H^+] = 8.3 \times 10^{-3} \, M$

pH = -log(8.3 x 10^{-3}) = 2.08

54. *Refer to Section 14.3.*

a. 2 mol. of H$^+$ will react with 2 mol. of X$^-$ to give 2 mol. of HX.

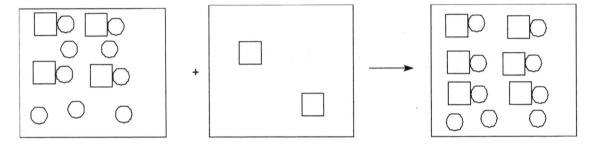

b. 2 mol. OH$^-$ will react with 2 mol. HX forming 2 mol. X$^-$ and 2 mol. H$_2$O (not shown).

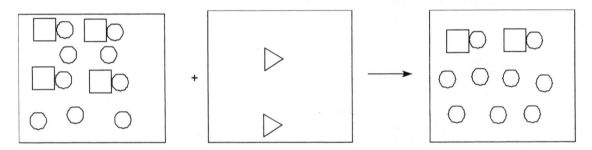

c. 5 mol. OH$^-$ will react with the 4 mol. HX forming 4 mol. X$^-$ and 4 mol. H$_2$O (not shown). There will be one mol. OH$^-$ left over, unreacted.

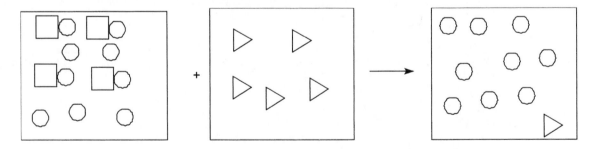

Consider the system in each beaker

Beaker A   [HA] = [NaA]. The Henderson-Hasselbalch equation shows that $pH = pK_a$.

Beaker B   The same system as Beaker A, only more dilute; $pH = pK_a$.

Beaker C   HCl reacts with NaA forming HA. 0.150 mol. HA; 0.050 mol. A⁻; 0 mol. HCl.

Beaker D   NaOH reacts with HA forming NaA. 0 mol. HA; 0.200 mol. NaA; 0 mol. NaOH.

Beaker E   HCl and NaOH in equimolar amounts react to form water and pH = 7.00.

a. **EQ**   See above.

b. **GT**   The pH in Beaker B is the pKa. The pH in Beaker C is the $pK_a + \log[A^-]/[HA]$. Since $[A^-]/[HA]$ is less than 1, the log of this ratio will be negative; pH will be reduced by that amount making the pH in Beaker C smaller than the pH in Beaker B.

c. **GT**   Beaker D has the salt of a weak acid present. These salts are basic and have a pH > 7.00.

d. **MI**   The pK_a of a weak acid is varied (see Table 14.1). One needs to know the value of the pK_a before a comparison to a known pH can be made.

e. **LT**   Beaker A has $pH = pK_a$. Beaker D is more basic than the buffer system formed from HA and A⁻.

a. **False**.   The formate ion concentration in 0.10 M HCHO₂ is much less than the formate ion concentration in 0.10 M NaCHO₂.
HCHO₂ is a weak acid, dissociating very little, while NaCHO₂ is a salt and dissociates completely.

b. **True**.   Excessive amounts of acid will convert all the B⁻ to HB, changing the buffer solution to an acid solution.

c. **False**.   A buffer is a mixture of roughly equal amounts of a weak acid and the conjugate base of that acid.

d. **False**.   Since K_a for HCO₃⁻ is 4.7×10^{-11}, K_b for CO₃²⁻ is 2.1×10^{-4}.
$K_b = K_w / K_a$ for the conjugate base (B⁻) of the weak acid HB.

a.   The pH of the solution before titration is high. This indicates that there is only base present. As more titrating agent is added, the pH decreases, indicating that the titrating agent is an **acid**.

b.   The weakest base will have the lowest [OH⁻] and the lowest pH. Thus, **C** is the weakest base.

c. Halfway to the equivalence point pK_a = pH. It takes 52 mL to reach the equivalence point, so it takes 26 mL to reach the half-equivalence point. The pH at 26 mL for curve B is 8.0. Thus pK_a = 8.0 and K_a = $10^{-8.0}$ = **1 x 10^{-8}**.

d. It takes 52 mL of 0.1 M titrating agent to reach the equivalence point. The volume of the base titrated is 50 mL. Since the volumes are so close, the molarities must be close. Thus, the concentration of the base is approximately **0.1 M**.

e. The equivalence point is the middle of the vertical portion of the curve. The pH at this point is **5**.

---

61. *Refer to Section 14.3.*

Write the equation for the reaction. Calculate $[H^+]$ (from pH) and moles of acetate (from the reaction stoichiometry) and from that, moles of HB still in solution. Then calculate the moles of acetic acid added and finally, the volume of acetic acid.

$$HC_2H_3O_2(aq) + OH^-(aq) \rightleftharpoons C_2H_3O_2^-(aq) + H_2O(l)$$

$$[H^+] = 10^{-pH} = 10^{-4.20} = 6.3 \times 10^{-5}\ M$$

$$0.1000\ L \times \frac{1.25\ mol.\ NaOH}{1\ L} \times \frac{1\ mol.\ C_2H_3O_2^-}{1\ mol.\ NaOH} = 0.125\ mol.\ C_2H_3O_2^-$$

$$[H^+] = K_a \times \frac{n_{HC_2H_3O_2}}{n_{C_2H_3O_2^-}} \quad \Rightarrow \quad 6.3 \times 10^{-5} = 1.8 \times 10^{-5} \times \frac{n_{HC_2H_3O_2}}{0.125}$$

$$n_{HC_2H_3O_2} = 0.44\ mol.$$

This is the moles of acid present *after* reaction has occurred. The total moles of acid that must have been present *before* reaction is:

$$n_{total} = n_{HC_2H_3O_2} + n_{C_2H_3O_2^-} = 0.44\ mol. + 0.125\ mol. = 0.56\ mol.$$

Note that this solution contains roughly equal amounts of HB and B⁻, and is therefore a buffer.

$$0.56\ mol. \times \frac{60.05\ g\ HC_2H_3O_2}{1\ mol.} \times \frac{100\ g\ solution}{98\ g\ HC_2H_3O_2} \times \frac{1\ mL}{1.0542\ g} = 33\ mL\ solution$$

---

62. *Refer to Sections 14.1 and 14.3.*

After solution A is titrated to its equivalence point, all the acid will have been converted to the conjugate base. Also, when solutions A and B are combined, [HB] = [B⁻], thus [HB] / [B⁻] = 1. Calculate the $[H^+]$ of this solution, and the K_a for the acid. From K_a and pH of the original solution, one can calculate [HB] and thus the molar mass.

From the final solution,

$$[H^+] = 10^{-pH} = 10^{-4.26} = 5.5 \times 10^{-5} \ M$$

$$[H^+] = K_a \times \frac{[HB]}{[B^-]} \quad \Rightarrow \quad 5.5 \times 10^{-5} = K_a \times 1$$

$$K_a = 5.5 \times 10^{-5}$$

From the initial solution,

$$[H^+] = 10^{-pH} = 10^{-2.56} = 2.8 \times 10^{-3} \ M$$

$$K_a = \frac{[B^-][H^+]}{[HB]} \quad \Rightarrow \quad 5.5 \times 10^{-5} = \frac{(2.8 \times 10^{-3})^2}{[HB]}$$

$$[HB] = 0.14 \ M$$

$$0.2500 \ L \times \frac{0.14 \ mol. \ HB}{1 \ L} = 0.035 \ mol. \ HB$$

$$\frac{4.00 \ g}{0.035 \ mol. \ HB} = 1.1 \times 10^2 \ g/mol. \ HB$$

---

63. *Refer to Section 14.3.*

Calculate the mole ratio of NH_4^+ to NH_3 needed to make a solution with the requisite pH. Consider the mole ratio and the definition of a buffer.

$$[H^+] = 10^{-pH} = 10^{-6.50} = 3.2 \times 10^{-7} \ M$$

$$[H^+] = K_a \times \frac{n_{NH_4^+}}{n_{NH_3}} \quad \Rightarrow \quad 3.2 \times 10^{-7} = 5.6 \times 10^{-10} \times \frac{n_{NH_4^+}}{n_{NH_3}}$$

$$\frac{n_{NH_4^+}}{n_{NH_3}} = 5.7 \times 10^2$$

A buffer requires roughly equal amounts of a weak acid and its conjugate base. A 570:1 ratio does **not** meet this requirement.

---

64. *Refer to Sections 14.1 and 14.3 and Example 14.6.*

---

a. $\quad K_a = \dfrac{[NO_2^-][H^+]}{[HNO_2]} \quad \Rightarrow \quad 6.0 \times 10^{-4} = \dfrac{x^2}{1.000 \ M}$

$x^2 = 6.0 \times 10^{-4}$

$x = [H^+] = 0.024 \ M$

$pH = -\log(0.024) = 1.62$

b. $0.05000 \text{ L} \times \dfrac{1.000 \text{ mol.}}{1 \text{ L}} = 0.05000 \text{ mol. HNO}_2$

   At half-neutralization, half of the HNO_2 will be converted to NO_2^-, thus:
   $$[NO_2^-] = [HNO_2] = 0.02500 \ M.$$

   $$[H^+] = K_a \times \dfrac{n_{HNO_2}}{n_{NO_2^-}} = 6.0 \times 10^{-4} \times \dfrac{0.02500}{0.02500} = 6.0 \times 10^{-4} \ M$$

   $$pH = -\log (6.0 \times 10^{-4}) = 3.22$$

c. At the equivalence point, all the nitrous acid has been converted to nitrite ion and the solution is weakly basic.
   $$NO_2^-(aq) + H_2O(l) \rightleftharpoons HNO_2(aq) + OH^-(aq)$$
   Calculate the volume of base added and the total volume of the final solution. Then calculate $[NO_2^-]$, $[OH^-]$, pOH and finally pH.

   $$0.0500 \text{ mol. HNO}_2 \times \dfrac{1 \text{ mol. NaOH}}{1 \text{ mol. HNO}_2} \times \dfrac{1 \text{ L}}{0.850 \text{ mol. NaOH}} = 0.0588 \text{ L}$$

   $$V_{total} = 0.05000 \text{ L (HNO}_2) + 0.0588 \text{ L (NaOH)} = 0.1088 \text{ L}$$

   $$[NO_2^-] = \dfrac{0.05000 \text{ mol.}}{0.1088 \text{ L}} = 0.4596 \ M$$

   $$K_b = \dfrac{[HNO_2][OH^-]}{[NO_2^-]} \quad \Rightarrow \quad 1.7 \times 10^{-11} = \dfrac{x^2}{0.4596 \ M}$$

   $$x^2 = 7.8 \times 10^{-12}$$
   $$x = [OH^-] = 2.8 \times 10^{-6} \ M$$

   $$pOH = -\log (2.8 \times 10^{-6}) = 5.55$$
   $$pH = 14.0 - pOH = 14.00 - 5.55 = 8.45$$

d. As calculated in part (c) above, 0.0588 L (58.8 mL) of NaOH solution were needed to reach equivalence. 0.10 mL less that volume is 0.0587 L (58.7 mL).

   $$0.0587 \text{ L} \times \dfrac{0.850 \text{ mol. NaOH}}{1 \text{ L}} \times \dfrac{1 \text{ mol. HNO}_2}{1 \text{ mol. NaOH}} = 0.0499 \text{ mol. HNO}_2$$

   $$0.05000 \text{ mol. HNO}_2 - 0.0499 \text{ mol. HNO}_2 = 0.0001 \text{ mol. HNO}_2$$

   $$[H^+] = K_a \times \dfrac{n_{HNO_2}}{n_{NO_2^-}} = 6.0 \times 10^{-4} \times \dfrac{0.0001}{0.0499} = 1 \times 10^{-6} \ M$$

   $$pH = -\log (1 \times 10^{-6}) = 6.0$$

e. At this point, any NaOH added in addition to the 58.8 mL (from part (c) above) is excess. Calculate the moles of excess NaOH and [OH⁻] in the resulting solution.

$$n_{NaOH} = 0.10 \text{ mL} \times \frac{1 \text{ L}}{1000 \text{ mL}} \times \frac{0.850 \text{ mol.}}{1 \text{ L}} = 8.5 \times 10^{-5} \text{ mol.} \quad \text{(in excess)}$$

$$[OH^-] = \frac{8.5 \times 10^{-5} \text{ mol.}}{0.1089 \text{ L}} = 7.8 \times 10^{-4} \ M$$

Note that we can neglect the OH⁻ formed by the nitrate (see part (c) above) since its contribution is so small, considering that K_b has only two significant figures.

pOH = -log (7.8 x 10⁻⁴) = 3.11
pH = 14.00 - pOH = 14.00 - 3.11 = 10.89

f. Consider one more data point, with 10.00 mL NaOH in excess, giving a total volume of 68.8 mL.

$$n_{NaOH} = 10.00 \text{ mL} \times \frac{1 \text{ L}}{1000 \text{ mL}} \times \frac{0.850 \text{ mol.}}{1 \text{ L}} = 8.50 \times 10^{-3} \text{ mol.} \quad \text{(in excess)}$$

$$[OH^-] = \frac{8.50 \times 10^{-3} \text{ mol.}}{0.0688 \text{ L}} = 0.124 \ M$$

pOH = -log (0.124) = 0.907
pH = 14.00 - pOH = 14.00 - 0.907 = 13.09

pH	NaOH (mL)
1.62	0.00
3.22	29.4
5.92	58.7
8.45	58.8
10.89	58.9
13.09	68.8

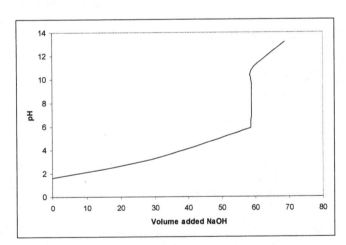

65. Refer to Section 14.2.

The indicator will be in the midst of changing color when [HIn] = [In⁻]. Calculate the [H⁺] at this point. Determine the amount of $C_2H_3O_2^-$ that is present, and thus the amount of NaOH used to produce this amount of $C_2H_3O_2^-$.

$$[H^+] = K_a \times \frac{[HIn]}{[In^-]} \quad \Rightarrow \quad [H^+] = 1.0 \times 10^{-5} \times 1 = 1.0 \times 10^{-5} \ M$$

$$[H^+] = K_a \times \frac{n_{HC_2H_3O_2}}{n_{C_2H_3O_2^-}} \quad \Rightarrow \quad 1.0 \times 10^{-5} = 1.8 \times 10^{-5} \times \frac{n_{HC_2H_3O_2}}{n_{C_2H_3O_2^-}}$$

$$\frac{n_{HC_2H_3O_2}}{n_{C_2H_3O_2^-}} = 0.56 \quad \Rightarrow \quad n_{HC_2H_3O_2} = 0.56 n_{C_2H_3O_2^-} \qquad\qquad \text{eq. 1}$$

Before titration:

$$0.05000 \ \text{L} \times \frac{1.00 \ \text{mol.}}{1 \ \text{L}} \ = \ 0.0500 \ \text{mol. } HC_2H_3O_2$$

After titration:

$$n_{HC_2H_3O_2} = x, \quad \text{and} \quad n_{C_2H_3O_2^-} = 0.0500 - x \qquad\qquad \text{Substituting into eq. 1,}$$

$$x = 0.56(0.0500 - x)$$
$$x = [HC_2H_3O_2] = 0.018 \ M$$

$$n_{C_2H_3O_2^-} = 0.0500 - x = 0.032 \ \text{mol.}$$

$$0.032 \ \text{mol.} \ C_2H_3O_2^- \times \frac{1 \ \text{mol. NaOH}}{1 \ \text{mol.} \ C_2H_3O_2^-} \times \frac{1 \ \text{L}}{1.00 \ \text{mol. NaOH}} = 0.032 \ \text{L NaOH}$$

A better indicator would be phenolphthalein.

---

66. *Refer to Section 13.4 and Example 13.9.*

---

a. $H_2SO_4(aq) \rightleftharpoons H^+(aq) + HSO_4^-(aq)$
   This is a strong acid and ionizes completely.

   $[H^+] = 0.1500 \ M$

   $pH = -\log (0.1500) = 0.8239$

b. For the second ionization: $HSO_4^-(aq) \rightleftharpoons H^+(aq) + SO_4^{2-}(aq)$
   $[SO_4^{2-}] = x$
   $[HSO_4^-] = 0.1500 - x$
   $[H^+] = 0.1500 + x$

$$K_a = \frac{[SO_4^{2-}][H^+]}{[HSO_4^-]} \quad \Rightarrow \quad 1.1 \times 10^{-2} = \frac{(0.1500 + x)(x)}{0.1500 - x}$$

$K_a/a = 1.1 \times 10^{-2} / 0.1500 = 7.3 \times 10^{-2}$
$K_a/a = 7.3 \times 10^{-2} > 2.5 \times 10^{-3}$ (The 5% rule does **not** apply. See Chapter 13, Problem 38.)

$$K_a = \frac{[SO_4^{2-}][H^+]}{[HSO_4^-]} \Rightarrow 1.1 \times 10^{-2} = \frac{(0.1500 + x)(x)}{0.1500 - x} = \frac{0.1500x + x^2}{0.1500 - x}$$

$0.00165 - 0.011x = 0.1500x + x^2$

$x^2 + 0.161x - 0.00165 = 0$

$$x = \frac{-b \pm \sqrt{b^2 - 4ac}}{2a} = \frac{-0.161 \pm \sqrt{(0.161)^2 - 4(1)(-0.00165)}}{2(1)} = 0.0097$$

$[H^+] = 0.1500 + 0.0097 = 0.1597\ M$

$pH = -\log(0.1597) = 0.7968$

---

67. *Refer to Section 14.1 and Appendix 3.*

$$[H^+] = K_a \times \frac{[HB]}{[B^-]}$$

Take the -log of both sides.

$$-\log[H^+] = -\log\left(K_a \times \frac{[HB]}{[B^-]}\right) = -\log K_a - \log\left(\frac{[HB]}{[B^-]}\right)$$

$$pH = pK_a - \log\left(\frac{[HB]}{[B^-]}\right) \Rightarrow pH = pK_a + \log\left(\frac{[B^-]}{[HB]}\right)$$

---

68. *Refer to Section 14.3.*

Since 21.0 mL is needed for neutralization, the titration with 7.00 mL represents ⅓ (7/21) neutralization. Thus ⅓ of the base (B) has been converted to the conjugate acid (HB$^+$), with ⅔ remaining as B.

Consequently, $\dfrac{n_{HB^+}}{n_B} = \dfrac{\frac{1}{3}}{\frac{2}{3}} = 0.5$

$[H^+] = 10^{-pH} = 10^{-8.95} = 1.1 \times 10^{-9}\ M$

$$[H^+] = K_a \times \frac{[HB^+]}{[B]} \Rightarrow 1.1 \times 10^{-9} = K_a \times 0.5$$

$K_a = 2.2 \times 10^{-9}$

$$K_b = \frac{K_w}{K_a} = \frac{1.0 \times 10^{-14}}{2.2 \times 10^{-9}} = 4.5 \times 10^{-6}$$

Chapter 15: Complex Ions

---

2. Refer to Section 15.1 and Examples 15.1 and 15.2.

a. $[Co(en)_2(SCN)Cl]^+$ has three different ligands:

   en: ethylenediamine is a molecule and has no charge.

   SCN^-:   is an anion with a –1 charge.

   Cl^-:   chloride is an anion with a -1 charge.

b. charge of complex = oxid. no. central metal + charges of ligands

   let x = oxid. no. central metal

   $+1 = x + 2(0) + 1(-1) + 1(-1)$

   $x = +3$

c. Sulfide is an anion with a –2 charge. Thus two $[Co(en)_2(SCN)Cl]^+$ are needed to balance the charge of the sulfide: **$[Co(en)_2(SCN)Cl]_2S$.**

---

4. Refer to Section 15.1.

Put the metal and the ligands together and add up the charges to get the overall charge.

a. $Pt^{2+} + 2NH_3 + C_2O_4^{2-} \rightarrow [Pt(NH_3)_2(C_2O_4)]$

b. $Pt^{2+} + 2NH_3 + SCN^- + Br^- \rightarrow [Pt(NH_3)_2(SCN)Br]$

c. $Pt^{2+} + en + 2NO_2^- \rightarrow [Pt(en)(NO_2)_2]$

   * *en* = ethylenediamine

---

6. Refer to Section 15.1.

The coordination number is the total number of atoms bonded to the metal.

a. $[Mn(NH_3)_2(OH)_4]^{2+}$

   $2(NH_3) + 4(OH^-) = 6$, thus the coordination number is **6**.

b. $[CuCl_4]^{3-}$

   $4Cl^- = 4$, thus the coordination number is **4**.

c. $[Cr(H_2O)_2(ox)_2]^{2-}$ (recall that (*ox*) is a chelating ligand, taking up 2 coordination sites on the Cr)

   $2(H_2O) + 2(2ox) = 6$, thus the coordination number is **6**.

d. $[Au(CN)_2]^+$

   $2(CN^-) = 2$, thus the coordination number is **2**.

8. *Refer to Section 15.1 and Example 15.1.*

Use the equation given in Section 15.1 and used in Example 15.1:
charge of complex = oxid. no. central metal + charges of ligands
let x = oxid. no. central metal

a. Each of the two NH_3 ligands has a charge of 0 and each of the four OH^- ligands has a charge of –1.
$+2 = x + 2(0) + 4(-1)$
$x = +6$

b. Each of the 4 Cl^- ligands has a charge of –1.
$-3 = x + 4(-1)$
$x = +1$

c. H_2O has a charge of 0 and *ox* has a charge of –2.
$-2 = x + 2(0) + 2(-2)$
$x = +2$

d. CN^- has a charge of –1.
$+1 = x + 2(-1)$
$x = +3$

10. *Refer to Section 15.1 and Example 15.1.*

The phosphate ion has a –3 charge; the aluminum ion has a +3 charge. Treat the complexes in #6 as polyatomic ions and construct formulas as in Section 2.5.

a. $[Mn(NH_3)_2(OH)_4]_3(PO_4)_2$

b. $Al[CuCl_4]$

c. $Al_2[Cr(H_2O)_2(ox)_2]_3$

d. $[Au(CN)_2]_3PO_4$

12. *Refer to Section 15.1 and Table 15.2 and Problem 8(above).*

Use the coordination number of the metal ion to determine the number of ligands. Add the charges of the metal ion and the ligands to get the charge of the complex.

a. Pt^{4+} has a coordination number of 6, and NH_3 has a 0 charge, thus
$[Pt(NH_3)_6]^{4+}$

b. Ag^+ has a coordination number of 2, and CN^- has a -1 charge, thus
$[Ag(CN)_2]^-$

c. Zn^{2+} has a coordination number of 4, and $C_2O_4^{2-}$ has a -2 charge and is chelating (each oxalate occupying 2 coordination sites), thus

$[Zn(C_2O_4)_2]^{2-}$

d. Cd^{2+} has a coordination number of 4, and CN^- has a -1 charge, thus

$[Cd(CN)_4]^{2-}$

---

14. Refer to Section 15.1 and Chapter 3.

Since chloride has a -1 charge, 2 moles of the Cl^- associate with one mole of complex cation:

$[Cr(H_2O)_5(OH)]Cl_2$

MM = 51.996 + 5(18.02) + 1(17.01)] + 2(35.45) = 230.01 g/mol.

$$\text{mass \%} = \frac{\text{mass of Cr}}{\text{mass of salt}} \times 100\% = \frac{51.996\,g}{230.01\,g} \times 100\% = 22.606\%$$

---

16. Refer to Section 15.1, Problem 3.87 and Problem 14(above).

$$\text{mass \%} = \frac{\text{mass of Co}}{\text{mass of B-12}} \times 100\%$$

mass of Co = (# Co atoms)(molar mass of Co)

let x = number of Co atoms

$$4.4\% = \frac{x(58.93\,g/mol.)}{1.3 \times 10^3\,g/mol.} \times 100\% \Rightarrow x \approx 1$$

There is 1 Co atom.

---

18. Refer to Sections 15.1 and 15.2.

When a polyatomic ligand bonds to a metal, the atom in the ligand which bonds is the one capable of acting as a Lewis base. $OH_2 = H_2O$; water is written in this fashion to emphasize that it is the oxygen that is bonded to the metal.

a.

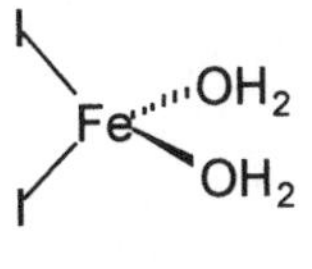

$$Fe \begin{matrix} OH_2 \\ OH_2 \end{matrix}$$

b. $[NCS-Ag-SCN]^-$

c.

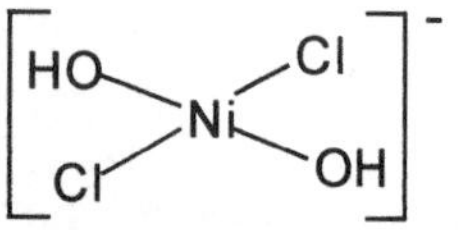

$$\left[\begin{array}{c} \text{HO} \quad\quad \text{Cl} \\ \text{Ni} \\ \text{Cl} \quad\quad \text{OH} \end{array}\right]^{-}$$

d.

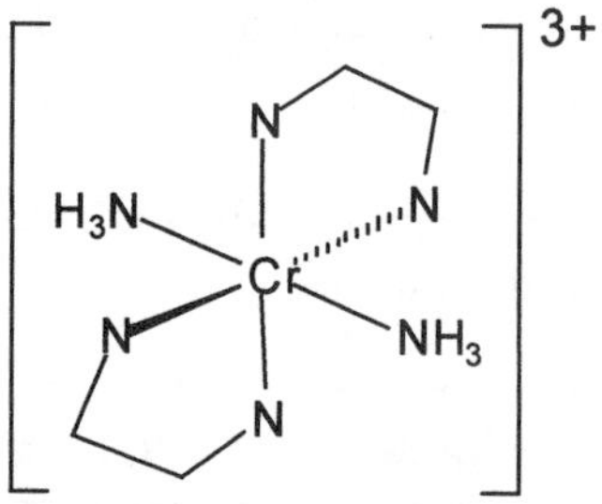

$$\left[\text{Cr(en) structure}\right]^{3+}$$

e.

$$\left[\text{Co(oxalate)(OH}_2)_2 \text{ structure}\right]^{+}$$

$en = $ N⌣N structure

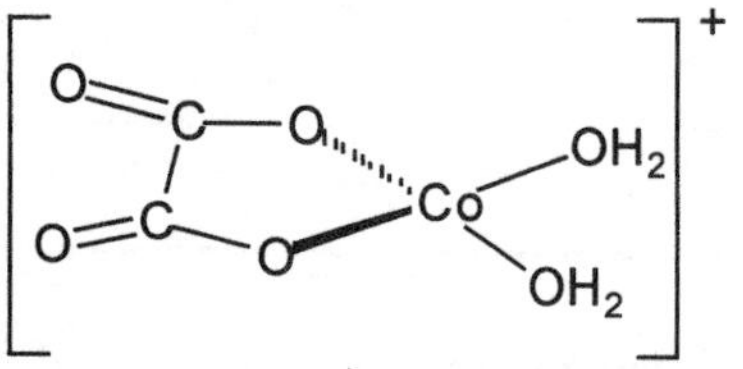

20. Refer to Sections 15.1 and 15.2.

The acac ligand will chelate as a bidentate ligand to the Fe in much the same manner as *en*.

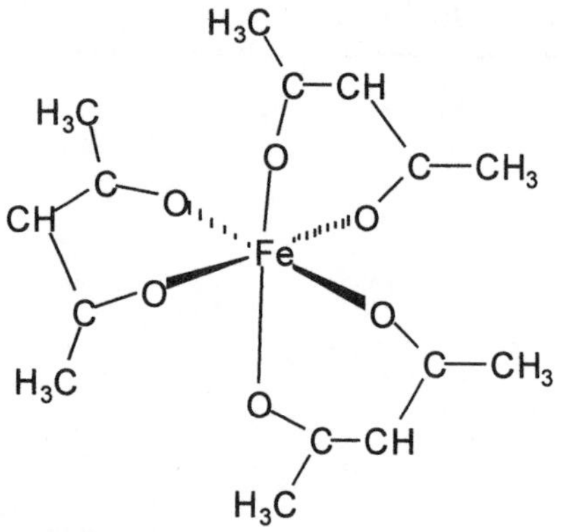

---

22. Refer to Section 15.2, Example 15.3 and Figures 15.5 and 15.6.

---

a. As in Figure 15.5, there are four of one ligand (SCN⁻) and two of another (NH₃). There are two possible geometric isomers (NH₃ *cis* and NH₃ *trans*).

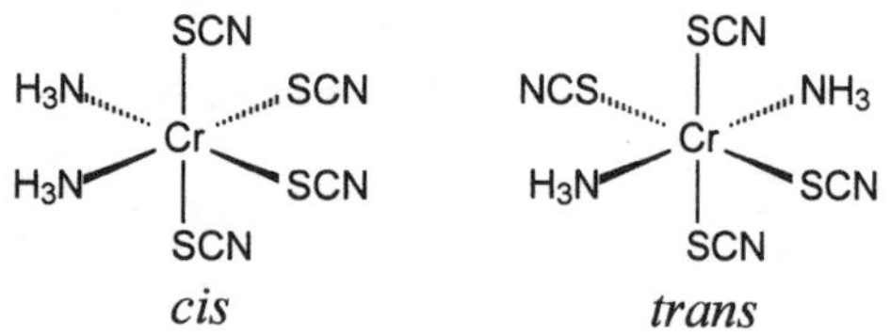

cis   trans

b. As in Figure 15.6, there are three of one ligand (NO_2^-), and three of another (NH_3). There are two possible geometric isomers (NH_3 in a *facial* and NH_3 in an *meridial* arrangement).

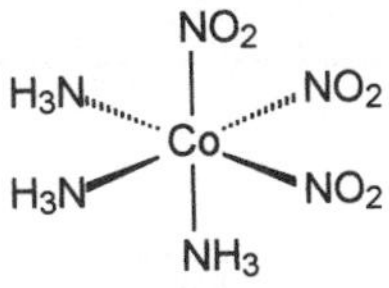

facial 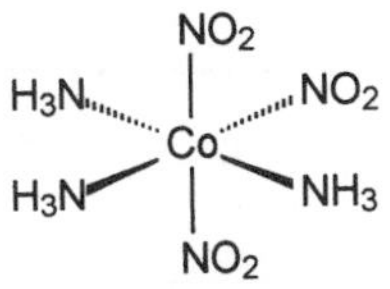   *meridial*

c. This complex has three NH_3, two H_2O and one OH^-. The NH_3 can be distributed in either *facial* or *meridial* fashion. For the *meridial* isomer, the H_2O can be either *cis* or *trans*.

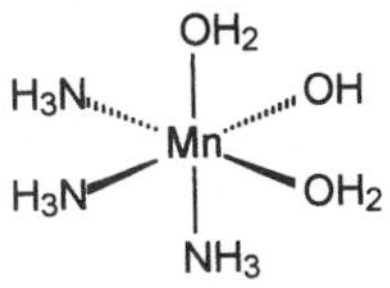

facial 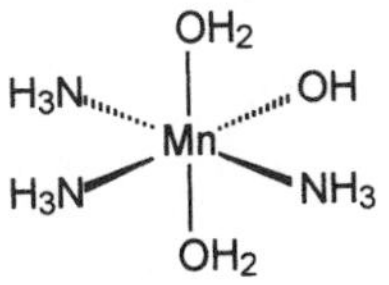   *meridial, trans* H_2O 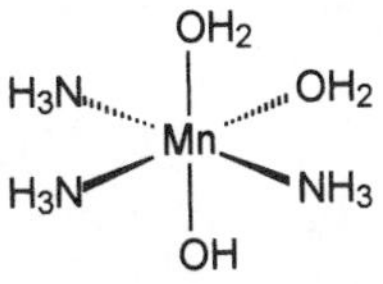   *meridial, cis* H_2O

---

24. Refer to Section 15.2, Figure 15.6, and Example 15.3.

This complex has three H_2O, two OH^-, and one Cl^-. The H_2O can be distributed in either *facial* (the ligands on the face of the octahedron) or *meridial* (the ligands in the same plane) fashion. For the *meridial* isomer, the OH^- can be either *cis* or *trans*.

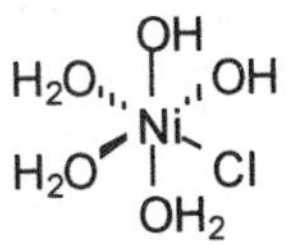

facial 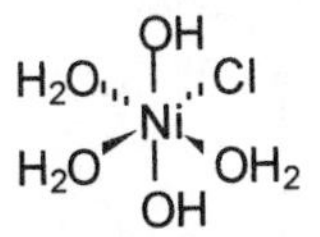   *meridial, trans* OH 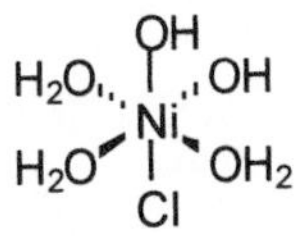   *meridial, cis* OH

---

26. Refer to Section 15.3 and Chapter 6, (p. 146 and p. 149).

When forming a cation, the electrons are removed first from the valence **s** orbital, then the **d** orbital.

a. Fe: $[Ar]4s^2 3d^6$
   Fe^{3+}: $[Ar]3d^5$ see p. 149

b. V: $[Ar]4s^2 3d^3$
   V^{2+}: $[Ar]3d^3$

c. Zn: $[Ar]4s^2 3d^{10}$
   Zn^{2+}: $[Ar]3d^{10}$

d. Cu: $[Ar]4s^1 3d^{10}$ see p. 146
   Cu^+: $[Ar]3d^{10}$

e. Mn: $[Ar]4s^2 3d^5$
    Mn^{4+}: $[Ar]3d^3$

---

28. Refer to Section 15.3 and Chapter 6.

a. $[Ar]3d^5$:   (↑)(↑)(↑)(↑)(↑)
                    5 unpaired electrons

b. $[Ar]3d^3$:   (↑)(↑)(↑)( )( )
                    3 unpaired electrons

c. $[Ar]3d^{10}$: (↑↓)(↑↓)(↑↓)(↑↓)(↑↓)
                    0 unpaired electrons

d. $[Ar]3d^{10}$: (↑↓)(↑↓)(↑↓)(↑↓)(↑↓)
                    0 unpaired electrons

e. $[Ar]3d^3$:   (↑)(↑)(↑)( )( )
                    3 unpaired electrons

---

30. Refer to Section 15.3 and Example 15.5.

a. Mo: $[Kr]5s^2 4d^4$
    Mo^{3+}: $[Kr]4d^3$

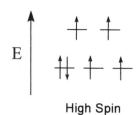

High Spin

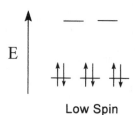

Low Spin

Note that the electron distribution does not change.

b. Pd: $[Kr]5s^2 4d^8$
    Pd^{4+}: $[Kr]4d^6$

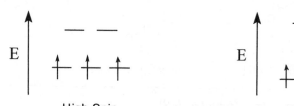

High Spin   Low Spin

Mn: [Ar]$4s^2 3d^5$
Mn^{3+}:   [Ar]$3d^4$

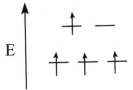

High Spin 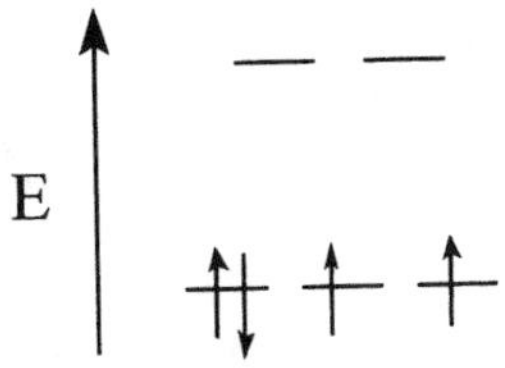   Low Spin

The high and low spin systems have different electron distributions, thus the ligand field strength dictates which complex will be formed.

Mn^{4+}:   [Ar]$3d^3$

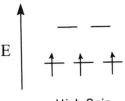

High Spin 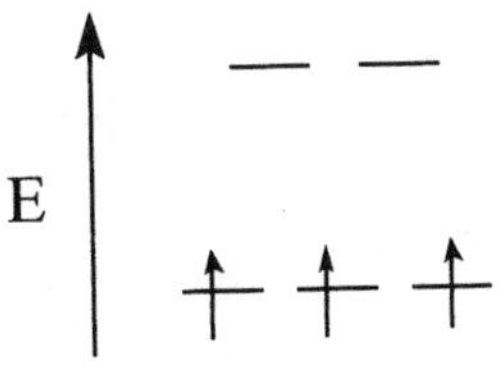   Low Spin

The high and low spin systems have the same electron distribution.

34. Refer to Section 15.3.

Both complexes contain Co^{3+} ([Ar]$3d^6$). To be diamagnetic, [Co(NH$_3$)$_6$]$^{3+}$ must have all electrons paired, which would make this a low spin complex. To be paramagnetic, [CoF$_6$]$^{3-}$ must have unpaired electrons, which would make this a high spin complex. This is consistent with the fact that NH$_3$ is a stronger field ligand than F$^-$.

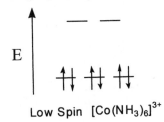

Low Spin [Co(NH$_3$)$_6$]$^{3+}$ 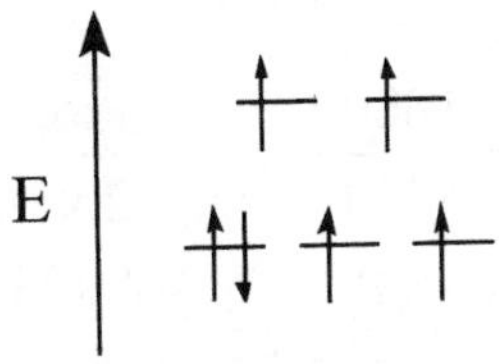   High Spin   [CoF$_6$]$^{3-}$

For weak field ligands, Δ_0 is small, resulting in high spin complexes.

a. Rh^{3+}: $[Kr]4d^6$ 4 unpaired electrons

b. Mn^{3+}: $[Ar]3d^4$ 4 unpaired electrons

c. Ag^+: $[Kr]4d^{10}$ 0 unpaired electrons

d. Pt^{4+}: $[Xe]5d^6$ 4 unpaired electrons

e. Au^{3+}: $[Xe]5d^8$ 2 unpaired electrons

$$E = \frac{2.60 \times 10^{-2} \text{ kJ}}{1 \text{ mol.}} \times \frac{1000 \text{ J}}{1 \text{ kJ}} \times \frac{1 \text{ mol.}}{6.022 \times 10^{23}} = 4.32 \times 10^{-19} \text{ J}$$

$$E = \frac{hc}{\lambda} \Rightarrow \lambda = \frac{(6.626 \times 10^{-34} \text{ J} \cdot \text{s})(2.998 \times 10^8 \text{ m/s})}{4.32 \times 10^{-19} \text{ J}} = 4.60 \times 10^{-7} \text{ m}$$

$\lambda = 460$ nm

The color observed is the complementary color of that absorbed by the complex. Since red is observed, the complementary color (from the color wheel) is green-blue. Thus the wavelength of maximum absorption is approximately **500 nm**.

42. Refer to Section 15.4 and Table 15.4.

a. $K_f = \dfrac{[Cd(CN)_4{}^{2-}]}{[Cd^{2+}][CN^-]^4} = 2 \times 10^{18}$

Rearrange the equation to solve for $[CN^-]^4$

$[CN^-]^4 = \dfrac{[Cd(CN)_4{}^{2-}]}{[Cd^{2+}] \times 2 \times 10^{18}}$

Substitute $[Cd^{2+}] = 10^{-8} \times [Cd(CN)_4{}^{2-}]$

$[CN^-]^4 = \dfrac{[Cd(CN)_4{}^{2-}]}{10^{-8} \times [Cd(CN)_4{}^{2-}] \times 2 \times 10^{18}}$

$[CN^-]^4 = \dfrac{1}{2 \times 10^{10}}$

$[CN^-]^4 = 5 \times 10^{-11}$

$\sqrt[4]{[CN^-]^4} = \sqrt[4]{5 \times 10^{-11}}$

$[CN^-] = 3 \times 10^{-3} \, M$

b. $K_f = \dfrac{[Fe(CN)_6{}^{4-}]}{[Fe^{2+}][CN^-]^6} = 4 \times 10^{45}$

$[CN^-]^6 = \dfrac{[Fe(CN)_6{}^{4-}]}{[Fe^{2+}] \times 4 \times 10^{45}}$

$[CN^-]^6 = \dfrac{[Fe(CN)_6{}^{4-}]}{10^{-20} \times [Fe(CN)_6{}^{4-}] \times 4 \times 10^{45}}$

$[CN^-]^6 = \dfrac{1}{4 \times 10^{25}}$

$[CN^-]^6 = 2.5 \times 10^{-26}$

$\sqrt[6]{[CN^-]^6} = \sqrt[6]{2.5 \times 10^{-26}}$

$[CN^-] = 5 \times 10^{-5} \, M$

44. Refer to Section 15.4 and Example 15.6.

$Cr^{3+}(aq) + 4OH^-(aq) \rightleftharpoons Cr(OH)_4^-(aq)$

When 85% of the Cr^{3+} is converted to $Cr(OH)_4^-$, the amount left is 15% of the original amount. Thus the ratio

$$\frac{[Cr(OH)_4^-]}{[Cr^{3+}]} = \frac{85}{15}$$

$$K_f = \frac{[Cr(OH)_4^-]}{[Cr^{3+}][OH^-]^4} = 8 \times 10^{29} = \frac{85}{15[OH^-]^4}$$

$[OH^-]^4 = 1 \times 10^{-29}$
$[OH^-] = 6 \times 10^{-8}$
$pOH = -\log(6 \times 10^{-8}) = 7.2$
$pH = 14.00 - 7.2 = 6.8$

46. Refer to Section 15.1 and Chapter 3.

a. Assume 100 g of sample.

$$22.0\text{ g Co} \times \frac{1\text{ mol.}}{58.93\text{ g}} = 0.373\text{ mol. Co}$$

$$31.4\text{ g N} \times \frac{1\text{ mol.}}{14.01\text{ g}} = 2.24\text{ mol. N}$$

$$6.78\text{ g H} \times \frac{1\text{ mol.}}{1.008\text{ g}} = 6.73\text{ mol. H}$$

$$39.8\text{ g Cl} \times \frac{1\text{ mol.}}{35.45\text{ g}} = 1.12\text{ mol. Cl}$$

0.373 / 0.373 = 1 mol. Co
2.24 / 0.373 = 6 mol. N
6.73 / 0.373 = 18 mol. H
1.12 / 0.373 = 3 mol. Cl

Thus the simplest formula is: $CoN_6H_{18}Cl_3$.

b. The salt most likely contains six NH_3 molecules and three Cl^- ions.

$[Co(NH_3)_6]Cl_3(s) \rightleftharpoons [Co(NH_3)_6]^{3+}(aq) + 3Cl^-(aq)$

In venous blood, oxygen is not coordinated to iron (hemoglobin, blue), while in arterial blood, oxygen is coordinated to iron (oxyhemoglobin, red). When there is a lot of oxygen around (as in the lungs), oxyhemoglobin is formed. When there is a deficiency of oxygen, the oxygen of the oxyhemoglobin dissociates from the iron, forming hemoglobin.

$$hemoglobin + O_2 \rightleftharpoons oxyhemoglobin + H_2O$$

It is because of this equilibrium that hemoglobin is able to transport oxygen from the lungs to the cells.

50. Refer to Section 15.4.

$$K_f = \frac{[Fe(bipy)_3^{2+}]}{[Fe^{2+}][bipy]^3}$$

$$0.17 \, mol. \, bipy \times \frac{1 \, mol. \, Fe(bipy)_3^{2+}}{3 \, mol. \, bipy} = 5.7 \times 10^{-2} \, mol. \, Fe(bipy)_3^{2+}$$

$$2.4 \, L \times \frac{0.52 \, mol. \, Fe^{2+}}{1 \, L} \times \frac{1 \, mol. \, Fe(bipy)_3^{2+}}{1 \, mol. \, Fe^{2+}} = 0.12 \, mol. \, Fe(bipy)_3^{2+}$$

Thus, bipy is the limiting reagent.
Bipy will be used up. Thus, [bipy] $\approx$ 0
The theoretical yield is 5.7×10^{-2} mol. $Fe(bipy)_3^{2+}$

$$[Fe(bipy)_3^{2+}] = \frac{5.7 \times 10^{-2} \, mol.}{2.4 \, L} = 0.024M$$

The amount of Fe^{2+} at equilibrium will be the initial amount minus that which reacted with bipy.

$$0.17 \, mol. \, bipy \times \frac{1 \, mol. \, Fe^{2+}}{3 \, mol. \, bipy} = 5.7 \times 10^{-2} \, mol. \, Fe^{2+}$$

$$\frac{5.7 \times 10^{-2} \, mol. \, Fe^{2+}}{2.4 \, L} = 0.024 \, M \, Fe^{2+}$$

$$0.052 \, M - 0.024 \, M = 0.028 \, M \, Fe^{2+}$$

52. Refer to Chapter 15.

a. **False**. *en* occupies two coordination sites. Thus, the coordination number is 6.

b. **False**. $[Ni(CN)_6]^{4-}$ is a low spin complex with a large Δ_o, thus it absorbs at a short wavelength. $[Ni(NH_3)_6]^{2+}$ is a high spin complex with a small Δ_o, thus it absorbs at a long wavelength.

c. **True**. Cr^{3+} has a partially filled d orbital. Thus there will be crystal field splitting energy and thus its complexes will be colored. Zn^{2+} has a d^{10} configuration. Thus there is no crystal field splitting energy and its complexes are not colored.

d. **False**. Ions with eight or more d electrons cannot form both high- and low-spin octahedral complexes.

---

54. *Refer to Section 15.1 and "Chemistry Beyond the Classroom: Chelates."*

Calculate the amount of Pb that was consumed, then the amount of EDTA needed to bind the lead. Recall that EDTA forms a 1:1 complex with Pb.

$$10.0 \text{ g paint} \times \frac{5.0 \text{ g Pb}}{100 \text{ g paint}} \times \frac{1 \text{ mol. Pb}}{207.2 \text{ g}} = 2.4 \times 10^{-3} \text{ mol. Pb}$$

$Na_4(EDTA) = Na_4(C_{10}H_{12}N_2O_8)$

MM = 380.2 g/mol.

$$2.4 \times 10^{-3} \text{ mol. Pb} \times \frac{1 \text{ mol. EDTA}}{1 \text{ mol. Pb}} \times \frac{1 \text{ mol. Na}_4\text{EDTA}}{1 \text{ mol. EDTA}} \times \frac{380.2 \text{ g}}{1 \text{ mol. Na}_4\text{EDTA}} = 0.91 \text{ g EDTA}$$

---

55. *Refer to Section 15.1 and Chapter 3.*

Calculate the molar mass of the simplest formula, then determine the molecular formula. From that, determine the ligand(s) present, the cation and the anion.

$MM(PtN_2H_6Cl_2) = 300.2$ g/mol.

600 is 2 x 300, thus the molecular formula is $(Pt_{1x2}N_{2x2}H_{6x2}Cl_{2x2}) = Pt_2N_4H_{12}Cl_4$

Since we have a complex cation and a complex anion, it is likely one Pt is with the cation, and the other is with the anion. Also N_4H_{12} is likely to be four NH_3. According to Table 15.2, Pt has a +2 charge and a coordination number of 4. Thus a likely candidate is:

$[Pt(NH_3)_4]^{2+}$ $[PtCl_4]^{2-}$ $\Rightarrow$ $[Pt(NH_3)_4][PtCl_4]$

A second candidate is:

$[Pt(NH_3)_3Cl]^{+}$ $[Pt(NH_3)Cl_3]^{-}$ $\Rightarrow$ $[Pt(NH_3)_3Cl][Pt(NH_3)Cl_3]$

Assume a 100 g sample.

$$20.25 \text{ g Cu} \times \frac{1 \text{ mol.}}{63.55 \text{ g}} = 0.3186 \text{ mol. Cu}$$

$$15.29 \text{ g C} \times \frac{1 \text{ mol.}}{12.01 \text{ g}} = 1.273 \text{ mol. C}$$

$$7.07 \text{ g H} \times \frac{1 \text{ mol.}}{1.008 \text{ g}} = 7.01 \text{ mol. H}$$

$$26.86 \text{ g N} \times \frac{1 \text{ mol.}}{14.01 \text{ g}} = 1.917 \text{ mol. N}$$

$$20.39 \text{ g O} \times \frac{1 \text{ mol.}}{16.00 \text{ g}} = 1.274 \text{ mol. O}$$

$$10.23 \text{ g S} \times \frac{1 \text{ mol.}}{32.07 \text{ g}} = 0.3190 \text{ mol. S}$$

$0.3186 / 0.3186 = 1$ mol. Cu
$1.273 / 0.3186 = 4$ mol. C
$7.07 / 0.3186 = 22$ mol. H
$1.917 / 0.3186 = 6$ mol. N
$1.274 / 0.3186 = 4$ mol. O
$0.3190 / 0.3186 = 1$ mol. S

Thus the simplest formula is: $CuC_4H_{22}N_6O_4S$.

b. To deduce the structural formula, one must determine the possible ligands that are present. The O_4S could be the sulfate anion. The -2 charge of the sulfate anion would balance the +2 charge that accompanies most copper cations, which would indicate that the remaining ligands are neutral. A ligand containing C, H and N mentioned in this chapter is ethylenediamine (*en*). If we subtract two *en* units from the simplest formula, we are left with H_6N_2, which is likely two NH_3. This gives: $[Cu(C_2H_8N_2)_2(NH_3)_2]SO_4$

NH₃ *trans*   NH₃ *cis*

57. *Refer to Section 15.3 and Problem 34 above.*

Calculate the energy per atom (or photon) in joules and then calculate the wavelength of light with that energy. The observed color will be the complement of the absorbed color.

$\Delta_0 = E = 55$ kcal/mol.

$$\frac{55\,\text{kcal}}{1\,\text{mol.}} \times \frac{4.184\,\text{kJ}}{1\,\text{kcal}} \times \frac{1000\,\text{J}}{1\,\text{kJ}} \times \frac{1\,\text{mol.}}{6.022 \times 10^{23}\,\text{photons}} = 3.8 \times 10^{-19}\,\text{J/photon}$$

$$E = \frac{hc}{\lambda} \quad \Rightarrow \quad 3.8 \times 10^{-19}\,\text{J} = \frac{(6.626 \times 10^{-34}\,\text{J} \cdot \text{s})(2.998 \times 10^{8}\,\text{m/s})}{\lambda}$$

$\lambda = 5.2 \times 10^{-7}$ m = 520 nm

This corresponds to green. If the absorbed color is green, the observed color is **red-violet**.

Chapter 16: Precipitation Equilibria

---

2. Refer to Section 16.1 and Example 16.1.

a. $AgCl(s) \rightleftharpoons Ag^+(aq) + Cl^-(aq)$ $\qquad$ $K_{sp} = [Ag^+][Cl^-]$

b. $Al_2(CO_3)_3(s) \rightleftharpoons 2Al^{3+}(aq) + 3CO_3^{2-}(aq)$ $\qquad$ $K_{sp} = [Al^{3+}]^2[CO_3^{2-}]^3$

c. $MnS_2(s) \rightleftharpoons Mn^{4+}(aq) + 2S^{2-}(aq)$ $\qquad$ $K_{sp} = [Mn^{4+}][S^{2-}]^2$

d. $Mg(OH)_2(s) \rightleftharpoons Mg^{2+}(aq) + 2OH^-(aq)$ $\qquad$ $K_{sp} = [Mg^{2+}][OH^-]^2$

---

4. Refer to Section 16.1 and Example 16.1.

The ions are those of the product side of the reaction, and the exponents in the K_{sp} expression are the coefficients in the equilibrium expression.

a. $CaCO_3(s) \rightleftharpoons Ca^{2+}(aq) + CO_3^{2-}(aq)$

b. $Co(OH)_3(s) \rightleftharpoons Co^{3+}(aq) + 3OH^-(aq)$

c. $Ag_2S(s) \rightleftharpoons 2Ag^+(aq) + S^{2-}(aq)$

d. $PbCl_2(s) \rightleftharpoons Pb^{2+}(aq) + 2Cl^-(aq)$

---

6. Refer to Section 16.1 and Example 16.2.

Write the balanced net ionic equation and the expression for K_{sp}. Solve for the missing concentration.

a. $Li_3PO_4(s) \rightleftharpoons 3Li^+(aq) + PO_4^{3-}(aq)$ $\qquad$ $K_{sp} = [Li^+]^3[PO_4^{3-}]$
$3.2 \times 10^{-9} = [Li^+]^3(7.5 \times 10^{-4})$
$4.3 \times 10^{-6} = [Li^+]^3$
$[Li^+] = 1.6 \times 10^{-2}\ M$

b. $AgNO_3(s) \rightleftharpoons Ag^+(aq) + NO_3^-(aq)$ $\qquad$ $K_{sp} = [Ag^+][NO_3^-]$
$6.0 \times 10^{-4} = (0.025)[NO_3^-]$
$[NO_3^-] = 2.4 \times 10^{-2}\ M$

c. $Sn(OH)_2(s) \rightleftharpoons Sn^{2+}(aq) + 2OH^-(aq)$ $\qquad$ $K_{sp} = [Sn^{2+}][OH^-]^2$
$pOH = 14.00 - 9.35 = 4.65$
$[OH^-] = 10^{-4.65} = 2.2 \times 10^{-5}\ M$
$1.4 \times 10^{-28} = [Sn^{2+}](2.2 \times 10^{-5})^2$
$[Sn^{2+}] = 2.9 \times 10^{-19}\ M$

8. Refer to Section 16.1, Table 16.1, and Example 16.3.

	K_{sp}
BaF_2	1.8×10^{-7}
$BaSO_4$	1.1×10^{-10}
$BaCrO_4$	1.2×10^{-10}

a. $K_{sp} = [Ba^{2+}][F^-]^2$
$1.8 \times 10^{-7} = (0.0034)[F^-]^2$
$[F^-]^2 = 5.3 \times 10^{-5}$
$[F^-] = 7.3 \times 10^{-3}\ M$

b. $K_{sp} = [Ba^{2+}][SO_4^{2-}]$
$1.1 \times 10^{-10} = (0.0034)[SO_4^{2-}]$
$[SO_4^{2-}] = 3.2 \times 10^{-8}\ M$

c. $K_{sp} = [Ba^{2+}][CrO_4^{2-}]$
$1.2 \times 10^{-10} = (0.0034)[CrO_4^{2-}]$
$[CrO_4^{2-}] = 3.5 \times 10^{-8}\ M$

10. Refer to Section 16.1 and Example 16.3.

a. A precipitate will begin to form when $Q = K_{sp}$.
$Cd(OH)_2(s) \rightleftharpoons Cd^{2+}(aq) + 2OH^-(aq)$ $\qquad K_{sp} = [Cd^{2+}][OH^-]^2$
$pOH = 14.00 - 9.62 = 4.38$
$[OH^-] = 10^{-4.38} = 4.2 \times 10^{-5}\ M$
$2.5 \times 10^{-14} = [Cd^{2+}](4.2 \times 10^{-5})^2$
$[Cd^{2+}] = 1.4 \times 10^{-5}\ M$

b. $2.5 \times 10^{-14} = (0.0013)[OH^-]^2$
$[OH^-]^2 = 1.9 \times 10^{-11}$
$[OH^-] = 4.4 \times 10^{-6}\ M$
$pOH = -\log(4.4 \times 10^{-6}) = 5.36$
$pH = 14.00 - 5.36 = 8.64$

c. $\%\ \text{left} = \dfrac{4.4 \times 10^{-6}\ M}{4.2 \times 10^{-5}\ M} \times 100 = 10\%$

12. Refer to Section 16.1 and Examples 16.3 and 16.4.

Calculate the molar concentrations of Pb^{2+} and CrO_4^{2-} in solution, then calculate Q.

$$\frac{0.50\,\text{mg Pb(NO}_3)_2}{1\,\text{L}} \times \frac{1\,\text{g}}{1000\,\text{mg}} \times \frac{1\,\text{mol. Pb(NO}_3)_2}{331.22\,\text{g}} \times \frac{1\,\text{mol. Pb}^{2+}}{1\,\text{mol. Pb(NO}_3)_2} = 1.5 \times 10^{-6}\ M\,Pb^{2+}$$

$$\frac{0.020\,\text{mg K}_2\text{CrO}_4}{1\,\text{L}} \times \frac{1\,\text{g}}{1000\,\text{mg}} \times \frac{1\,\text{mol. K}_2\text{CrO}_4}{194.2\,\text{g K}_2\text{CrO}_4} \times \frac{1\,\text{mol. CrO}_4{}^{2-}}{1\,\text{mol. K}_2\text{CrO}_4} = 1.0 \times 10^{-7}\,M\,\text{CrO}_4{}^{2-}$$

$Q = [\text{Pb}^{2+}][\text{CrO}_4{}^{2-}] = (1.5 \times 10^{-6})(1.0 \times 10^{-7}) = 1.5 \times 10^{-13}$

$Q > K_{sp} (= 2 \times 10^{-14})$, therefore a **precipitate will form**.

To just start to precipitate:

$K_{sp} = [\text{Pb}^{2+}][\text{CrO}_4{}^{2-}] \implies 2 \times 10^{-14} = [\text{Pb}^{2+}](1.0 \times 10^{-7})$

$[\text{Pb}^{2+}] = 2 \times 10^{-7}\,M$

---

14. *Refer to Section 16.1 and Example 16.3.*

---

a. Calculate Q for Zn(OH)_2 and compare to K_{sp}

$\text{Zn(NO}_3)_2(aq) + 2\text{KOH}(aq) \rightarrow \text{Zn(OH)}_2(s) + 2\text{KNO}_3(aq)$

$\text{Zn(OH)}_2(s) \rightleftharpoons \text{Zn}^{2+}(aq) + 2\text{OH}^-(aq)$

$K_{sp} = [\text{Zn}^{2+}][\text{OH}^-]^2$

$\text{pOH} = 14.00 - 9.00 = 5.00$

$[\text{OH}^-]\,10^{-5.00} = 1.0 \times 10^{-5}$

$$0.0200\,\text{L} \times \frac{1.0 \times 10^{-5}\,\text{mol. OH}^-}{1\,\text{L}} = 2.0 \times 10^{-7}\,\text{mol. OH}^-$$

total volume $= 35.00\,\text{mL} + 20.0\,\text{mL} = 55.0\,\text{mL}$

$$[\text{OH}^-] = \frac{2.0 \times 10^{-7}\,\text{mol. OH}^-}{0.0550\,\text{L}} = 3.6 \times 10^{-6}\,M\,\text{OH}^-$$

$$[\text{Zn}^{2+}] = 0.0350\,\text{L} \times \frac{0.061\,\text{mol. Zn}^{2+}}{1\,\text{L}} \times \frac{1}{0.0550\,\text{L}} = 3.9 \times 10^{-2}\,M\,\text{Zn}^{2+}$$

$Q = [\text{Zn}^{2+}][\text{OH}^-]^2 = (3.9 \times 10^{-2})(3.6 \times 10^{-6})^2 = 5.0 \times 10^{-13}$

$Q > K_{sp} (= 4 \times 10^{-17})$, therefore a **precipitate will form**.

b. NO_3^- is a spectator ion, so all the NO_3^- introduced initially will remain.

$$[\text{NO}_3^-] = \frac{3.9 \times 10^{-2}\,\text{mol. Zn(NO}_3)_2}{1\,\text{L}} \times \frac{2\,\text{mol. NO}_3^-}{1\,\text{mol. Zn(NO}_3)_2} = 7.8 \times 10^{-2}\,M\,\text{NO}_3^-$$

$[\text{K}^+] = [\text{OH}^-] = 3.6 \times 10^{-6}\,M$

Since $[\text{Zn}^{2+}]$ is more than twice that of $[\text{OH}^-]$, it is safe to assume (given the very small K_{sp}) that all the OH$^-$ is removed from solution as $\text{Zn(OH)}_2(s)$ and that the only Zn^{2+} left in solution is the excess Zn^{2+}.

$$\frac{3.6\times10^{-6}\text{ mol. OH}^-}{1\text{ L}}\times\frac{1\text{ mol. Zn}^{2+}}{2\text{ mol. OH}^-}=1.8\times10^{-6}\ M\ \text{Zn}^{2+}\text{ (precipitated as Zn(OH)}_2)$$

$[\text{Zn}^{2+}] = 3.9 \times 10^{-2}\ M$ (initial) - $1.8 \times 10^{-6}\ M$ (precipitated)
$[\text{Zn}^{2+}] = 3.9 \times 10^{-2}\ M$ (remaining in solution)
The $[\text{OH}^-]$ is calculated from the K_{sp} and $[\text{Zn}^{2+}]$.

$K_{sp} = [\text{Zn}^{2+}][\text{OH}^-]^2 \Rightarrow 4 \times 10^{-17} = (3.9 \times 10^{-2})[\text{OH}^-]^2$

$[\text{OH}^-]^2 = 1 \times 10^{-15}$
$[\text{OH}^-]^2 = 3 \times 10^{-8}\ M$
$\text{pOH} = -\log(3 \times 10^{-8}) = 7.5$
$\text{pH} = 14.00 - 7.5 = 6.5$

---

16. Refer to Section 16.1.

Write the balanced net ionic equation, calculate the concentration of the ions and then determine K_{sp}.

$$\text{Ag}_2\text{SO}_4(s) \rightleftharpoons 2\text{Ag}^+(aq) + \text{SO}_4^{2-}(aq)$$

$$\frac{1.2\text{ g Ag}_2\text{SO}_4}{0.2500\text{ L}}\times\frac{1\text{ mol. Ag}_2\text{SO}_4}{311.87\text{ g Ag}_2\text{SO}_4}=0.015\ M\ \text{Ag}_2\text{SO}_4$$

$$\frac{1.2\text{ g Ag}_2\text{SO}_4}{0.2500\text{ L}}\times\frac{1\text{ mol. Ag}_2\text{SO}_4}{311.87\text{ g Ag}_2\text{SO}_4}\times\frac{2\text{ mol. Ag}^+}{1\text{ mol. Ag}_2\text{SO}_4}=0.031\ M\ \text{Ag}^+$$

$$\frac{1.2\text{ g Ag}_2\text{SO}_4}{0.2500\text{ L}}\times\frac{1\text{ mol. Ag}_2\text{SO}_4}{311.87\text{ g Ag}_2\text{SO}_4}\times\frac{1\text{ mol. SO}_4^{2-}}{1\text{ mol. Ag}_2\text{SO}_4}=0.015\ M\ \text{SO}_4^{2-}$$

$$K_{sp} = [\text{Ag}^+]^2[\text{SO}_4^{2-}] = (0.031)^2(0.015) = 1.4 \times 10^{-5}$$

---

18. Refer to Section 16.1, Examples 16.5 and 16.6 and Table 16.1.

Write the balanced net ionic equation, then set up the expression for K_{sp}. Calculate the concentration of Mg^{2+}, and from that the mass of Mg(OH)_2.

$$\text{Mg(OH)}_2(s) \rightleftharpoons \text{Mg}^{2+}(aq) + 2\text{OH}^-(aq) \qquad\qquad K_{sp} = [\text{Mg}^{2+}][\text{OH}^-]^2$$

a. $K_{sp} = 6 \times 10^{-12} = (x)(2x)^2 = 4x^3$
   $x = 1 \times 10^{-4}\ M\ \text{Mg}^{2+}$

$$\frac{1\times10^{-4}\text{ mol. Mg}^{2+}}{1\text{ L}}\times\frac{1\text{ mol. Mg(OH)}_2}{1\text{ mol. Mg}^{2+}}\times\frac{58.3\text{ g Mg(OH)}_2}{1\text{ mol. Mg(OH)}_2}=6\times10^{-3}\text{ g/L Mg(OH)}_2$$

b. $[\text{OH}^-] = 2(0.041)\ M = 0.082\ M$ (We can ignore the small contribution from Mg(OH)_2)

   $K_{sp} = 6 \times 10^{-12} = (x)(0.082)^2$
   $x = 9 \times 10^{-10}\ M\ \text{Mg}^{2+}$

$$\frac{9 \times 10^{-10} \text{ mol. Mg}^{2+}}{1 \text{ L}} \times \frac{1 \text{ mol. Mg(OH)}_2}{1 \text{ mol. Mg}^{2+}} \times \frac{58.3 \text{ g Mg(OH)}_2}{1 \text{ mol. Mg(OH)}_2} = 5 \times 10^{-8} \text{ g/L Mg(OH)}_2$$

c. $[Mg^{2+}] = 0.0050 \, M$ (We can ignore the small contribution from $Mg(OH)_2$)
$K_{sp} = 6 \times 10^{-12} = (0.0050)(2x)^2$
$x^2 = 3 \times 10^{-10} \, M \, OH^-$
$x = 2 \times 10^{-5} \, M \, OH^-$

$$\frac{2 \times 10^{-5} \text{ mol. OH}^-}{1 \text{ L}} \times \frac{1 \text{ mol. Mg(OH)}_2}{2 \text{ mol. (OH)}^-} \times \frac{58.3 \text{ g Mg(OH)}_2}{1 \text{ mol. Mg(OH)}_2} = 5 \times 10^{-4} \text{ g/L Mg(OH)}_2$$

---

20. Refer to Section 16.1, Example 16.5 and Table 16.1.

$AgC_2H_3O_2(aq) \rightleftharpoons Ag^+(aq) + C_2H_3O_2^-(aq)$ $K_{sp} = [Ag^+][C_2H_3O_2^-]$

$$\frac{10.0 \text{ g AgC}_2\text{H}_3\text{O}_2}{1.0 \text{ L}} \times \frac{1 \text{ mol. AgC}_2\text{H}_3\text{O}_2}{166.94 \text{ g AgC}_2\text{H}_3\text{O}_2} = 0.060 \, M \, AgC_2H_3O_2$$

a. at 25°C: $K_{sp} = [Ag^+][C_2H_3O_2^-] \Rightarrow 1.9 \times 10^{-3} = x^2$
$x = 0.044 \, M$
$0.060 \, M > 0.044 \, M$, Thus, not all the $AgC_2H_3O_2$ will dissolve.

b. at 25°C: $K_{sp} = [Ag^+][C_2H_3O_2^-] \Rightarrow 2 \times 10^{-2} = x^2$
$x = 0.141 \, M$
$0.141 \, M > 0.060 \, M$, Thus, all the $AgC_2H_3O_2$ will dissolve.

---

22. Refer to Section 16.2 and Example 16.6.

---

for $PbSO_4$: $K_{sp} = [Pb^{2+}][SO_4^{2-}] = 1.8 \times 10^{-8}$
for $Pb(OH)_2$: $K_{sp} = [Pb^{2+}][OH^-]^2 = 2.8 \times 10^{-16}$

a. $[Pb^{2+}]$ needed to precipitate $PbSO_4$:
$1.8 \times 10^{-8} = [Pb^{2+}](0.020)$
$[Pb^{2+}] = 9.0 \times 10^{-7} \, M$

$[Pb^{2+}]$ needed to precipitate $Pb(OH)_2$:
$2.8 \times 10^{-16} = [Pb^{2+}](0.020)^2$
$[Pb^{2+}] = 7.0 \times 10^{-13} \, M$

It takes less Pb^{2+} to precipitate $Pb(OH)_2$ than to precipitate $PbSO_4$. $Pb(OH)_2$ will precipitate 1st.

b. When $PbSO_4$ first starts to precipitate, $[Pb^{2+}] = 9.0 \times 10^{-7}$
$(9.0 \times 10^{-7})[OH^-]^2 = 2.8 \times 10^{-16}$
$[OH^-]^2 = 3.1 \times 10^{-10}$
$[OH^-] = 1.8 \times 10^{-5} \, M$
$pOH = -\log(1.8 \times 10^{-5}) = 4.74$
$pH = 14.00 - 4.74 = 9.26$

24. Refer to Section 16.2.

$$\frac{0.925\,g\,AgNO_3}{0.375\,L} \times \frac{1\,mol.\,AgNO_3}{169.91\,g\,AgNO_3} \times \frac{1\,mol.\,Ag^+}{1\,mol.\,AgNO_3} = 0.0145\,M\,Ag^+$$

$$\frac{6.25\,g\,Mg(NO_3)_2}{0.375\,L} \times \frac{1\,mol.\,Mg(NO_3)_2}{148.32\,g\,Mg(NO_3)_2} \times \frac{1\,mol.\,Mg^{2+}}{1\,mol.\,Mg(NO_3)_2} = 0.112\,M\,Mg^{2+}$$

a. for Ag_2CO_3: $K_{sp} = [Ag^+]^2[CO_3^{2-}] = 8 \times 10^{-12}$
    $8 \times 10^{-12} = (0.0145)^2[CO_3^{2-}]$
    $[CO_3^{2-}] = 4 \times 10^{-8}\,M$

    for $MgCO_3$: $K_{sp} = [Mg^{2+}][CO_3^{2-}] = 6.8 \times 10^{-6}$
    $6.8 \times 10^{-6} = (0.112)[CO_3^{2-}]$
    $[CO_3^{2-}] = 6.1 \times 10^{-5}\,M$

    Ag_2CO_3 requires less carbonate, thus, it precipitates first.

b. When Ag_2CO_3 starts to precipitate, $[CO_3^{2-}] = 4 \times 10^{-8}\,M$

---

26. Refer to Section 16.2.

$$Ba(NO_3)_2(aq) + Na_2CO_3(aq) \rightleftharpoons BaCO_3(s) + 2\,NaNO_3(aq)$$
$$Ca(NO_3)_2(aq) + Na_2CO_3(aq) \rightleftharpoons CaCO_3(s) + 2\,NaNO_3(aq)$$

$$\frac{95\,mg\,Ba(NO_3)_2}{0.5000\,L} \times \frac{1\,g}{1000\,mg} \times \frac{1\,mol.\,Ba(NO_3)_2}{261.35\,g} \times \frac{1\,mol.\,Ba^{2+}}{1\,mol.\,Ba(NO_3)_2} = 7.3 \times 10^{-4}\,M\,Ba^{2+}$$

$$\frac{95\,mg\,Ca(NO_3)_2}{0.5000\,L} \times \frac{1\,g}{1000\,mg} \times \frac{1\,mol.\,Ca(NO_3)_2}{164.1\,g} \times \frac{1\,mol.\,Ca^{2+}}{1\,mol.\,Ca(NO_3)_2} = 1.2 \times 10^{-3}\,M\,Ca^{2+}$$

$$\frac{100.0\,mg\,Na_2CO_3}{0.5000\,L} \times \frac{1\,g}{1000\,mg} \times \frac{1\,mol.\,Na_2CO_3}{105.99\,g} \times \frac{1\,mol.\,CO_3^{2-}}{1\,mol.\,Na_2CO_3} = 1.89 \times 10^{-3}\,M\,CO_3^{2-}$$

$BaCO_3$: $K_{sp} = [Ba^{2+}][CO_3^{2-}] = 2.6 \times 10^{-9}$
$Q = (7.3 \times 10^{-4})(1.89 \times 10^{-3}) = 1.4 \times 10^{-6}$
$Q > K$, Thus, $BaCO_3$ will precipitate.

$CaCO_3$: $K_{sp} = [Ca^{2+}][CO_3^{2-}] = 4.9 \times 10^{-9}$
$Q = (1.2 \times 10^{-3})(1.89 \times 10^{-3}) = 2.3 \times 10^{-6}$
$Q > K$, Thus, $CaCO_3$ will precipitate.

There will be a precipitate of both $CaCO_3$ and $BaCO_3$.

28. Refer to Section 16.2 and Example 16.8.

a. $CaCO_3(s) + 2H^+(aq) \rightarrow Ca^{2+}(aq) + H_2CO_3(aq)$
b. $NiS(s) + 2H^+(aq) \rightarrow Ni^{2+}(aq) + H_2S(aq)$
c. $Al(OH)_3(s) + 3H^+(aq) \rightarrow Al^{3+}(aq) + 3H_2O(l)$
d. $Sb(OH)_4^-(aq) + 4H^+(aq) \rightarrow Sb^{3+}(aq) + 4H_2O(l)$
e. no reaction

30. Refer to Section 16.2, Table 16.2, and Example 16.9.

a. $Cu(OH)_2(s) + 4NH_3(aq) \rightarrow Cu(NH_3)_4^{2+}(aq) + 2OH^-(aq)$
b. $Cd^{2+}(aq) + 4NH_3(aq) \rightarrow Cd(NH_3)_4^{2+}(aq)$
c. The lead must be combining with an anion to give an electrically neutral precipitate. A likely candidate is OH^- formed by the reaction of ammonia and water.
   $Pb^{2+}(aq) + 2NH_3(aq) + 2H_2O(l) \rightarrow Pb(OH)_2(s) + 2NH_4^+(aq)$

32. Refer to Section 16.2 and Example 16.9.

Determine the products from Tables 16.1 (for part a) and 16.2 (for parts b and c). From the product, work backwards to balance the net ionic equation.

a. $Ni^{2+}(aq) + 2OH^-(aq) \rightarrow Ni(OH)_2(s)$
b. $Sn^{4+}(aq) + 6OH^-(aq) \rightarrow Sn(OH)_6^{2-}(aq)$
c. $Al(OH)_3(s) + OH^-(aq) \rightarrow Al(OH)_4^-(aq)$

34. Refer to Section 16.2 and Example 16.8.

a. $Co(OH)_2(s) + 2H^+(aq) \rightleftharpoons Co^{2+}(aq) + 2H_2O(l)$
b. Determine the equilibrium equations which sum to give the equation in part a.

$Co(OH)_2(s) \rightleftharpoons Co^{2+}(aq) + 2OH^-(aq)$	$K_{sp} = 2 \times 10^{-16}$
$\underline{2H^+(aq) + 2OH^-(aq) \rightleftharpoons 2H_2O(l)}$	$\underline{1/(K_w)^2 = 1 \times 10^{28}}$
$Co(OH)_2(s) + 2H^+(aq) \rightleftharpoons Co^{2+}(aq) + 2H_2O(l)$	

$$K = (2 \times 10^{-16})(1 \times 10^{28}) = 2 \times 10^{12}$$

36. Refer to Section 16.2 and Example 16.9.

a. Calculate K for the reaction, recalling that when you add two equilibrium equations, you must multiply the equilibrium constants.

$$AgCl(s) \rightleftharpoons Ag^+(aq) + Cl^-(aq) \qquad K_{sp} = 1.8 \times 10^{-10}$$
$$I^-(aq) + Ag^+(aq) \rightleftharpoons AgI(s) \qquad 1/K_{sp} = 1 \times 10^{16}$$
$$\overline{AgCl(s) + I^-(aq) \rightleftharpoons Cl^-(aq) + AgI(s)} \qquad K = K_{sp} \times 1/K_{sp} = 2 \times 10^6$$

b. Saturated AgCl has some Ag^+ ions in solution. Since AgI has a lower K_{sp} than AgCl, the addition of NaI will cause AgI to precipitate.

---

38. *Refer to Section 16.2 and Example 16.9.*

a.
$$Cu(OH)_2(s) \rightleftharpoons Cu^{2+}(aq) + 2OH^-(aq) \qquad K_{sp} = 2 \times 10^{-19}$$
$$Cu^{2+}(aq) + 4NH_3(aq) \rightleftharpoons Cu(NH_3)_4^{2+}(aq) \qquad K_f = 2 \times 10^{12}$$
$$\overline{Cu(OH)_2(s) + 4NH_3(aq) \rightleftharpoons Cu(NH_3)_4^{2+}(aq) + 2OH^-(aq)} \qquad K = K_{sp} \times K_f = 4 \times 10^{-7}$$

b. $K = \dfrac{[Cu(NH_3)_4^{2+}][OH^-]^2}{[NH_3]^4}$ and $2[Cu(NH_3)_4^{2+}] = [OH^-]$

$$K = 4 \times 10^{-7} = \frac{(x)(2x)^2}{(4.5\,M)^4} = \frac{4x^3}{410}$$

$$x^3 = 4 \times 10^{-5}$$
$$x = 3 \times 10^{-2}\,M$$

---

40. *Refer to Sections 16.1 and 16.2 and Example 16.9.*

a. Calculate K for the reaction. Then calculate $[Zn(OH)_4^{2-}]$ from the equilibrium expression. Bear in mind that solids do not factor into the equation.

$$Zn(OH)_2(s) \rightleftharpoons Zn^{2+}(aq) + 2OH^-(aq) \qquad K_{sp} = 4 \times 10^{-17}$$
$$Zn^{2+}(aq) + 4OH^-(aq) \rightleftharpoons [Zn(OH)_4]^{2-}(aq) \qquad K_f = 3 \times 10^{14}$$
$$\overline{Zn(OH)_2(s) + 2OH^-(aq) \rightleftharpoons [Zn(OH)_4]^{2-}(aq)} \qquad K = K_{sp} \times K_f = 1.2 \times 10^{-2}$$

$$K = \frac{[Zn(OH)_4^{2-}]}{[OH^-]^2} \Rightarrow 1.2 \times 10^{-2} = \frac{[Zn(OH)_4^{2-}]}{(0.1)^2} \Rightarrow [Zn(OH)_4^{2-}] = 0.0001M$$

One mol. of $[Zn(OH)_4]^{2-}$ is formed for each mole of $Zn(OH)_2$ that dissolves, thus, the molar solubility is 0.0001 M.

b. Molar solubility in pure water: $K_{sp} = 4 \times 10^{-17} = (x)(2x)^2 = 4x^3$
    x = molar solubility = 2×10^{-6}

a. Sum equations to get the goal equation and multiply K values to get the overall K.

$$Fe(OH)_2(s) \rightleftharpoons Fe^{2+}(aq) + 2OH^-(aq) \qquad\qquad K_{sp} = 5 \times 10^{-17}$$
$$Fe^{2+}(aq) + 6CN^-(aq) \rightleftharpoons [Fe(CN)_6]^{4-}(aq) \qquad\qquad K_f = 4 \times 10^{45}$$
$$\overline{Fe(OH)_2(s) + 6CN^-(aq) \rightleftharpoons [Fe(CN)_6]^{4-}(aq) + 2OH^-(aq) \qquad K = K_{sp} \times K_f = 2 \times 10^{29}}$$

b. Calculate concentrations of $[Fe(CN)_6]^{4-}$ and of OH^-. Put these into the equilibrium expression and solve for $[CN^-]$.

$$\frac{10.0\,g\,Fe(OH)_2}{1.00\,L} \times \frac{1\,mol.\,Fe(OH)_2}{89.87\,g} \times \frac{1\,mol.\,[Fe(CN)_6]^{4-}}{1\,mol.\,Fe(OH)_2} = 0.111\,M\,[Fe(CN)_6]^{4-}$$

$$\frac{10.0\,g\,Fe(OH)_2}{1.00\,L} \times \frac{1\,mol.\,Fe(OH)_2}{89.87\,g} \times \frac{2\,mol.\,OH^-}{1\,mol.\,Fe(OH)_2} = 0.222\,M\,OH^-$$

$$K = \frac{[Fe(CN)_6^{4-}][OH^-]^2}{[CN^-]^6} \Rightarrow 2 \times 10^{29} = \frac{(0.111)(0.222)^2}{[CN^-]^6}$$

$$[CN^-] = 5 \times 10^{-6}\,M$$

---

44. *Refer to Chapter 16.*

a. $Pb(NO_3)_2$ is soluble in water, thus addition of $Pb(NO_3)_2$ is addition of a product, Pb^{2+}. Adding a product shifts the equilibrium towards the reactants. More $PbCl_2$ precipitates.

b. ΔH is positive, thus the reaction is endothermic. One can view this as heat being a reactant. Increasing the temperature for an endothermic reaction is like adding a reactant; the equilibrium is shifted to form more products. Thus, more $PbCl_2$ will dissolve.

c. Ag^+ will react with Cl^- causing the Cl^- to be removed as $AgCl$ precipitates. When a product is removed, the equilibrium will shift to form more products, thus more $PbCl_2$ will dissolve.

d. HCl (a strong acid) will dissociate to form H^+ and Cl^-. Since Cl^- is a product, adding HCl is adding a product. Adding a product shifts the equilibrium towards the reactants. More $PbCl_2$ precipitates.

---

46. *Refer to Sections 16.1 and 13.5.*

a. $$\frac{17.5\,g\,NH_4Cl}{1.50\,L} \times \frac{1\,mol.\,NH_4Cl}{53.49\,g\,NH_4Cl} = 0.218\,M\,NH_4Cl$$

	$NH_3(aq)$	+ $H_2O(l)$	$\rightleftharpoons$	$NH_4^+(aq)$	+ $OH^-(aq)$
$[\ ]_0$	3.75	---		0.218	0
$\Delta[\ ]$	$-x$	---		$+x$	$+x$
$[\ ]_{eq}$	$3.75 - x$	---		$0.218 + x$	x

$$K_b = 1.8 \times 10^{-5} = \frac{[NH_4^+][OH^-]}{[NH_3]} = \frac{(0.218 + x)(x)}{(3.75 - x)}$$

Assuming the 5% rule is valid for both $[NH_4^+]$ and $[OH^-]$, then

$$K_b = 1.8 \times 10^{-5} = \frac{(0.218)(x)}{3.75}$$

$$x = [OH^-] = 3.1 \times 10^{-4}\ M$$

b. $Mg^{2+}(aq) + 2OH^-(aq) \rightleftharpoons Mg(OH)_2(s)$ $\qquad K_{sp} = 6 \times 10^{-12}$

$$[Mg^{2+}] = \frac{5.0\ g\ MgCl_2}{1.50\ L} \times \frac{1\ mol.\ MgCl_2}{95.2\ g\ MgCl_2} \times \frac{1\ mol.\ Mg^{2+}}{1\ mol.\ MgCl_2} = 0.0350\ M\ Mg^{2+}$$

$Q = [Mg^{2+}][OH^-]^2 = (0.0350)(3.1 \times 10^{-4})^2 = 3.4 \times 10^{-9}$

$K_{sp} < Q$, Therefore a **precipitate will form**.

c. $K_{sp} = [Mg^{2+}][OH^-]^2$

$6 \times 10^{-12} = [Mg^{2+}](3.1 \times 10^{-4})^2$

$[Mg^{2+}] = 6 \times 10^{-5}\ M$

---

48. Refer to Section 16.1.

---

at 95°C: $K_{sp} = [Ca^{2+}][C_2O_4^{2-}] \Rightarrow 1 \times 10^{-8} = x^2 \Rightarrow x = 1 \times 10^{-4}\ M$
at 25°C: $K_{sp} = [Ca^{2+}][C_2O_4^{2-}] \Rightarrow 4 \times 10^{-9} = x^2 \Rightarrow x = 6 \times 10^{-5}\ M$

difference $= 1 \times 10^{-4} - 6 \times 10^{-5} = 4 \times 10^{-5}\ M$

$$0.500\ L \times \frac{4 \times 10^{-5}\ mol.}{L} \times \frac{128\ g}{1\ mol.} \times \frac{1000\ mg}{1\ g} = 2\ mg$$

Since the top box in the problem represents a saturated solution, we know that four moles of MX_2 will dissolve in one liter. For the subsequent boxes, dissolve up to four MX_2's, and leave any remainder undissolved.

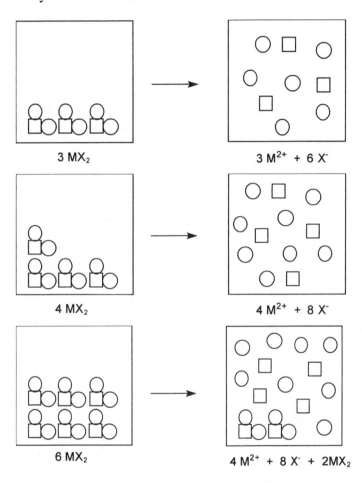

3 MX_2 3 M^{2+} + 6 X^-

4 MX_2 4 M^{2+} + 8 X^-

6 MX_2 4 M^{2+} + 8 X^- + 2MX_2

a. **True** If the substance is insoluble it means that the equilibrium lies towards reactants and there will be more reactants than products at equilibrium making the ratio of products to reactants less than one.

b. **False** If more $PbCl_2$ dissolves at a higher temperature than at a lower temperature it means that heat is a reactant (since adding heat causes dissolution). Thus, the reaction is endothermic.

c. **False** $Cu(s)$ is not an ion in the dissociation of $Cu(OH)_2$.
$$Cu(OH)_2(s) \rightleftharpoons Cu^{2+}(aq) + 2OH^-(aq)$$

Consider the equilibrium equations with "heat" as a product (exothermic) or reactant (endothermic) and apply Le Chatelier's principle.

exothermic: $M(s) \rightleftharpoons M(aq) + heat$
endothermic: $M(s) + heat \rightleftharpoons M(aq)$

In the former case, adding heat to the system would shift the reaction to the left (more reactants), which would produce more undissolved material. This corresponds to a decrease in solubility with increasing temperature.
In the latter case, adding heat to the system would shift the reaction to the right (more products), which would produce more dissolved material. This corresponds to an increase in solubility with increasing temperature.
Dissolving this material is an endothermic process.

Calculate the solubility of $Mg(OH)_2$ in water using K_{sp}. Then calculate K for the reaction given. Using K, calculate the solubility of $Mg(OH)_2$ in 0.2 M NH_4Cl and compare the two solubilities.

Solubility in water:
$$K_{sp} = 6 \times 10^{-12} = [Mg^{2+}][OH^-]^2 = (x)(2x)^2 = 4x^3$$
$$x^3 = 1.5 \times 10^{-12}$$
$$x = 1 \times 10^{-4}\ M$$

Solubility in 0.2 M NH_4Cl:

$Mg(OH)_2(s) \rightleftharpoons Mg^{2+}(aq) + 2OH^-(aq)$	$K_{sp} = 6 \times 10^{-12}$
$2[\ NH_4^+(aq) \rightleftharpoons H^+(aq) + NH_3(aq)\ ]$	$K_1 = (K_a)^2 = 3.1 \times 10^{-19}$
$2[\ H^+(aq) + OH^-(aq) \rightleftharpoons H_2O(l)\ ]$	$K_2 = (1/K_w)^2 = 1 \times 10^{28}$
$Mg(OH)_2(s) + 2NH_4^+(aq) \rightleftharpoons Mg^{2+}(aq) + 2NH_3(aq) + 2H_2O(l)$	$K = K_{sp} \times K_1 \times K_2 = 0.02$

$$K = \frac{[Mg^{2+}][NH_3]^2}{[NH_4^+]^2} \qquad \text{set } [Mg^{2+}] = x, \text{ then } [NH_3] = 2x \text{ and } [NH_4^+] = 0.2 - x$$

$$K = 0.02 = \frac{(x)(2x)^2}{(0.2 - x)^2} \quad \Rightarrow \quad 0.0008 - 0.008\,x + x^2 = 4x^3$$

Using successive approximations: $\left(x \approx \sqrt[3]{0.0008/4}\right)$
$$x = 0.06\ M$$

$Mg(OH)_2$ is approximately 600 times more soluble in 0.2 M NH_4Cl than in "pure" water.

57. *Refer to Section 16.2 and Chapter 14.*

Calculate K for the given reaction. Determine $[H^+]$ using the equation for buffers; with that, calculate $[Ca^{2+}]$.

$CaF_2(s) \rightleftharpoons Ca^{2+}(aq) + 2F^-(aq)$	$K_{sp} = 1.5 \times 10^{-10}$
$2[\,H^+(aq) + F^-(aq) \rightleftharpoons HF(aq)\,]$	$K_1 = (1/K_a)^2 = 2.1 \times 10^6$
$CaF_2(s) + 2H^+(aq) \rightleftharpoons Ca^{2+}(aq) + 2HF(aq)$	$K = K_{sp} \times K_1 = 3.2 \times 10^{-4}$

$$[H^+] = K_a \times \frac{[HCHO_2]}{[CHO_2^-]} = 1.9 \times 10^{-4} \times \frac{0.30}{0.20} = 2.9 \times 10^{-4}\ M$$

$$K = \frac{[Ca^{2+}][HF]^2}{[H^+]^2} \Rightarrow 3.2 \times 10^{-4} = \frac{(x)(2x)^2}{(2.9 \times 10^{-4})^2} = \frac{4x^3}{8.4 \times 10^{-8}}$$

$x^3 = 6.8 \times 10^{-12}$
$x = 1.9 \times 10^{-4}\ M$

58. *Refer to Section 16.1.*

Since AgI has the smaller K_{sp}, it will precipitate first. Thereafter, this is a common ion problem. Calculate the $[Ag^+]$ in solution at the point at which AgCl will just begin to precipitate. Using that concentration and K_{sp} (AgI), calculate $[I^-]$.

K_{sp} (AgCl) $= 1.8 \times 10^{-10} = [Ag^+][Cl^-] = (x)(0.020\ M)$
$x = 9.0 \times 10^{-9}\ M\ Ag^+$

K_{sp} (AgI) $= 1 \times 10^{-16} = [Ag^+][I^-] = (9.0 \times 10^{-9}\ M)(x)$
$x = 1 \times 10^{-8}\ M\ I^-$

59. *Refer to Section 16.1.*

a. $K_{sp} = 6 \times 10^{-12} = [Mg^{2+}][OH^-]^2 = (0.056)(x)^2$
$x^2 = 1 \times 10^{-10}$
$x = 1 \times 10^{-5}\ M$

b. Na^+: No. NaOH is extremely soluble.

Ca^{2+}: $Q = [Ca^{2+}][OH^-]^2 = (0.01)(1 \times 10^{-5})^2 = 1 \times 10^{-12}$
$Q < K_{sp}$, therefore **Ca^{2+} does not precipitate.**

Al^{3+}: $Q = [Al^{3+}][OH^-]^3 = (4 \times 10^{-7})(1 \times 10^{-5})^3 = 4 \times 10^{-22}$
$Q > K_{sp}$, therefore **Al^{3+} does precipitate.**

Fe^{3+}: $Q = [Fe^{3+}][OH^-]^3 = (2 \times 10^{-7})(1 \times 10^{-5})^3 = 2 \times 10^{-22}$
$Q > K_{sp}$, therefore **Fe^{3+} does precipitate.**

c. Calculate [OH⁻] when half the Mg^{2+} is precipitated, then calculate $[Al^{3+}]$ and $[Fe^{3+}]$ in solution and the percent precipitated.

$$K_{sp} = 6 \times 10^{-12} = [Mg^{2+}][OH^-]^2 = (0.028)(x)^2$$
$$x = 1.5 \times 10^{-5} \, M$$

$$K_{sp} (Al(OH)_3) = [Al^{3+}][OH^-]^3 \Rightarrow 2 \times 10^{-31} = (x)(1.5 \times 10^{-5})^3$$
$$x = 6 \times 10^{-17} \, M$$

$$\text{percent in solution} = \frac{6 \times 10^{-17}}{4 \times 10^{-7}} \times 100\% = 1 \times 10^{-8} \, \%$$

Essentially all the Al^{3+} will have precipitated.

$$K_{sp} (Fe(OH)_3) = [Fe^{3+}][OH^-]^3 \Rightarrow 3 \times 10^{-39} = (x)(1.5 \times 10^{-5})^3$$
$$x = 9 \times 10^{-25} \, M$$

$$\text{percent in solution} = \frac{9 \times 10^{-25}}{2 \times 10^{-7}} \times 100\% = 4 \times 10^{-16} \, \%$$

Essentially all the Fe^{3+} will have precipitated.

d. Calculate and add up the masses of the species that precipitate.

$$0.028 \text{ mol. } Mg(OH)_2 \times \frac{58.32 \text{ g}}{1 \text{ mol.}} = 1.6 \text{ g}$$

$$4 \times 10^{-7} \text{ mol. } Al(OH)_3 \times \frac{70.00 \text{ g}}{1 \text{ mol.}} = 3 \times 10^{-5} \text{ g}$$

$$2 \times 10^{-7} \text{ mol. } Fe(OH)_3 \times \frac{106.9 \text{ g}}{1 \text{ mol.}} = 2 \times 10^{-5} \text{ g}$$

Total mass = 1.6 g.

---

60. Refer to Section 16.2 and Example 16.9.

---

a. $Zn(NH_3)_4^{2+}(aq) \rightleftharpoons Zn^{2+}(aq) + 4NH_3(aq)$ $K_1 = 1/K_f = 2.8 \times 10^{-9}$

    $Zn^{2+}(aq) + 4OH^-(aq) \rightleftharpoons Zn(OH)_4^{2-}(aq)$ $K_f = 3 \times 10^{14}$

$Zn(NH_3)_4^{2+}(aq) + 4OH^-(aq) \rightleftharpoons Zn(OH)_4^{2-}(aq) + 4NH_3(aq)$ $K = K_1 \times K_f = 8 \times 10^5$

b. Calculate the [OH⁻] in a 1.0 M NH_3 solution, then calculate the ratio of the two complexes using the equilibrium concentration expression.

$$NH_3(aq) + H_2O \rightleftharpoons NH_4^+(aq) + OH^-(aq)$$

$$K_b = \frac{[NH_4^+][OH^-]}{[NH_3]} \Rightarrow 1.8 \times 10^{-5} = \frac{x^2}{1.0}$$

$$x^2 = 1.8 \times 10^{-5}$$
$$x = 0.0042 \; M \; OH^-$$

$$K = \frac{[Zn(OH)_4^{2-}][NH_3]^4}{[Zn(NH_3)_4^{2+}][OH^-]^4} \quad \Rightarrow \quad 8 \times 10^5 = \frac{[Zn(OH)_4^{2-}]}{[Zn(NH_3)_4^{2+}]} \times \frac{(1.0)^4}{(0.0042)^4}$$

$$\frac{[Zn(OH)_4^{2-}]}{[Zn(NH_3)_4^{2+}]} = 2 \times 10^{-4}$$

$$\frac{[Zn(NH_3)_4^{2+}]}{[Zn(OH)_4^{2-}]} = 4 \times 10^3$$

61. Refer to Section 16.2 and Example 16.9.

$Ag(NH_3)_2^+(aq) \rightleftharpoons Ag^+(aq) + 2NH_3(aq)$	$K_1 = 1/K_f = 5.9 \times 10^{-8}$
$2[\, H^+(aq) + NH_3(aq) \rightleftharpoons NH_4^+(aq)\,]$	$K_2 = (1/K_a)^2 = 3.2 \times 10^{18}$
$Ag^+(aq) + Cl^-(aq) \rightleftharpoons AgCl(s)$	$K_3 = 1/K_{sp} = 5.6 \times 10^9$

$$Ag(NH_3)_2^+(aq) + 2H^+(aq) + Cl^-(aq) \rightleftharpoons AgCl(s) + NH_4^+(aq)$$

$$K = K_1 \times K_2 \times K_3$$
$$K = 1.0 \times 10^{21}$$

Chapter 17: Spontaneity of Reaction

---

2. *Refer to Section 17.1.*

    a. Non-spontaneous. Wax does not melt spontaneously at 25°C.

    b. Non-spontaneous. Apples do not sort themselves, this task requires human intervention.

    c. Non-spontaneous. Chemistry notes do not outline themselves, unfortunately.

    d. Spontaneous. Wind scattering leaves increases disorder, thus ΔS is positive and the process is spontaneous.

---

4. *Refer to Section 17.1.*

Consider physically what is happening in the chemical equation and determine if the process occurs spontaneously.

    a. Spontaneous. Solid carbon dioxide readily sublimes at room temperature.

    b. Non-spontaneous. Table salt does not melt at 25°C.

    c. Non-spontaneous. Table salt does not decompose at 25°C.

    d. Non-spontaneous. Carbon dioxide does not decompose at 25°C.

The only spontaneous process is (a).

---

6. *Refer to Section 17.2 and Example 17.1.*

    a. ΔS (-) Entropy generally decreases changing from a liquid to a solid (freezing).

    b. ΔS (-) Precipitating $PbCl_2$ from Pb^{2+} and $2Cl^-$ involves converting 3 moles of aqueous ions to 1 mole of solid, thus entropy decreases.

    c. ΔS (+) During the process of combustion, a liquid (candle wax) is converted to gases (water and CO_2). Entropy increases as liquids are converted to gases.

    d. ΔS (-) Weeding a garden results in a neater, more ordered garden.

---

8. *Refer to Section 17.2 and Example 17.1.*

    a. ΔS (?) Because the moles of gaseous product equals the moles of gaseous reactant, rhe sign of ΔS cannot be predicted.

    b. ΔS (+) There are twelve moles of gaseous products and no moles of gaseous reactants. Thus, the entropy increases.

c. ΔS (-)   One mole of liquid forms one mole of solid. Liquids have more entropy than solids, thus the entropy decreases.

d. ΔS (+)   Two moles of gas are converted to four moles of gas. The total number of moles of gases increases, therefore entropy increases.

---

10. Refer to Section 17.2 and Example 17.1.

a. ΔS (-)   In this reaction, the total moles of gas decreases from reactants to products, thus the entropy decreases.

b. ΔS (+)   In this reaction, the total moles of gas increases from reactants to products, thus the entropy increases.

c. ΔS (+)   In this reaction, the total moles of gas increases from reactants to products, thus the entropy increases.

---

12. Refer to Section 17.2 and Example 17.2.

$\Delta S° = \Sigma \, S°_{(products)} - \Sigma \, S°_{(reactants)}$

a. $\Delta S° = [(4 \text{ mol.})(240.0 \text{ J/mol.·K}) + (6 \text{ mol.})(188.7 \text{ J/mol.·K})]$
$- [(4 \text{ mol.})(192.3 \text{ J/mol.·K}) + (7 \text{ mol.})(205.0 \text{ J/mol.·K})]$
$\Delta S° = -112.0 \text{ J/K}$

b. $\Delta S° = [(1 \text{ mol.})(191.5 \text{ J/mol.·K}) + (4 \text{ mol.})(188.7 \text{ J/mol.·K})]$
$- [(2 \text{ mol.})(109.6 \text{ J/mol.·K}) + (1 \text{ mol.})(121.2 \text{ J/mol.·K})]$
$\Delta S° = 605.9 \text{ J/K}$

c. $\Delta S° = [(1 \text{ mol.})(213.6 \text{ J/mol.·K})] - [(1 \text{ mol.})(5.7 \text{ J/mol.·K}) + (1 \text{ mol.})(205.0 \text{ J/mol.·K})]$
$\Delta S° = 2.9 \text{ J/K}$

d. $\Delta S° = [(1 \text{ mol.})(201.7 \text{ J/mol.·K}) + (3 \text{ mol.})(186.8 \text{ J/mol.·K})]$
$- [(1 \text{ mol.})(186.2 \text{ J/mol.·K}) + (3 \text{ mol.})(223.0 \text{ J/mol.·K})]$
$\Delta S° = -93.1 \text{ J/K}$

---

14. Refer to Section 17.2 and Example 17.3.

$\Delta S° = \Sigma \, S°_{(products)} - \Sigma \, S°_{(reactants)}$

a. $\Delta S° = [(5 \text{ mol.})(146.4 \text{ J/mol.·K}) + (3 \text{ mol.})(-73.6 \text{ J/mol.·K}) + (2 \text{ mol.})(69.9 \text{ J/mol.·K})]$
$- [(5 \text{ mol.})(210.7 \text{ J/mol.·K}) + (3 \text{ mol.})(191.2 \text{ J/mol.·K}) + (4 \text{ mol.})(0.0 \text{ J/mol.·K})]$
$\Delta S° = -976.1 \text{ J/K}$

b. $\Delta S° = [(1 \text{ mol.})(23.8 \text{ J/mol.·K}) + (6 \text{ mol.})(-315.9 \text{ J/mol.·K}) + (4 \text{ mol.})(69.9 \text{ J/mol.·K})]$
$- [(6 \text{ mol.})(137.7 \text{ J/mol.·K}) + (1 \text{ mol.})(50.2 \text{ J/mol.·K}) + (8 \text{ mol.})(0.0 \text{ J/mol.·K})]$
$\Delta S° = -816.0 \text{ J/K}$

c. $\Delta S° = [(2 \text{ mol.})(213.6 \text{ J/mol.·K}) + (1 \text{ mol.})(188.7 \text{ J/mol.·K})]$
$- [(1 \text{ mol.})(200.8 \text{ J/mol.·K}) + (^5/_2 \text{ mol.})(205.0 \text{ J/mol.·K})]$
$\Delta S° = -97.1 \text{ J/K}$

---

16. Refer to Section 17.2 and Example 17.3.

$\Delta S° = \Sigma S°_{(products)} - \Sigma S°_{(reactants)}$

a. $\Delta S° = [(4 \text{ mol.})(69.9 \text{ J/mol.·K}) + (2 \text{ mol.})(210.7 \text{ J/mol.·K}) + (3 \text{ mol.})(31.8 \text{ J/mol.·K})]$
$- [(2 \text{ mol.})(155.6 \text{ J/mol.·K}) + (3 \text{ mol.})(205.7 \text{ J/mol.·K})]$
$\Delta S° = -131.9 \text{ J/K} = -0.1319 \text{ kJ/K}$

b. $\Delta S° = [(5 \text{ mol.})(56.5 \text{ J/mol.·K}) + (1 \text{ mol.})(90.4 \text{ J/mol.·K}) + (6 \text{ mol.})(0.0 \text{ J/mol.·K})]$
$- [(1 \text{ mol.})(364.5 \text{ J/mol.·K}) + (4 \text{ mol.})(69.9 \text{ J/mol.·K})]$
$\Delta S° = -271.2 \text{ J/K} = -0.2712 \text{ kJ/K}$

c. $\Delta S° = [(3 \text{ mol.})(-315.9 \text{ J/mol.·K}) + (1 \text{ mol.})(53.0 \text{ J/mol.·K}) + (2 \text{ mol.})(69.9 \text{ J/mol.·K})]$
$- [(1 \text{ mol.})(191.2 \text{ J/mol.·K}) + (3 \text{ mol.})(-137.7 \text{ J/mol.·K}) + (4 \text{ mol.})(0.0 \text{ J/mol.·K})]$
$\Delta S° = -533.0 \text{ J/K} = -0.5330 \text{ kJ/K}$

---

18. Refer to Section 17.3.

$\Delta G° = \Delta H° - T\Delta S°$ (Convert T to Kelvin and $\Delta S°$ to kJ/K)

a. $\Delta G° = 293 \text{ kJ} - (318 \text{ K})(0.695 \text{ kJ/K})$
$\Delta G° = 514 \text{ kJ}$

b. $\Delta G° = -1137 \text{ kJ} - (318 \text{ K})(0.496 \text{ kJ/K})$
$\Delta G° = -1295 \text{ kJ}$

c. $\Delta G° = -86.6 \text{ kJ} - (318 \text{ K})(-0.382 \text{ kJ/K})$
$\Delta G° = +34.9 \text{ kJ}$

---

20. Refer to Section 17.4 and Example 17.3.

Calculate $\Delta H°$ and $\Delta S°$ (see Problem 12 above) for the reactions. Then calculate $\Delta G°$. If $\Delta G°$ is negative, the reaction is spontaneous.

a. $\Delta H° = \Sigma \Delta H_f°_{(products)} - \Sigma \Delta H_f°_{(reactants)}$
$\Delta H° = [(4 \text{ mol.})(33.2 \text{ kJ/mol.}) + (6 \text{ mol.})(-241.8 \text{ kJ/mol.})]$
$- [(4 \text{ mol.})(-46.1 \text{ kJ/mol.}) + (7 \text{ mol.})(0.0 \text{ kJ/mol.})]$
$\Delta H° = -1133.6 \text{ kJ}$

$\Delta G° = \Delta H° - T\Delta S°$
$\Delta G° = -1133.6 \text{ kJ} - (415 \text{ K})(-0.1120 \text{ kJ/K})$
$\Delta G° = -1087.1 \text{ kJ};$ **Spontaneous.**

b. $\Delta H^\circ = \Sigma \Delta H_f^\circ {}_{(products)} - \Sigma \Delta H_f^\circ {}_{(reactants)}$

   $\Delta H^\circ = [(1 \text{ mol.})(0.0 \text{ kJ/mol.}) + (4 \text{ mol.})(-241.8 \text{ kJ/mol.})]$
   $\qquad - [(2 \text{ mol.})(-187.8 \text{ kJ/mol.}) + (1 \text{ mol.})(50.6 \text{ kJ/mol.})]$

   $\Delta H^\circ = -642.2 \text{ kJ}$

   $\Delta G^\circ = \Delta H^\circ - T\Delta S^\circ$
   $\Delta G^\circ = -642.2 \text{ kJ} - (415 \text{ K})(0.6059 \text{ kJ/K})$
   $\Delta G^\circ = -894 \text{ kJ};$ **Spontaneous.**

c. $\Delta H^\circ = \Sigma \Delta H_f^\circ {}_{(products)} - \Sigma \Delta H_f^\circ {}_{(reactants)}$

   $\Delta H^\circ = [(1 \text{ mol.})(-393.5 \text{ kJ/mol.})] - [(1 \text{ mol.})(0.0 \text{ kJ/mol.}) + (1 \text{ mol.})(0.0 \text{ kJ/mol.})]$

   $\Delta H^\circ = -393.5 \text{ kJ}$

   $\Delta G^\circ = \Delta H^\circ - T\Delta S^\circ$
   $\Delta G^\circ = -393.5 \text{ kJ} - (415 \text{ K})(0.0029 \text{ kJ/K})$
   $\Delta G^\circ = -394.7 \text{ kJ};$ **Spontaneous.**

d. $\Delta H^\circ = \Sigma \Delta H_f^\circ {}_{(products)} - \Sigma \Delta H_f^\circ {}_{(reactants)}$

   $\Delta H^\circ = [(1 \text{ mol.})(-134.5 \text{ kJ/mol.}) + (3 \text{ mol.})(-92.3 \text{ kJ/mol.})]$
   $\qquad - [(1 \text{ mol.})(-74.8 \text{ kJ/mol.}) + (3 \text{ mol.})(0.0 \text{ kJ/mol.})]$

   $\Delta H^\circ = -336.6 \text{ kJ}$

   $\Delta G^\circ = \Delta H^\circ - T\Delta S^\circ$
   $\Delta G^\circ = -336.6 \text{ kJ} - (415 \text{ K})(-0.0931 \text{ kJ/K})$
   $\Delta G^\circ = -298.0 \text{ kJ};$ **Spontaneous.**

---

22. Refer to Section 17.4.

$\Delta G^\circ = \Sigma \Delta G_f^\circ {}_{(products)} - \Sigma \Delta G_f^\circ {}_{(reactants)}$

a. $\Delta G^\circ = [(5 \text{ mol.})(-108.7 \text{ kJ/mol.}) + (3 \text{ mol.})(-228.1 \text{ kJ/mol.}) + (2 \text{ mol.})(-237.2 \text{ kJ/mol.})]$
   $\qquad - [(5 \text{ mol.})(86.6 \text{ kJ/mol.}) + (3 \text{ mol.})(-447.2 \text{ kJ/mol.}) + (4 \text{ mol.})(0.0 \text{ kJ/mol.})]$
   $\Delta G^\circ = -793.6 \text{ kJ}$

b. $\Delta G^\circ = [(1 \text{ mol.})(0.0 \text{ kJ/mol.}) + (6 \text{ mol.})(-4.7 \text{ kJ/mol.}) + (4 \text{ mol.})(-237.2 \text{ kJ/mol.})]$
   $\qquad - [(6 \text{ mol.})(-78.9 \text{ kJ/mol.}) + (1 \text{ mol.})(-727.8 \text{ kJ/mol.}) + (8 \text{ mol.})(0.0 \text{ kJ/mol.})]$
   $\Delta G^\circ = 224.2 \text{ kJ}$

c. $\Delta G^\circ = [(2 \text{ mol.})(-394.4 \text{ kJ/mol.}) + (1 \text{ mol.})(-228.6 \text{ kJ/mol.})]$
   $\qquad - [(1 \text{ mol.})(-209.2 \text{ kJ/mol.}) + (^5/_2 \text{ mol.})(0.0 \text{ kJ/mol.})]$
   $\Delta G^\circ = -1226.6 \text{ kJ}$

---

24. Refer to Section 17.4 and Example 17.5.

Determine ΔH_f° from the table in Appendix 1. Write the equation for the formation of the compounds from their elements. Calculate ΔS° and then ΔG_f°.

a. $N_2(g) + 2H_2(g) + \frac{3}{2} O_2(g) \rightarrow NH_4NO_3(s)$

$\Delta H_{\text{reaction}} = \Delta H_f^\circ = -365.6 \text{ kJ}$

$\Delta S^\circ = \Sigma\, S^\circ_{\text{products}} - \Sigma\, S^\circ_{\text{reactants}}$
$\Delta S^\circ = [(1 \text{ mol.})(0.1511 \text{ kJ/mol.·K})] - [(1 \text{ mol.})(0.1915 \text{ kJ/mol.·K})$
$\qquad + (2 \text{ mol.})(0.1306 \text{ kJ/mol.·K}) + (\frac{3}{2} \text{ mol.})(0.2050 \text{ kJ/mol.·K})]$

$\Delta S^\circ = -0.6091 \text{ kJ/K}$

$\Delta G_{\text{reaction}} = \Delta G_f^\circ = \Delta H^\circ - T\Delta S^\circ$

$\Delta G_f^\circ = -365.6 \text{ kJ} - (298 \text{ K})(-0.6091 \text{ kJ/K}) = -184.1 \text{ kJ}$

Note that ΔG_f° is defined for 1 mole, thus $\Delta G_f^\circ = -184.1 \text{ kJ/mol}$.

b. $C(s) + \frac{1}{2} H_2(g) + \frac{3}{2} Cl_2(g) \rightarrow CHCl_3(l)$

$\Delta H_{\text{reaction}} = \Delta H_f^\circ = -134.5 \text{ kJ}$

$\Delta S^\circ = \Sigma\, S^\circ_{\text{products}} - \Sigma\, S^\circ_{\text{reactants}}$
$\Delta S^\circ = [(1 \text{ mol.})(0.2017 \text{ kJ/mol.·K})] - [(1 \text{ mol.})(0.0057 \text{ kJ/mol.·K})$
$\qquad + (\frac{1}{2} \text{ mol.})(0.1306 \text{ kJ/mol.·K}) + (\frac{3}{2} \text{ mol.})(0.2230 \text{ kJ/mol.·K})]$

$\Delta S^\circ = -0.2038 \text{ kJ/K}$

$\Delta G_{\text{reaction}} = \Delta G_f^\circ = \Delta H^\circ - T\Delta S^\circ$

$\Delta G_f^\circ = -134.5 \text{ kJ} - (298 \text{ K})(-0.2038 \text{ kJ/K})$

$\Delta G_f^\circ = -73.8 \text{ kJ}$

Note that ΔG_f° is defined for 1 mole, thus $\Delta G_f^\circ = -73.8 \text{ kJ/mol}$.

c. $K(s) + \frac{1}{2} Cl_2(g) \rightarrow KCl(s)$

$\Delta H_{\text{reaction}} = \Delta H_f^\circ = -436.7 \text{ kJ}$

$\Delta S^\circ = \Sigma\, S^\circ_{(\text{prod})} - \Sigma\, S^\circ_{(\text{react.})}$
$\Delta S^\circ = [(1 \text{ mol.})(0.0826 \text{ kJ/mol.·K})]$
$\qquad - [(1 \text{ mol.})(0.0642 \text{ kJ/mol.·K}) + (\frac{1}{2} \text{ mol.})(0.2230 \text{ kJ/mol.·K})]$

$\Delta S^\circ = -0.0931 \text{ kJ/K}$

$\Delta G_{\text{reaction}} = \Delta G_f^\circ = \Delta H^\circ - T\Delta S^\circ$

$\Delta G_f^\circ = -436.7 \text{ kJ} - (298 \text{ K})(-0.0931 \text{ kJ/K})$

$\Delta G_f^\circ = -409.0 \text{ kJ}$

Note that ΔG_f° is defined for 1 mole, thus $\Delta G_f^\circ = -409.0 \text{ kJ/mol}$.

$O_3 (g) + H_2O(l) \rightarrow H_2O_2(aq) + O_2(g)$

$\Delta G° = \Sigma \ \Delta G_f°_{(products)} - \Sigma \ \Delta G_f°_{(reactants)}$

$\Delta G° = [(1 \text{ mol.})(-134 \text{ kJ/mol.}) + (1 \text{ mol.})(0.0 \text{ kJ/mol.})]$
$\qquad\qquad - [(1 \text{ mol.})(-163.2 \text{ kJ/mol.}) + (1 \text{ mol.})(-237.0 \text{ kJ/mol.})]$
$\Delta G° = -60 \text{ kJ}$

$\Delta G°$ is negative, so the reaction is spontaneous and the reaction is **plausible**.

---

28. *Refer to Section 17.4.*

$Mg(s) + 2H_2O(l) \rightarrow Mg(OH)_2(s) + H_2(g)$

$\Delta G° = \Sigma \ \Delta G_f°_{(products)} - \Sigma \ \Delta G_f°_{(reactants)}$

$\Delta G° = [(1 \text{ mol.})(-833.6 \text{ kJ/mol.}) + (1 \text{ mol.})(0.0 \text{ kJ/mol.})]$
$\qquad\qquad - [(1 \text{ mol.})(0.0 \text{ kJ/mol.}) + (2 \text{ mol.})(-237.2 \text{ kJ/mol.})]$

$\Delta G° = -359.2 \text{ kJ} \qquad\qquad (\text{at } 25°C)$

To calculate $\Delta G°$ at 15°C, calculate $\Delta H°$ and $\Delta S°$, then use $\Delta G° = \Delta H° - T\Delta S°$

$\Delta H° = \Sigma \ \Delta H_f°_{(products)} - \Sigma \ \Delta H_f°_{(reactants)}$
$\Delta H° = [(1 \text{ mol.})(-924.5 \text{ kJ/mol.}) + (1 \text{ mol.})(0.0 \text{ kJ/mol.})]$
$\qquad - [(1 \text{ mol.})(0.0 \text{ kJ/mol.}) + (2 \text{ mol.})(-285.5 \text{ kJ/mol.})]$
$\Delta H° = -352.9 \text{ kJ}$

$\Delta S° = \Sigma \ S°_{(products)} - \Sigma \ S°_{(reactants)}$
$\Delta S° = [(1 \text{ mol.})(0.0632 \text{ kJ/mol.·K}) + (1 \text{ mol.})(0.1306 \text{ kJ/mol.·K})]$
$\qquad - [(1 \text{ mol.})(0.0327 \text{ kJ/mol.·K}) + (2 \text{ mol.})(0.0699 \text{ kJ/mol.·K})]$
$\Delta S° = 0.0213 \text{ kJ/K}$

$\Delta G° = \Delta H° - T\Delta S°$
$\Delta G° = -352.9 \text{ kJ} - (288 \text{ K})(0.0213 \text{ kJ/K})$
$\Delta G° = -359.0 \text{ kJ} \qquad\qquad (\text{at } 15°C)$

---

30. *Refer to Sections 17.2 and 17.3.*

a. $\Delta G° = \Delta H° - T\Delta S°$
$-148.4 \text{ kJ} = -109.0 \text{ kJ} - (298 \text{ K})(\Delta S°)$
$\Delta S° = 0.132 \text{ kJ/K}$
The sign of $\Delta S°$ is reasonable since Δn_{gas} is positive.

b. $\Delta S° = \Sigma\ S°_{(products)} - \Sigma\ S°_{(reactants)}$

0.132 kJ/K = [(4 mol.)(0.0645 kJ/mol.·K) + (1 mol.)(0.2050 kJ/mol.·K)]
$\qquad$ - [(2 mol.)$S°$(N$_2$O$_2$(s)) + (2 mol.)(0.0699 kJ/mol.·K)]

(2 mol.)$S°$(N$_2$O$_2$(s)) = 0.191 kJ/K

$S°$(N$_2$O$_2$(s)) = 0.0955 kJ/mol.·K

c. $\Delta H°_{(reaction)} = \Sigma\ \Delta H_f°_{(products)} - \Sigma\ \Delta H_f°_{(reactants)}$

-109.0 kJ = [(4 mol.)(-425.6 kJ/mol.) + (1 mol.)(0.0 kJ/mol.)]
$\qquad$ - [(2 mol.)$\Delta H_f°$(N$_2$O$_2$(s)) + (2 mol.)(-285.8 kJ/mol.)]

(2 mol.)$\Delta H_f°$(N$_2$O$_2$(s)) = -1021.8 kJ

$\Delta H_f°$(N$_2$O$_2$(s)) = -510.9 kJ/mol.

---

32. Refer to Sections 17.2 and 17.3.

Calculate $\Delta S°_{(reaction)}$ using the Gibbs-Helmholtz equation. Then use $\Delta S°_{(reaction)}$ to solve for $S°$(Co^{2+})

a. $\Delta G° = \Delta H° - T\Delta S°$

-1750.9 kJ = -2024.6 kJ - (298 K)($\Delta S°$)

$\Delta S° = -0.918$ kJ/K = - 918 J/K

b. $\Delta S° = \Sigma\ S°_{(products)} - \Sigma\ S°_{(reactants)}$

-0.918 kJ/K = [(2 mol.)(- 0.0736 kJ/mol.·K) + (5 mol.)$S°$(Co^{2+}(aq))
$\qquad$ + (8 mol.)(0.0699 kJ/mol.·K)] - [(2 mol.)(0.1912 kJ/mol.·K)
$\qquad$ + (16 mol.)(0.0 kJ/mol.·K)+ (5 mol.)(0.03004 kJ/mol.·K)]

$S°$(Co^{2+}) = - 0.159 kJ/mol.·K = - 159 J/mol.·K

---

34. Refer to Section 17.5.

Set up the equation for calculating $\Delta G°$. Then examine the impact changing T would have on the sign of $\Delta G°$. Specifically, would raising or lowering T change the sign of $\Delta G°$?

a. $\Delta G° = \Delta H° - T\Delta S°$

$\Delta G° = 851.5$ kJ - (T)(0.0385 kJ/K)

$\Delta G° = 0$ when $T = 2.21 \times 10^4$ K,

$\Delta G°$ is (+) when $T < 2.21 \times 10^4$ K and

$\Delta G°$ is (-) when $T > 2.21 \times 10^4$ K.

Therefore, varying T can alter the spontaneity of the reaction.

b. $\Delta G° = \Delta H° - T\Delta S°$

$\Delta G° = -50.6$ kJ - (T)(0.3315 kJ/K)

$\Delta G° = 0$ when $T = -153$ K,

Therefore, $\Delta G°$ is always negative and changing T will not alter the spontaneity of the reaction.

c. $\Delta G° = \Delta H° - T\Delta S°$
$\Delta G° = 98.9 \text{ kJ} - (T)(0.0939 \text{ kJ/K})$
$\Delta G° = 0$ when $T = 1050$ K,
$\Delta G°$ is (+) when $T < 1050$ K and
$\Delta G°$ is (-) when $T > 1050$ K.
Therefore, varying T can alter the spontaneity of the reaction.

---

36. Refer to Section 17.5 and Example 17.6.

Write the Gibbs-Helmholtz equation for the reaction, set $\Delta G° = 0$ and solve for T. Those temperatures at which $\Delta G° < 0$ represent the temperatures at which the reaction is spontaneous.

a. Spontaneous for $T > 2.21 \times 10^4$ K.

b. Spontaneous at all temperatures.

c. Spontaneous for $T > 1050$ K.

---

38. Refer to Sections 17.4 and 17.5 and Example 17.6.

Calculate $\Delta H°_{reaction}$ and $\Delta S°_{reaction}$. Using the Gibbs-Helmholtz equation, set $\Delta G° = 0$, and solve for T.

$\Delta H° = \Sigma \, \Delta H_f°_{products} - \Sigma \, \Delta H_f°_{reactants}$
$\Delta H° = [(1 \text{ mol.})(0 \text{ kJ/mol.}) + (2 \text{ mol.})(-393.5 \text{ kJ/mol.})]$
$- [(1 \text{ mol.})(-580.7 \text{ kJ/mol.}) + (2 \text{ mol.})(-110.5 \text{ kJ/mol.})]$
$\Delta H° = +14.7 \text{ kJ}$

$\Delta S° = \Sigma \, S°_{products} - \Sigma \, S°_{reactants}$
$\Delta S° = [(1 \text{ mol.})(0.0516 \text{ kJ/mol.·K}) + (2 \text{ mol.})(0.2136 \text{ kJ/mol.·K})]$
$- [(1 \text{ mol.})(0.0523 \text{ kJ/mol.·K}) + (2 \text{ mol.})(0.1976 \text{ kJ/mol.·K})]$
$\Delta S° = 0.0313 \text{ kJ/K}$

$\Delta G° = \Delta H° - T\Delta S°$
$0 = 14.7 \text{ kJ} - (T)(0.0313 \text{ kJ/K})$
$T = 470 \text{ K}$
$470 \text{ K} - 273 = 197°C$

---

40. Refer to Sections 17.4 and 17.5 and Figure 17.6.

a. Calculate $\Delta H°_{reaction}$ and $\Delta S°_{reaction}$. Use these values to calculate $\Delta G°$.

$\Delta H° = \Sigma \, \Delta H_f°_{products} - \Sigma \, \Delta H_f°_{reactants}$
$\Delta H° = (4 \text{ mol.})(0.0 \text{ kJ/mol.}) - (2 \text{ mol.})(-31.0 \text{ kJ/mol.})$
$\Delta H° = 62.0 \text{ kJ}$

$\Delta S° = \Sigma \ S°_{products} - \Sigma \ S°_{reactants}$

$\Delta S° = [(4 \ mol.)(0.0426 \ kJ/mol.\cdot K) + (1 \ mol.)(0.205 \ kJ/mol.\cdot K)]$
$\qquad\qquad - [(2 \ mol.)(0.1213 \ kJ/mol.\cdot K)]$

$\Delta S° = 0.1328 \ kJ/K$

$\Delta G° = \Delta H° - T\Delta S°$

$\Delta G° = 62.0 \ kJ - (T)(0.1328 \ kJ/K)$

T (K)	100	200	300	400	500
$\Delta G°$ (kJ)	48.7	35.4	22.2	8.9	-4.4

b. $\Delta G° = 0 = 62.0 \ kJ - (T)(0.1328 \ kJ/K)$

$T = 467 \ K$

$\Delta G°$ becomes negative (and the reaction becomes spontaneous) at temperatures above 467 K, therefore this is the lowest temperature at which the decomposition is possible.

---

42. Refer to Section 17.5.

Calculate $\Delta H°_{reaction}$ and $\Delta S°_{reaction}$. Then calculate T for $\Delta G° = 0$ to determine the lowest temperature at which the reaction is spontaneous.

a. $SnO_2(s) \rightarrow Sn(s) + O_2(g)$

$\Delta H° = \Sigma \ \Delta H_f°_{products} - \Sigma \ \Delta H_f°_{reactants}$
$\Delta H° = [(1 \ mol.)(0.0 \ kJ/mol.) + (1 \ mol.)(0.0 \ kJ/mol.)] - [(1 \ mol.)(-580.7 \ kJ/mol.)]$
$\Delta H° = 580.7 \ kJ$

$\Delta S° = \Sigma \ S°_{products} - \Sigma \ S°_{reactants}$
$\Delta S° = [(1 \ mol.)(0.0516 \ kJ/mol.\cdot K) + (1 \ mol.)(0.2050 \ kJ/mol.\cdot K)]$
$\qquad\quad - [(1 \ mol.)(0.0523 \ kJ/mol.\cdot K)]$
$\Delta S° = 0.2043 \ kJ/K$

$\Delta G° = \Delta H° - T\Delta S°$
$\Delta G° = 0 = 580.7 \ kJ - (T)(0.2043 \ kJ/K)$
$\Delta G° = 0$ when $T = 2842 \ K$

b. $SnO_2(s) + 2H_2(g) \rightarrow Sn(s) + 2H_2O(g)$

$\Delta H° = \Sigma \ \Delta H_f°_{products} - \Sigma \ \Delta H_f°_{reactants}$
$\Delta H° = [(1 \ mol.)(0.0 \ kJ/mol.) + (2 \ mol.)(-241.8 \ kJ/mol.)]$
$\qquad\quad - [(1 \ mol.)(-580.7 \ kJ/mol.) + (2 \ mol.)(0.0 \ kJ/mol.)]$
$\Delta H° = 97.1 \ kJ$

$\Delta S° = \Sigma \ S°_{products} - \Sigma \ S°_{reactants}$
$\Delta S° = [(1 \ mol.)(0.0516 \ kJ/mol.\cdot K) + (2 \ mol.)(0.1887 \ kJ/mol.\cdot K)]$
$\qquad\quad - [(1 \ mol.)(0.0523 \ kJ/mol.\cdot K) + (2 \ mol.)(0.1306 \ kJ/mol.\cdot K)]$
$\Delta S° = 0.1155 \ kJ/K$

$$\Delta G^\circ = \Delta H^\circ - T\Delta S^\circ$$
$$\Delta G^\circ = 0 = 97.1 \text{ kJ} - (T)(0.1155 \text{ kJ/K})$$
$$\Delta G^\circ = 0 \text{ when } T = 841 \text{ K}$$

c. $SnO_2(s) + C(s) \rightarrow Sn(s) + CO_2(g)$

$$\Delta H^\circ = \Sigma \Delta H_f^\circ{}_{\text{products}} - \Sigma \Delta H_f^\circ{}_{\text{reactants}}$$
$$\Delta H^\circ = [(1 \text{ mol.})(0.0 \text{ kJ/mol.}) + (1 \text{ mol.})(-393.5 \text{ kJ/mol.})]$$
$$\qquad - [(1 \text{ mol.})(-580.7 \text{ kJ/mol.}) + (1 \text{ mol.})(0.0 \text{ kJ/mol.})]$$
$$\Delta H^\circ = 187.2 \text{ kJ}$$

$$\Delta S^\circ = \Sigma S^\circ{}_{\text{products}} - \Sigma S^\circ{}_{\text{reactants}}$$
$$\Delta S^\circ = [(1 \text{ mol.})(0.0516 \text{ kJ/mol.}\cdot\text{K}) + (1 \text{ mol.})(0.2136 \text{ kJ/mol.}\cdot\text{K})]$$
$$\qquad - [(1 \text{ mol.})(0.0523 \text{ kJ/mol.}\cdot\text{K}) + (1 \text{ mol.})(0.0057 \text{ kJ/mol.}\cdot\text{K})]$$
$$\Delta S^\circ = 0.2072 \text{ kJ/K}$$

$$\Delta G^\circ = \Delta H^\circ - T\Delta S^\circ$$
$$\Delta G^\circ = 0 = 187.2 \text{ kJ} - (T)(0.2072 \text{ kJ/K})$$

$$\Delta G^\circ = 0 \text{ when } T = 903.5 \text{ K}$$

Reaction (b) is spontaneous at the lowest temperature.

---

44. *Refer to Section 17.5.*

Calculate $\Delta H^\circ{}_{\text{reaction}}$ and $\Delta S^\circ{}_{\text{reaction}}$. $\Delta G^\circ = 0$ at equilibrium, so calculate T for $\Delta G^\circ = 0$ to determine the temperature at which the reaction is at equilibrium.

$Sn_{\text{white}}(s) \rightleftharpoons Sn_{\text{grey}}(s)$

$$\Delta H^\circ = \Sigma \Delta H_f^\circ{}_{\text{products}} - \Sigma \Delta H_f^\circ{}_{\text{reactants}}$$
$$\Delta H^\circ = [(1 \text{ mol.})(-2.09 \text{ kJ/mol.})] - [(1 \text{ mol.})(0.0 \text{ kJ/mol.})]$$
$$\Delta H^\circ = -2.09 \text{ kJ}$$

$$\Delta S^\circ = \Sigma S^\circ{}_{\text{products}} - \Sigma S^\circ{}_{\text{reactants}}$$
$$\Delta S^\circ = [(1 \text{ mol.})(0.04414 \text{ kJ/mol.}\cdot\text{K})] - [(1 \text{ mol.})(0.05155 \text{ kJ/mol.}\cdot\text{K})]$$
$$\Delta S^\circ = -0.00741 \text{ kJ/K}$$

$$\Delta G^\circ = \Delta H^\circ - T\Delta S^\circ$$
$$\Delta G^\circ = 0 = -2.09 \text{ kJ} - (T)(-0.00741 \text{ kJ/K})$$
$$T = 282 \text{ K}$$

The system is at equilibrium when $T = 282$ K

---

46. *Refer to Section 17.5.*

Calculate $\Delta H^\circ{}_{\text{reaction}}$ and $\Delta S^\circ{}_{\text{reaction}}$. $\Delta G^\circ = 0$ at equilibrium, so calculate T for $\Delta G^\circ = 0$ to determine the temperature at which the reaction is at equilibrium.

$$C_{graphite}(s) \rightleftharpoons C_{diamond}(s)$$

$\Delta H° = \Sigma \Delta H_f°_{products} - \Sigma \Delta H_f°_{reactants}$
$\Delta H° = [(1 \text{ mol.})(1.9 \text{ kJ/mol.})] - [(1 \text{ mol.})(0 \text{ kJ/mol.})]$
$\Delta H° = 1.9 \text{ kJ}$

$\Delta S° = \Sigma S°_{products} - \Sigma S°_{reactants}$
$\Delta S° = [(1 \text{ mol.})(0.0024 \text{ kJ/mol.·K})] - [(1 \text{ mol.})(0.0057 \text{ kJ/mol.·K})]$
$\Delta S° = -0.0033 \text{ kJ/K}$

$\Delta G° = \Delta H° - T\Delta S°$
$\Delta G° = 0 = 1.9 \text{ kJ} - (T)(-0.0033 \text{ kJ/K})$
$T = -576 \text{ K}$

The system is at equilibrium when $T = -576$ K. Since the temperature can never be negative, there is no temperature at which graphite and diamond are in equilibrium (for $P = 1$ atm).

---

48. *Refer to Section 17.5.*

Calculate $\Delta H°_{reaction}$ and $\Delta S°_{reaction}$. The boiling point is the temperature at which the solid and the liquid are at equilibrium. $\Delta G° = 0$ at equilibrium, so calculate T for $\Delta G° = 0$ to determine the boiling point.

$\Delta H° = \Sigma \Delta H_f°_{products} - \Sigma \Delta H_f°_{reactants}$
$\Delta H° = [(1 \text{ mol.})(30.91 \text{ kJ/mol.})] - [(1 \text{ mol.})(0.0 \text{ kJ/mol.})] = 30.91 \text{ kJ}$

$\Delta S° = \Sigma S°_{products} - \Sigma S°_{reactants}$
$\Delta S° = [(1 \text{ mol.})(0.2454 \text{ kJ/mol.·K})] - [(1 \text{ mol.})(0.1522 \text{ kJ/mol.·K})] = 0.0932 \text{ kJ/K}$

$\Delta G° = \Delta H° - T\Delta S°$
$\Delta G° = 0 = 30.91 \text{ kJ} - (T)(0.0932 \text{ kJ/K})$
$T = 332 \text{ K} = 59°C$

The boiling point is 59°C.

---

50. *Refer to Section 17.5.*

Calculate Q for the concentrations listed, calculate $\Delta G°$ using the values in Appendix 1, then calculate $\Delta G_{reaction}$.

a. $Q = \dfrac{[H^+][C_2H_3O_2^-]}{[HC_2H_3O_2]} = \dfrac{(0.85)(0.85)}{(0.15)} = 4.8$

$\Delta G = \Delta G° + RT \ln Q$
$\Delta G = 27.2 \text{ kJ} + (0.00831 \text{ kJ/mol.·K})(298 \text{ K})(\ln 4.8)$
$\Delta G = 31.1 \text{ kJ}$
ΔG is positive, so the reaction is **non-spontaneous**.

b. $Q = \dfrac{[H^+][C_2H_3O_2^-]}{[HC_2H_3O_2]} = \dfrac{(2.0 \times 10^{-3})(2.0 \times 10^{-3})}{(1)} = 4.0 \times 10^{-6}$

$\Delta G = \Delta G^\circ + RT \ln Q$

$\Delta G = 27.2 \text{ kJ} + (0.00831 \text{ kJ/mol·K})(298 \text{ K})(\ln 4.0 \times 10^{-6})$

$\Delta G = -3.6 \text{ kJ}$

ΔG is negative, so the reaction is **spontaneous**.

---

52. Refer to Sections 17.4 and 17.5 and Example 17.7.

Calculate ΔG° using the values in Appendix 1, calculate Q for the concentrations listed, then calculate $\Delta G_{\text{reaction}}$.

a. $\Delta G^\circ = \Sigma \Delta G_f^\circ{}_{\text{products}} - \Sigma \Delta G_f^\circ{}_{\text{reactants}}$

$\Delta G^\circ = [(2 \text{ mol.})(-237.2 \text{ kJ/mol.}) + (4 \text{ mol.})(-4.7 \text{ kJ/mol.})]$
$\qquad - [(1 \text{ mol.})(0.0 \text{ kJ/mol.}) + (4 \text{ mol.})(0.0 \text{ kJ/mol.}) + (4 \text{ mol.})(-78.9 \text{ kJ/mol.})]$

$\Delta G^\circ = -177.6 \text{ kJ}$

b. $[H^+] = 10^{-\text{pH}} = 10^{-3.12} = 7.6 \times 10^{-4} \, M$

$Q = \dfrac{[Fe^{3+}]^4}{P_{O_2}[H^+]^4[Fe^{2+}]^4} = \dfrac{(0.250)^4}{(0.755)(7.6 \times 10^{-4})^4(0.250)^4} = 3.97 \times 10^{12}$

$\Delta G = \Delta G^\circ + RT \ln Q$

$\Delta G = -177.6 \text{ kJ} + (0.00831 \text{ kJ/mol·K})(298 \text{ K})(\ln 3.97 \times 10^{12})$

$\Delta G = -105.6 \text{ kJ}$

---

54. Refer to Sections 17.4 and 17.5 and Example 17.7.

a. $\Delta G^\circ = \Sigma \Delta G_f^\circ{}_{\text{products}} - \Sigma \Delta G_f^\circ{}_{\text{reactants}}$

$\Delta G^\circ = [(1 \text{ mol.})(77.1 \text{ kJ/mol.}) + (1 \text{ mol.})(-131.2 \text{ kJ/mol.})] - [(1 \text{ mol.})(-109.8 \text{ kJ/mol.})]$

$\Delta G^\circ = 55.7 \text{ kJ}$

b. Assume 25°C as in part (a) above. Set $\Delta G = 0$, solve the equation for $[Ag^+]$ and $[Cl^-]$. Note that $[Ag^+] = [Cl^-] = x$.

$\Delta G = \Delta G^\circ + RT \ln Q \qquad\qquad Q = [Ag^+][Cl^-]$

$-1.0 \text{ kJ} = 55.7 \text{ kJ} + (8.31 \text{ J/mol·K})(0.001 \text{ kJ/J})(298 \text{ K})(\ln [Ag^+][Cl^-])$

$\ln [Ag^+][Cl^-] = -22.9$

$[Ag^+][Cl^-] = x^2 = 1.15 \times 10^{-10}$

$x = [Ag^+] = [Cl^-] = 1.07 \times 10^{-5} \, M$

c. $K_{sp} = [Ag^+][Cl^-] = 1.15 \times 10^{-10}$

Actual $K_{sp} = 1.8 \times 10^{-10}$

Since the calculated value and K_{sp} are approximately equal, the answer in (b) is reasonable.

Manipulate the equations so that their sum is the desired equation. Remember that reversing the equation changes the sign of $\Delta G°$, and any factor applied to the equation must also be applied to $\Delta G°$.

$-1[N_2O_5(g) \rightarrow 2NO(g) + {}^3/_2O_2(g)]$ $\Delta G° = (-1)(-59.2 \text{ kJ})$

$\underline{-2[NO(g) + {}^1/_2O_2(g) \rightarrow NO_2(g)]}$ $\Delta G° = (-2)(-35.6 \text{ kJ})$

$2NO(g) + {}^3/_2O_2(g) + 2NO_2(g) \rightarrow N_2O_5(g) + 2NO(g) + O_2(g)$

${}^1/_2O_2(g) + 2NO_2(g) \rightarrow N_2O_5(g)$ $\Delta G° = 130.4 \text{ kJ}$

a. $\Delta G° = \Sigma \Delta G_f°{}_{\text{products}} - \Sigma \Delta G_f°{}_{\text{reactants}}$

     $\Delta G° = [(1 \text{ mol.})(0.0 \text{ kJ/mol.}) + (1 \text{ mol.})(0.0 \text{ kJ/mol.})] - [(1 \text{ mol.})(-201.3 \text{ kJ/mol.})]$

     $\Delta G° = 201.3 \text{ kJ}$

     $\Delta G°$ is positive, this reaction is not spontaneous and therefore **not feasible** (at 25°C).

b. $ZnS(s) \rightarrow Zn(s) + S(s)$ $\Delta G° = 201.3 \text{ kJ}$

     $\underline{S(s) + O_2(g) \rightarrow SO_2(g)}$ $\Delta G° = -300.2 \text{ kJ}$

     $ZnS(s) + S(s) + O_2(g) \rightarrow Zn(s) + S(s) + SO_2(g)$

     $ZnS(s) + O_2(g) \rightarrow Zn(s) + SO_2(g)$ $\Delta G° = -98.9 \text{ kJ}$

This reaction is spontaneous, and therefore **feasible** at 25°C

Calculate the moles of ADP required to give $\Delta G°{}_{\text{reaction}} = -390 \text{ kJ}$.

$\Delta G°{}_{\text{reaction}} = (x \text{ mol.})(\Delta G°{}_{(\text{ADP reaction})}) + \Delta G°{}_{(\text{glucose reaction})}$

$-390 \text{ kJ} = (x)(31 \text{ kJ}) + (-2870 \text{ kJ})$

$x = 80 \text{ mol.}$ (Multiply the ADP reaction by 80 and add to the $C_6H_{12}O_6$ reaction.)

$80[ADP(aq) + HPO_4{}^{2-}(aq) + 2H^+(aq) \rightarrow ATP(aq) + H_2O(l)]$ $\Delta G° = 80(31 \text{ kJ})$

$\underline{C_6H_{12}O_6(aq) + 6O_2(g) \rightarrow 6CO_2(g) + 6H_2O(l)}$ $\Delta G° = -2870 \text{ kJ}$

$80ADP(aq) + 80HPO_4{}^{2-}(aq) + 160H^+(aq) + C_6H_{12}O_6(aq) + 6O_2(g)$

     $\rightarrow 6CO_2(g) + 6H_2O(l) + 80ATP(aq) + 80H_2O(l)$

$80ADP(aq) + 80HPO_4{}^{2-}(aq) + 160H^+(aq) + C_6H_{12}O_6(aq) + 6O_2(g)$

     $\rightarrow 6CO_2(g) + 86H_2O(l) + 80ATP(aq)$ $\Delta G° = -390 \text{ kJ}$

a. $\Delta G° = \Sigma \Delta G_f°{}_{\text{products}} - \Sigma \Delta G_f°{}_{\text{reactants}}$

     $\Delta G° = [(1 \text{ mol.})(0.0 \text{ kJ/mol.}) + (1 \text{ mol.})(-26.7 \text{ kJ/mol.})] - [(1 \text{ mol.})(-79.3 \text{ kJ/mol.})]$

     $\Delta G° = +52.6 \text{ kJ}$

b. $\Delta G° = -RT\ln K_a$

$52.6 \text{ kJ} = -(0.00831 \text{ kJ/mol·K})(298 \text{ K})(\ln K_a)$

$\ln K_a = -21.2$

$K_a = e^{-21.2} = 6.2 \times 10^{-10}$

---

64. *Refer to Section 17.6 and Example 17.9.*

a. $\Delta G° = -RT \ln K$

$\Delta G° = -(0.00831 \text{ kJ/mol·K})(298 \text{ K})\ln (4.4 \times 10^{-19})$

$\Delta G° = 105 \text{ kJ}$

b. $\Delta G° = \Sigma \Delta G°_{f \text{ products}} - \Sigma \Delta G°_{f \text{ reactants}}$

$105 \text{ kJ} = (3 \text{ mol.})(86.6 \text{ kJ/mol.}) - [(1 \text{ mol.})(51.3 \text{ kJ/mol.}) + (1 \text{ mol.})(\Delta G°_f(N_2O))]$

$\Delta G°_f(N_2O) = 104 \text{ kJ/mol.}$

---

66. *Refer to Section 17.5.*

a. $\Delta G° = -RT \ln K$

$\Delta G° = -(0.00831 \text{ kJ/mol·K})(721 \text{ K})\ln(50.0)$

$\Delta G° = -23.4 \text{ kJ/mol.}$

b. $\Delta G° = \Sigma \Delta G°_{f \text{ products}} - \Sigma \Delta G°_{f \text{ reactants}}$

$\Delta G° = (2 \text{ mol.})(1.7 \text{ kJ/mol.}) - [(1 \text{ mol.})(0.0 \text{ kJ/mol.}) + (1 \text{ mol.})(19.4 \text{ kJ/mol.})]$

$\Delta G° = -16 \text{ kJ/mol.}$

$\Delta G° = -RT \ln K$

$-16 \text{ kJ} = -(0.00831 \text{ kJ/mol·K})(298 \text{ K})\ln K$

$\ln K = 6.46$

$K = 6.4 \times 10^2$

---

68. *Refer to Section 17.6.*

Calculate $\Delta H°$ and $\Delta S°$. Then calculate $\Delta G°$ and K_a at each of the temperatures.

$$HF(aq) \rightleftharpoons H^+(aq) + F^-(aq)$$

$\Delta H° = \Sigma \Delta H°_{f \text{ products}} - \Sigma \Delta H°_{f \text{ reactants}}$

$\Delta H° = [(1 \text{ mol.})(0.0 \text{ kJ/mol.}) + (1 \text{ mol.})(-332.6 \text{ kJ/mol.})] - [(1 \text{ mol.})(-320.1 \text{ kJ/mol.})]$

$\Delta H° = -12.5 \text{ kJ}$

$\Delta S° = \Sigma S°_{\text{products}} - \Sigma S°_{\text{reactants}}$

$\Delta S° = [(1 \text{ mol.})(0.0 \text{ kJ/mol·K}) + (1 \text{ mol.})(-0.0138 \text{ kJ/mol·K})] - [(1 \text{ mol.})(0.0887 \text{ kJ/mol·K.})]$

$\Delta S° = -0.1025 \text{ kJ/K}$

At 25°C:

$\Delta G° = \Delta H° - T\Delta S°$

$\Delta G° = -12.5 \text{ kJ} - (298 \text{ K})(-0.1025 \text{ kJ/K})$

$\Delta G° = 18.0 \text{ kJ} = 1.80 \times 10^4 \text{ J}$

$$\Delta G^\circ = -RT \ln K$$
$$1.80 \times 10^4 \text{ J} = -(8.31 \text{ J/mol.·K})(298 \text{ K}) \ln K$$
$$\ln K = -7.27$$
$$K = 7.0 \times 10^{-4}$$

---

70. Refer to Section 17.6 and Chapter 13.

Calculate $[H^+]$ (and $[B^-]$) from the pH. Then calculate K_a, and from that, ΔG°. Note that K_a is for the reverse of the desired reaction, therefore the sign of ΔG° must also be reversed.

$$[B^-] = [H^+] = 10^{-pH} = 10^{-3.71} = 2.0 \times 10^{-4}$$

	$HB(aq)$	$\rightleftharpoons$	$H^+(aq)$	$+$	$B^-(aq)$
$[\ ]_0$	0.13		0		0
$\Delta[\ ]$	-2.0×10^{-4}		$+2.0 \times 10^{-4}$		$+2.0 \times 10^{-4}$
$[\ ]_{eq}$	$0.31 - 2.0 \times 10^{-4}$		2.0×10^{-4}		2.0×10^{-4}

$$K_a = \frac{[H^+][B^-]}{[HB]} = \frac{(2.0 \times 10^{-4})(2.0 \times 10^{-4})}{0.13 - (2.0 \times 10^{-4})} = 3.1 \times 10^{-7}$$

$$\Delta G^\circ = -RT \ln K$$
$$\Delta G^\circ = -(8.314 \text{ J/mol.·K})(0.001 \text{ kJ/J})(298 \text{ K}) \ln(3.1 \times 10^{-7})$$
$$\Delta G^\circ = +37.1 \text{ kJ} \qquad [\text{for: } HB(aq) \rightleftharpoons H^+(aq) + B^-(aq)]$$
$$\Delta G^\circ = -37.1 \text{ kJ} \qquad [\text{for: } H^+(aq) + B^-(aq) \rightleftharpoons HB(aq)]$$

---

72. Refer to Section 17.7 and Example 17.9.

Calculate ΔG° for the reaction. Manipulate the equations so that their sum is the desired equation. Remember that reversing the equation changes the sign of ΔG°, and any factor applied to the equation must also be applied to ΔG°.

$$\Delta G^\circ = \Sigma \, \Delta G_f^\circ \text{products} - \Sigma \, \Delta G_f^\circ \text{reactants}$$
$$\Delta G^\circ = [(1 \text{ mol.})(-237.2 \text{ kJ/mol.}) + (1 \text{ mol.})(0.0 \text{ kJ/mol.})]$$
$$\qquad - [(1 \text{ mol.})(-33.6 \text{ kJ/mol.}) + (^1/_2 \text{ mol.})(0.0 \text{ kJ/mol.})]$$
$$\Delta G^\circ = 203.6 \text{ kJ}$$

$$24H_2S(g) + 12O_2(g) \rightarrow 24H_2O(l) + 24S(s) \qquad \Delta G^\circ = 24(-203.6 \text{ kJ})$$
$$\underline{6CO_2(g) + 6H_2O(l) \rightarrow C_6H_{12}O_6(aq) + 6O_2(g)} \qquad \underline{\Delta G^\circ = 2870 \text{ kJ}}$$
$$24H_2S(g) + 12O_2(g) + 6CO_2(g) + 6H_2O(l) \rightarrow C_6H_{12}O_6(aq) + 6O_2(g) + 24H_2O(l) + 24S(s)$$
$$\qquad \qquad \Delta G^\circ = 24(-203.6 \text{ kJ}) + 2870 \text{ kJ}$$

$$24H_2S(g) + 6O_2(g) + 6CO_2(g) \rightarrow C_6H_{12}O_6(aq) + 18H_2O(l) + 24S(s) \quad \Delta G^\circ = -2016 \text{ kJ}$$

74. Refer to Sections 17.1 - 17.3.

a. **False**. An exothermic reaction is *usually* spontaneous, but not always. One must also consider entropy changes, which could render an exothermic reaction non-spontaneous.

b. **False**. When $\Delta G°$ is positive, the reaction is not spontaneous under the standard conditions (25°C and 1 atm), but may be spontaneous under different conditions.

c. **False**. $\Delta S°$ is positive for a reaction in which there is an increase in the moles of *gas*.

d. **False**. If $\Delta H°$ and $\Delta S°$ are both negative, $\Delta G°$ will be negative at low temperatures and positive at high temperatures.

76. Refer to Sections 17.2, 17.4, 17.5 and 17.6.

a. $\Delta H°$ and $\Delta G°$ become equal at **0 K**.
$\Delta G° = \Delta H° + T\Delta S°$, if $T = 0$ K, then
$\Delta G° = \Delta H°$

b. $\Delta G°$ and ΔG are equal when $Q = 1$
$\Delta G = \Delta G° + RT \ln Q$, when $Q = 1$, then $\ln Q = 0$ and
$\Delta G = \Delta G°$

c. $S°$ for steam is **greater** than $S°$ for water. Entropy is always higher for gases than for liquids.

78. Refer to Section 17.2.

a. Gases have a higher entropy than solids and liquids. Therefore, if the moles of gas decrease in a reaction, the **entropy decreases** and $\Delta S°$ is negative.

b. **Entropy changes due to temperature are small** and may be ignored. The exception is when the temperature change also involves a phase change.

c. The entropy of solids is lower than the entropy of liquids because **solids have an ordered three dimensional structure**, which largely inhibit molecular motion, while liquids do not have an ordered three dimensional structure and have significant molecular motion.

80. Refer to Chapter 17.

a. The relationship is linear and the reaction becomes more spontaneous at higher temperatures.

b. The reaction is endothermic because $\Delta H°$ must be positive. See Table 17.2

c. $\Delta S°$ is positive. See Table 17.2

d. When $\Delta G° = 0$; approximately 330 K.

e. 27°C = 300K
$\Delta G° = -RT \ln K$
10 kJ = -(0.00831 kJ/mol.·K)(300 K) ln K
$K = 2 \times 10^{-2}$

---

82. Refer to Section 17.7 and Chapter 9.

For the equation $C_2H_5OH(l) \rightleftharpoons C_2H_5OH(g)$, $\Delta G° = 0$ since this is a phase change (see problem 48 above).

$\Delta G° = \Delta H° - T \Delta S° \Rightarrow \Delta H° = T \Delta S°$

Calculate ΔS, ΔH_{vap} and then use the Claysius-Clapeyron equation in Chapter 9.

$\Delta S° = \Sigma S°_{products} - \Sigma S°_{reactants}$
$\Delta S° = [(1 \text{ mol.})(0.2827 \text{ kJ/mol.·K})] - [(1 \text{ mol.})(0.1607 \text{ kJ/mol.·K})]$
$\Delta S° = 0.1220 \text{ kJ/K}$

$\Delta H° = T \Delta S°$
$\Delta H° = (351.6 \text{ K})(0.1220 \text{ kJ/K})$
$\Delta H° = \Delta H_{vap} = 42.89 \text{ kJ}$

$$\ln\left(\frac{P_2}{P_1}\right) = \frac{+\Delta H_{vap}}{R}\left(\frac{1}{T_1} - \frac{1}{T_2}\right)$$

$$\ln\left(\frac{760 \text{ mm Hg}}{357 \text{ mm Hg}}\right) = \frac{42890 \text{ J/mol.}}{8.31 \text{ J/mol.·K}}\left(\frac{1}{T_1} - \frac{1}{351.6 \text{ K}}\right)$$

$$0.756 = (5161 \text{ K})\left(\frac{1}{T_1} - 0.00285\right) \Rightarrow \frac{1}{T_1} = 0.00300 \text{ K}^{-1}$$

$T_1 = 333 \text{ K} = 60°C$

---

83. Refer to Sections 17.4, 17.5 and 17.6.

Calculate $\Delta S°$ and $\Delta H°$, then $\Delta G°$. Use $\Delta G°$ to calculate K. Finally, calculate the final pressures of the gases using the equilibrium expression.

$\Delta S° = \Sigma S°_{products} - \Sigma S°_{reactants}$
$\Delta S° = [(1 \text{ mol.})(0.1306 \text{ kJ/mol.·K}) + (1 \text{ mol.})(0.2607 \text{ kJ/mol.·K})]$
     $- [(2 \text{ mol.})(0.2065 \text{ kJ/mol.·K})]$
$\Delta S° = -0.0217 \text{ kJ/K}$

$\Delta H° = \Sigma \Delta H_f°_{products} - \Sigma \Delta H_f°_{reactants}$
$\Delta H° = [(1 \text{ mol.})(0.0 \text{ kJ/mol.}) + (1 \text{ mol.})(62.4 \text{ kJ/mol.})] - [(2 \text{ mol.})(26.5 \text{ kJ/mol.})]$
$\Delta H° = 9.4 \text{ kJ}$

$\Delta G° = \Delta H° - T \Delta S°$
$\Delta G° = 9.4$ kJ - (773 K)(-0.0217 kJ/K)
$\Delta G° = 26.2$ kJ

$\Delta G° = -RT \ln K$
26.2 kJ = -(8.31 J/mol.·K)(0.001 kJ/J)(773 K) ln K
$\ln K = -4.08$
$K = 0.0169$

Since $\Delta G°$ is positive, the reaction will proceed in the reverse direction, giving:
$\quad [H_2] = [I_2] = 0.200 - x$ and $[HI] = 0.200 + 2x$

$$K = \frac{(P_{H_2})(P_{I_2})}{(P_{HI})^2} \quad \Rightarrow \quad 0.0169 = \frac{(0.200 - x)^2}{(0.200 + 2x)^2}$$

$$\sqrt{0.0169} = \frac{0.200 - x}{0.200 + 2x} \quad \Rightarrow \quad 0.0260 + 0.260x = 0.200 - x$$

$x = 0.138$ atm

$[H_2] = [I_2] = 0.200 - x = 0.062$ atm
$[HI] = 0.200 + 2x = 0.476$ atm

---

84. *Refer to Sections 17.5 and 17.6.*

a. Assume the phase change is at equilibrium.

$$\Delta H° = \Delta H_{fusion} = \frac{333 \text{ J}}{1 \text{ g}} \times \frac{18.02 \text{ g}}{1 \text{ mol.}} = 6.00 \times 10^3 \text{ J/mol.} = 6.00 \text{ kJ/mol.}$$

b. $\Delta G° = -RT \ln K$ Since neither solids nor liquids appear in the expression for K (both are one), we get: $\Delta G° = -RT \ln 1 = 0$.

c. $\Delta G° = \Delta H° - T \Delta S°$
$0 = 6.00$ kJ - (273 K)($\Delta S°$)
$\Delta S° = 0.0220$ kJ/K

d. $\Delta G° = \Delta H° - T \Delta S°$
$\Delta G° = 6.00$ kJ - (253 K)(0.0220 kJ/K)
$\Delta G° = 0.43$ kJ $\qquad$ Thus, ice does not melt spontaneously at -20°C.

e. $\Delta G° = \Delta H° - T \Delta S°$
$\Delta G° = 6.00$ kJ - (293 K)(0.0220 kJ/K)
$\Delta G° = -0.45$ kJ $\qquad$ But ice does melt spontaneously at +20°C.

---

85. *Refer to Section 17.5.*

a. Calculate $\Delta S°$ at 25°C using the Gibbs-Helmholtz equation, then calculate $\Delta G°$ at 37°C. Calculate the amount of energy from one gram sugar with 25% efficiency.

$\Delta G° = \Delta H° - T\,\Delta S°$

-5790 kJ = -5650 kJ - (298 K)($\Delta S°$)

$\Delta S° = 0.470$ kJ/K

$\Delta G° = \Delta H° - T\,\Delta S°$

$\Delta G° = -5650$ kJ - (310 K)(0.470 kJ/K)

$\Delta G° = -5796$ kJ

$$\frac{-5796 \text{ kJ}}{\text{mol.}} \times \frac{1 \text{ mol.}}{342.0 \text{ g}} \times 1.00 \text{ g} \times 0.25 = -4.2 \text{ kJ}$$

Thus one gram of sugar provides 4.2 kJ of work.

b. $m = 120 \text{ lb} \times \dfrac{453.6 \text{ g}}{1 \text{ lb}} \times \dfrac{1 \text{ kg}}{1000 \text{ g}} = 54.4 \text{ kg}$

$w = (9.79 \times 10^{-3})(54.4 \text{ kg})(4158 \text{ m}) = 2210 \text{ kJ}$

$2210 \text{ kJ} \times \dfrac{1.00 \text{ g sugar}}{4.2 \text{ kJ}} = 530 \text{ g sugar}$

Thus, a little more than one pound of sugar will provide enough energy for a 120 lb person to climb a 2.5 mile high mountain.

---

86. *Refer to Section 17.5.*

$CaH_2(s) \rightarrow Ca(s) + H_2(g)$

$\Delta H° = -\Delta H_f° = +186.2$ kJ

$\Delta S° = \Sigma\, S°_{products} - \Sigma\, S°_{reactants}$
$\Delta S° = [(1 \text{ mol.})(0.0414 \text{ kJ/mol.·K}) + (1 \text{ mol.})(0.1306 \text{ kJ/mol.·K})]$
    $- [(1 \text{ mol.})(0.0420 \text{ kJ/mol.·K})]$
$\Delta S° = 0.130$ kJ/K

$\Delta G° = -RT \ln K = -RT \ln (1 \text{ atm } H_2) = 0$ kJ
$\Delta G° = \Delta H° - T\,\Delta S°$
0 kJ = 186.2 kJ - (T)(0.130 kJ/K)
$T = 1432$ K = 1159°C

This is not a feasible source of H_2 fuel.

---

87. *Refer to Sections 17.4, 17.5 and 17.7.*

$Cu(s) + \tfrac{1}{2}O_2(g) \rightarrow CuO(s)$

$\Delta G° = \Sigma\, \Delta G_f°{}_{products} - \Sigma\, \Delta G_f°{}_{reactants}$
$\Delta G° = [(1 \text{ mol.})(-129.7 \text{ kJ/mol.})] - [(1 \text{ mol.})(0.0 \text{ kJ/mol.}) + (\tfrac{1}{2} \text{ mol.})(0.0 \text{ kJ/mol.})]$
$\Delta G° = -129.7$ kJ
$\Delta G°$ is negative, so the black CuO forms spontaneously at room temperature.

$2CuO(s) \rightarrow Cu_2O(s) + \frac{1}{2}O_2(g)$

$\Delta H° = \Sigma \Delta H_f°_{products} - \Sigma \Delta H_f°_{reactants}$
$\Delta H° = [(1 \text{ mol.})(-168.6 \text{ kJ/mol.}) + (\frac{1}{2} \text{ mol.})(0.0 \text{ kJ/mol.})] - [(2 \text{ mol.})(-157.3 \text{ kJ/mol.})]$
$\Delta H° = 146.0 \text{ kJ}$

$\Delta S° = \Sigma S°_{products} - \Sigma S°_{reactants}$
$\Delta S° = [(1 \text{ mol.})(0.0931 \text{ kJ/mol.·K}) + (\frac{1}{2} \text{ mol.})(0.2050 \text{ kJ/mol.·K})]$
$\qquad - [(2 \text{ mol.})(0.0426 \text{ kJ/mol.·K})]$
$\Delta S° = 0.1104 \text{ kJ/K}$

The reaction becomes spontaneous when $\Delta G° < 0$, giving
$\Delta G° = \Delta H° - T\Delta S°$
$0 = 146.0 \text{ kJ} - (T)(0.1104 \text{ kJ/K})$
$T = 1322 \text{ K} = 1049°C$ (Reaction becomes spontaneous above 1049°C)

Thus the black CuO converts to red Cu$_2$O when the temperature exceeds 1049°C.

$Cu_2O(s) \rightarrow 2Cu(s) + \frac{1}{2}O_2(g)$

$\Delta H° = \Sigma \Delta H_f°_{products} - \Sigma \Delta H_f°_{reactants}$
$\Delta H° = [(2 \text{ mol.})(0.0 \text{ kJ/mol.}) + (\frac{1}{2} \text{ mol.})(0.0 \text{ kJ/mol.})] - [(1 \text{ mol.})(-168.6 \text{ kJ/mol.})]$
$\Delta H° = 168.6 \text{ kJ}$

$\Delta S° = \Sigma S°_{products} - \Sigma S°_{reactants}$
$\Delta S° = [(2 \text{ mol.})(0.0332 \text{ kJ/mol.}) + (\frac{1}{2} \text{ mol.})(0.2050 \text{ kJ/mol.})] - [(1 \text{ mol.})(0.0931 \text{ kJ/mol.})]$
$\Delta S° = 0.0758 \text{ kJ}$

$\Delta G° = \Delta H° - T\Delta S°$
The reaction becomes spontaneous when $\Delta G° < 0$, giving
$0 = 168.6 \text{ kJ} - (T)(0.0758 \text{ kJ/K})$
$T = 2220 \text{ K} = 1950°C$ (Reaction becomes spontaneous above 1950°C)

Thus the red Cu$_2$O reverts to metallic Cu when the temperature exceeds 1950°C.

Chapter 18: Electrochemistry

---

2. Refer to Sections 4.4 and 18.1.

Remember that the oxidation is always shown on the left side of the cell notation. For part (c), note that the Pt serves as an inert electrode and is not part of the chemical equation.

a. $Ag(s) \rightarrow Ag^+(aq) + e^-$ oxidation half-reaction
      $Sn^{4+}(aq) + 2e^- \rightarrow Sn^{2+}(aq)$ reduction half-reaction

Multiply the oxidation half-reaction by two, then add the two half-reactions together.

$2Ag(s) \rightarrow 2Ag^+(aq) + 2e^-$
$\underline{Sn^{4+}(aq) + 2e^- \rightarrow Sn^{2+}(aq)}$
$2Ag(s) + Sn^{4+}(aq) + 2e^- \rightarrow 2Ag^+(aq) + 2e^- + Sn^{2+}(aq)$

$2Ag(s) + Sn^{4+}(aq) \rightarrow 2Ag^+(aq) + Sn^{2+}(aq)$

b. $Al(s) \rightarrow Al^{3+}(aq) + 3e^-$ oxidation half-reaction
      $Cu^{2+}(aq) + 2e^- \rightarrow Cu(s)$ reduction half-reaction

Multiply the oxidation half-reaction by two and the reduction half-reaction by three, then add the two half-reactions together.

$2Al(s) \rightarrow 2Al^{3+}(aq) + 6e^-$
$\underline{3Cu^{2+}(aq) + 6e^- \rightarrow 3Cu(s)}$
$2Al(s) + 3Cu^{2+}(aq) + 6e^- \rightarrow 2Al^{3+}(aq) + 6e^- + 3Cu(s)$
$2Al(s) + 3Cu^{2+}(aq) \rightarrow 2Al^{3+}(aq) + 3Cu(s)$

c.   Note: The cell notation should be:

$Pt \mid Fe^{2+}, Fe^{3+} \parallel MnO_4^-, H^+ \mid Mn^{2+} \mid Pt$

$Fe^{2+}(aq) \rightarrow Fe^{3+}(aq) + e^-$ oxidation half-reaction
$MnO_4^-(aq) + 5e^- \rightarrow Mn^{2+}(aq)$ reduction half-reaction
$MnO_4^-(aq) + 5e^- + 8H^+(aq) \rightarrow Mn^{2+}(aq)$
$MnO_4^-(aq) + 5e^- + 8H^+(aq) \rightarrow Mn^{2+}(aq) + 4H_2O(l)$

Balance electrons by multiplying the oxidation equation by 5, then add the two half-reactions together.

$5Fe^{2+}(aq) \rightarrow 5Fe^{3+}(aq) + 5e^-$
$\underline{MnO_4^-(aq) + 5e^- + 8H^+(aq) \rightarrow Mn^{2+}(aq) + 4H_2O(l)}$
$5Fe^{2+}(aq) + MnO_4^-(aq) + 5e^- + 8H^+(aq) \rightarrow 5Fe^{3+}(aq) + 5e^- + Mn^{2+}(aq) + 4H_2O(l)$
$5Fe^{2+}(aq) + MnO_4^-(aq) + 8H^+(aq) \rightarrow 5Fe^{3+}(aq) + Mn^{2+}(aq) + 4H_2O(l)$

a. Sn(s) is oxidized at the anode to $Sn^{2+}(aq)$.
   $Ag^+(aq)$ is reduced at the cathode to Ag(s).

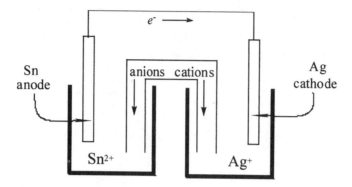

b. $H_2(g)$ is oxidized at the anode to $H^+(aq)$.
   $Hg_2^{2+}(Hg_2Cl_2)$ is reduced at the cathode to Hg(l).

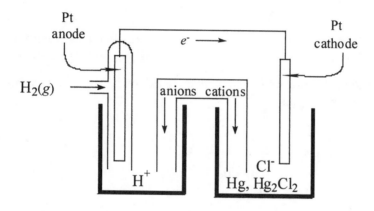

c. Pb(s) is oxidized at the anode to $Pb^{2+}(PbSO_4(s))$.
   $Pb^{4+}(PbO_2(s))$ is reduced at the cathode to $Pb^{2+}(PbSO_4(s))$.

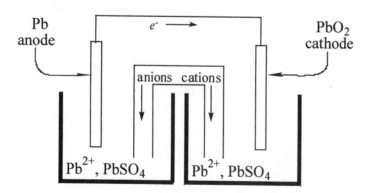

6. _Refer to Section 18.1 and Example 18.1._

$2I^-(aq) \rightarrow I_2(s) + 2e^-$ oxidation half-reaction: anode

$\underline{Br_2(l) + 2e^- \rightarrow 2Br^-(aq)}$ reduction half-reaction: cathode

$2I^-(aq) + Br_2(l) + 2e^- \rightarrow I_2(s) + 2e^- + 2Br^-(aq)$

$2I^-(aq) + Br_2(l) \rightarrow I_2(s) + 2Br^-(aq)$

$Pt \mid I^- \mid I_2 \parallel Br_2 \mid Br^- \mid Pt$

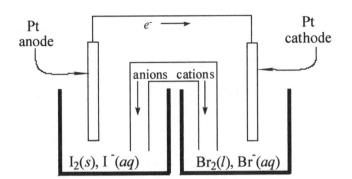

8. _Refer to Section 18.2 and Example 18.2._

A reducing agent is oxidized in the process of reducing another species. The more effective reducing agent is the one that is more easily oxidized; thus the species with the lower reduction potential is the better reducing agent.

a. $Cr^{3+}(aq) + 3e^- \rightarrow Cr(s)$ $E^\circ_{red} = -0.744$ V

       $Cd^{2+}(aq) + 2e^- \rightarrow Cd(s)$ $E^\circ_{red} = -0.402$ V

       Cr is the better reducing agent.

b. $I_2(s) + 2e^- \rightarrow 2I^-(aq)$ $E^\circ_{red} = 0.534$ V

       $Br_2(l) + 2e^- \rightarrow 2Br^-(aq)$ $E^\circ_{red} = 1.077$ V

       I^- is the better reducing agent.

c. $O_2(g) + 2H_2O(l) + 4e^- \rightarrow 4OH^-(aq)$ $E^\circ_{red} = 0.401$ V

       $NO_3^-(aq) + H_2O(l) + 2e^- \rightarrow NO_2^-(aq) + 2OH^-(aq)$ $E^\circ_{red} = 0.004$ V

       NO_2^- is the better reducing agent.

d. $NO_3^-(aq) + 4H^+(aq) + 3e^- \rightarrow NO(g) + 2H_2O(l)$ $E^\circ_{red} = 0.964$ V

       $NO_3^-(aq) + 2H_2O(l) + 3e^- \rightarrow NO(g) + 4OH^-(aq)$ $E^\circ_{red} = -0.140$ V

       NO in basic solution is the better reducing agent.

10. Refer to Section 18.2 and Example 18.2.

	Al^{3+}	Ni^{2+}	$AgBr$	ClO_3^- (acidic)	F_2
E_{ox}° (V)	-1.68	-0.236	0.073	1.458	2.889

The species with the highest E_{red}° is the strongest oxidizing agent, thus:

$Al^{3+} < Ni^{2+} < AgBr < ClO_3^-$ (acidic) $< F_2$

12. Refer to Section 18.2.

Cu^+ can be both an oxidizing and a reducing agent (E_{red}° = 0.518 V, E_{ox}° = -0.161 V).

Zn is a reducing agent (E_{ox}° = 0.762 V).

Ni^{2+} is an oxidizing agent (E_{red}° = -0.236 V).

Fe^{2+} can be both an oxidizing and a reducing agent (E_{red}° = -0.409 V, E_{ox}° = -0.769 V).

H^+ is an oxidizing agent (E_{red}° = 0.00 V).

The species with the highest E_{red}° is the strongest oxidizing agent, thus:
$Fe^{2+} < Ni^{2+} < H^+ < Cu^+$

The species with the highest E_{ox}° is the stronger reducing agent, thus:
$Fe^{2+} < Cu^+ < Zn$

Cu^+ and Fe^{2+} can be both oxidizing and reducing agents.

14. Refer to Section 18.2.

Write the half-reactions for the conversions listed, with the corresponding E° value. A species that will effect one transformation without the other will have an E° value between that of the two half-reactions.

a. $ClO_4^-(aq) + H_2O(l) + 2e^- \rightarrow ClO_3^-(aq) + 2OH^-(aq)$ E_{red}° = 0.398 V

    $ClO_3^-(aq) + 3H_2O(l) + 6e^- \rightarrow Cl^-(aq) + 6OH^-(aq)$ E_{red}° = 0.614 V

    Oxidizing agent is O_2

    $O_2(g) + 2H_2O(l) + 4e^- \rightarrow 4OH^-(aq)$ E_{red}° = 0.401 V

b. $Mg^{2+}(aq) + 2e^- \rightarrow Mg(s)$ E_{red}° = -2.357 V

    $Ba^{2+}(aq) + 2e^- \rightarrow Ba(s)$ E_{red}° = -2.906 V

    reducing agent: $Ca(s)$ (or $Na(s)$)

    $Ca^{2+}(aq) + 2e^- \rightarrow Ca(s)$ E_{red}° = -2.869 V

c. $Na^+(aq) + e^- \rightarrow Na(s)$ $E^\circ_{red} = -2.714$ V

    $Li^+(aq) + e^- \rightarrow Li(s)$ $E^\circ_{red} = -3.040$ V

    reducing agent: $Ca(s)$ (or $K(s)$ or $Ba(s)$)

    $Ca^{2+}(aq) + 2e^- \rightarrow Ca(s)$ $E^\circ_{red} = -2.869$ V

---

16. Refer to Section 18.2 and Example 18.3.

Write the half-reactions and the associated redox potentials. When combining the half-reactions, add the potentials to calculate E°. Bear in mind that **multiplying a half-reaction by a factor does not change that E° value**.

a. $Pb(s) \rightarrow Pb^{2+}(aq) + 2e^-$ $E^\circ_{ox} = 0.127$ V

    $2[\,Ag^+(aq) + e^- \rightarrow Ag(s)\,]$ $E^\circ_{red} = 0.799$ V

    _____

    $Pb(s) + 2Ag^+(aq) + 2e^- \rightarrow 2Ag(s) + Pb^{2+}(aq) + 2e^-$

    $Pb(s) + 2Ag^+(aq) \rightarrow 2Ag(s) + Pb^{2+}(aq)$ $E^\circ = 0.926$ V

b. $4[\,Fe^{2+}(aq) \rightarrow Fe^{3+}(aq) + e^-\,]$ $E^\circ_{ox} = -0.769$ V

    $O_2(g) + 4H^+(aq) + 4e^- \rightarrow 2H_2O(l)$ $E^\circ_{red} = 1.229$ V

    _____

    $4Fe^{2+}(aq) + O_2(g) + 4H^+(aq) + 4e^- \rightarrow 2H_2O(l) + 4Fe^{3+}(aq) + 4e^-$

    $4Fe^{2+}(aq) + O_2(g) + 4H^+(aq) \rightarrow 2H_2O(l) + 4Fe^{3+}(aq)$ $E^\circ = 0.460$ V

c.   Reverse the Zn-Zn^{2+} reaction and change the sign of E° to obtain a positive E°(cell).

    $Zn(s) \rightarrow Zn^{2+}(aq) + 2e^-$ $E^\circ_{ox} = 0.762$ V

    $Cd^{2+}(aq) + 2e^- \rightarrow Cd(s)$ $E^\circ_{red} = -0.402$ V

    _____

    $Zn(s) + Cd^{2+}(aq) + 2e^- \rightarrow Cd(s) + Zn^{2+}(aq) + 2e^-$

    $Zn(s) + Cd^{2+}(aq) \rightarrow Cd(s) + Zn^{2+}(aq)$ $E^\circ = 0.360$ V

---

18. Refer to Section 18.2, Table 18.1, and Example 18.3.

Write the half-reactions and the associated redox potentials. When combining the half-reactions, add the potentials to calculate E°. Bear in mind that **multiplying a half-reaction by a factor does not change that E° value**.

a. $2[Cr^{2+}(aq) \rightarrow Cr^{3+}(aq) + e^-]$ $E^\circ_{ox} = 0.408$ V

    $Sn^{4+}(aq) + 2e^- \rightarrow Sn^{2+}(aq)$ $E^\circ_{red} = 0.154$ V

    _____

    $Sn^{4+}(aq) + 2e^- + 2Cr^{2+}(aq) \rightarrow 2Cr^{3+}(aq) + 2e^- + Sn^{2+}(aq)$

    $Sn^{4+}(aq) + 2Cr^{2+}(aq) \rightarrow 2Cr^{3+}(aq) + Sn^{2+}(aq)$ $E^\circ = 0.562$ V

b. $Mn^{2+}(aq) + 2H_2O(l) \rightarrow MnO_2(s) + 4H^+(aq) + 2e^-$ $E^\circ_{ox} = -1.229$ V

    $H_2O_2(aq) + 2H^+(aq) + 2e^- \rightarrow 2H_2O(l)$ $E^\circ_{red} = 1.763$ V

    _____

    $Mn^{2+}(aq) + 2H_2O(l) + H_2O_2(aq) + 2H^+(aq) + 2e^-$

            $\rightarrow 2H_2O(l) + MnO_2(s) + 4H^+(aq) + 2e^-$

    $Mn^{2+}(aq) + H_2O_2(aq) \rightarrow MnO_2(s) + 2H^+(aq)$ $E^\circ = 0.534$ V

Write the half-reactions and the associated redox potentials. When combining the half-reactions, add the potentials to calculate $E°$. Bear in mind that **multiplying a half-reaction by a factor does not change that $E°$ value**.

a. $Pb(s) + SO_4^{2-}(aq) \rightarrow PbSO_4(s) + 2e^-$ $E°_{ox} = 0.356$ V

 $Pb^{2+}(aq) + 2e^- \rightarrow Pb(s)$ $E°_{red} = -0.127$ V

 $\overline{Pb(s) + SO_4^{2-}(aq) + Pb^{2+}(aq) + 2e^- \rightarrow PbSO_4(s) + 2e^- + Pb(s)}$

 $SO_4^{2-}(aq) + Pb^{2+}(aq) \rightarrow PbSO_4(s)$ $E° = 0.229$ V

b. $4[½Cl_2(g) + 3H_2O(l) \rightarrow ClO_3^-(aq) + 6H^+(aq) + 5e^-]$ $E°_{ox} = -1.458$ V

 $5[O_2(g) + 4H^+(aq) + 4e^- \rightarrow 2H_2O(l)]$ $E°_{red} = 1.229$ V

 $\overline{2Cl_2(g) + 12H_2O(l) + 5O_2(g) + 20H^+(aq) + 20e^-}$

 $\rightarrow 4ClO_3^-(aq) + 24H^+(aq) + 20e^- + 10H_2O(l)$

 $2Cl_2(g) + 2H_2O(l) + 5O_2(g) \rightarrow 4ClO_3^-(aq) + 4H^+(aq)$ $E° = -0.229$ V

c. $3[4OH^-(aq) \rightarrow O_2(g) + 2H_2O(l) + 4e^-]$ $E°_{ox} = -0.401$ V

 $2[ClO_3^-(aq) + 3H_2O(l) + 6e^- \rightarrow Cl^-(aq) + 6OH^-(aq)]$ $E°_{red} = 0.614$ V

 $\overline{12OH^-(aq) + 2ClO_3^-(aq) + 6H_2O(l) + 12e^-}$

 $\rightarrow 3O_2(g) + 6H_2O(l) + 12e^- + 2Cl^-(aq) + 12OH^-(aq)$

 $2ClO_3^-(aq) \rightarrow 3O_2(g) + 2Cl^-(aq)$ $E° = 0.213$ V

Readjust the reference potential by adding +0.300V to each $E°_{red}$ value.

a. $E°_{ox} = - E°_{red} = -0.300$ V

b. $E°_{red} = 1.077$ V $+ 0.300$ V $= 1.377$ V

c. $3[4OH^-(aq) \rightarrow O_2(g) + 2H_2O(l) + 4e^-]$ $E°_{ox} = -0.701$ V

 $2[ClO_3^-(aq) + 3H_2O(l) + 6e^- \rightarrow Cl^-(aq) + 6OH^-(aq)]$ $E°_{red} = 0.914$ V

 $\overline{12OH^-(aq) + 2ClO_3^-(aq) + 6H_2O(l) + 12e^-}$

 $\rightarrow 3O_2(g) + 6H_2O(l) + 12e^- + 2Cl^-(aq) + 12OH^-(aq)$

 $2ClO_3^-(aq) \rightarrow 3O_2(g) + 2Cl^-(aq)$ $E° = 0.213$ V

The two values are the same because $E°$ measures the difference between the two half-reactions.

A reaction is spontaneous when $\Delta G°$ is negative. Since $\Delta G° = -nFE°$, and n and F are positive constants, $\Delta G°$ is negative (and the reaction spontaneous) when $E°$ is positive.

a. $Zn(s) \rightarrow Zn^{2+}(aq) + 2e^-$ $E°_{ox} = 0.762$ V

$2[\, Fe^{3+}(aq) + e^- \rightarrow Fe^{2+}(aq)\,]$ $E°_{red} = 0.799$ V

$Zn(s) + 2Fe^{3+}(aq) + 2e^- \rightarrow 2Fe^{2+}(aq) + Zn^{2+}(aq) + 2e^-$

$Zn(s) + 2Fe^{3+}(aq) \rightarrow 2Fe^{2+}(aq) + Zn^{2+}(aq)$ $E° = 1.561$ V

$E°$ is positive, so the reaction is **spontaneous**.

b. $Cu(s) \rightarrow Cu^{2+}(aq) + 2e^-$ $E°_{ox} = -0.339$ V

$2H^+(aq) + 2e^- \rightarrow H_2(g)$ $E°_{red} = 0.0$ V

$Cu(s) + 2H^+(aq) + 2e^- \rightarrow Cu^{2+}(aq) + H_2(g) + 2e^-$

$Cu(s) + 2H^+(aq) \rightarrow Cu^{2+}(aq) + H_2(g)$ $E° = -0.339$ V

$E°$ is negative, so the reaction is **not** spontaneous.

c. $2Br^-(aq) \rightarrow Br_2(l) + 2e^-$ $E°_{ox} = -1.077$ V

$I_2(s) + 2e^- \rightarrow 2I^-(aq)$ $E°_{red} = 0.534$ V

$2Br^-(aq) + 2e^- + I_2(s) \rightarrow 2I^-(aq) + 2e^- + Br_2(l)$

$2Br^-(aq) + I_2(s) \rightarrow 2I^-(aq) + Br_2(l)$ $E° = -0.543$ V

$E°$ is negative, so the reaction is **not** spontaneous.

a. $ClO_3^-(aq) + 6H^+(aq) + 5e^- \rightarrow \frac{1}{2}Cl_2(g) + 3H_2O(l)$ $E°_{red} = 1.458$ V

$2Cl^-(aq) \rightarrow Cl_2(g) + 2e^-$ $E°_{ox} = -1.360$ V

Adding these two half-reactions gives a positive $E°$.

$ClO_3^-(aq) + 6H^+(aq) + 5e^- \rightarrow \frac{1}{2}Cl_2(g) + 3H_2O(l)$ $E°_{red} = 1.458$ V

$2F^-(aq) \rightarrow F_2(g) + 2e^-$ $E°_{ox} = -2.889$ V

Adding these two half-reactions gives a negative $E°$.
Thus, ClO_3^- will oxidize Cl^-, but not F^-.

b. $Fe(s) \rightarrow Fe^{2+}(aq) + 2e^-$ $E°_{ox} = 0.409$ V

$Cr^{3+}(aq) + 3e^- \rightarrow Cr(s)$ $E°_{red} = -0.744$ V

Adding these two half-reactions gives a negative $E°$.

$Fe(s) \rightarrow Fe^{2+}(aq) + 2e^-$ $E^\circ_{ox} = 0.409$ V

$Cr^{3+}(aq) + e^- \rightarrow Cr^{2+}(aq)$ $E^\circ_{red} = -0.408$ V

Adding these two half-reactions gives a positive E°.
Thus, Fe reduces Cr^{3+} to Cr^{2+}.

c. $ClO_3^-(aq) + H_2O(l) \rightarrow ClO_4^-(aq) + 2H^+(aq) + 2e^-$ $E^\circ_{ox} = -1.19$ V

    $O_2(g) + 4H^+(aq) + 4e^- \rightarrow 2H_2O(l)$ $E^\circ_{red} = 1.229$ V

    Adding these two half-reactions gives a positive E°.
    Thus, O_2 is reduced spontaneously in acidic media.

    $ClO_3^-(aq) + 2OH^-(aq) \rightarrow ClO_4^-(aq) + H_2O(l) + 2e^-$ $E^\circ_{ox} = -0.398$ V

    $O_2(g) + 2H_2O(l) + 4e^- \rightarrow 4OH^-(aq)$ $E^\circ_{red} = 0.401$ V

    Adding these two half-reactions gives a positive E°.
    Thus, O_2 is reduced spontaneously in basic media also.

---

28. Refer to Sections 18.2 and 18.3 and Example 18.4.

Consider the possible half-reactions and associated potentials. If the potentials of a pair of redox half-reactions give a positive E°, then a reaction will occur.

a. $S(s) + Hg(l) \rightarrow$?

    $S(s) + 2H^+(aq) + 2e^- \rightarrow H_2S(aq)$ $E^\circ_{red} = 0.144$ V

    $2Hg(l) \rightarrow Hg_2^{2+}(aq) + 2e^-$ $E^\circ_{ox} = -0.769$ V

    Adding E°_{ox} to E°_{red} gives a negative E°, thus no reaction will occur.

b. $MnO_2(s) + Hg(l) \rightarrow$?

    $MnO_2(s) + 4H^+(aq) + 2e^- \rightarrow Mn^{2+}(aq) + 2H_2O(l)$ $E^\circ_{red} = 1.229$ V

    $\underline{2Hg(l) \rightarrow Hg_2^{2+}(aq) + 2e^- \hspace{4cm} E^\circ_{ox} = -0.796 \text{ V}}$

    $MnO_2(s) + 4H^+(aq) + 2Hg(l) + 2e^- \rightarrow Mn^{2+}(aq) + 2H_2O(l) + Hg_2^{2+}(aq) + 2e^-$

    $MnO_2(s) + 4H^+(aq) + 2Hg(l) \rightarrow Mn^{2+}(aq) + 2H_2O(l) + Hg_2^{2+}(aq)$ $E^\circ = 0.433$ V

c. $Al(s) + K^+(aq) \rightarrow$?

    $Al(s) \rightarrow Al^{3+}(aq) + 3e^-$ $E^\circ_{ox} = +1.68$ V

    $K^+(aq) + e^- \rightarrow K(s)$ $E^\circ_{red} = -2.936$ V

    Adding E°_{ox} to E°_{red} gives a negative E°, thus no reaction will occur.

---

30. Refer to Sections 18.2 and 18.3 and Example 18.4.

If HCl is oxidizing one of the listed species, then the HCl itself must be reduced. Since the Cl^- is already in a reduced state, the likely species to be reduced is H^+.

a. $2H^+(aq) + 2e^- \rightarrow 2H_2(g)$ $E^\circ_{red} = 0.0$ V

     $Au(s) \rightarrow Au^{3+}(aq) + 3e^-$ $E^\circ_{ox} = -1.498$ V

     $Au(s) + 4Cl^- \rightarrow AuCl_4^-(aq) + 3e^-$ $E^\circ_{ox} = -1.001$ V

     Adding E°_{red} to either E°_{ox} gives a negative E°, thus HCl will **not** oxidize Au.

b. $2H^+(aq) + 2e^- \rightarrow 2H_2(g)$ $E^\circ_{red} = 0.0$ V

     $Mg(s) \rightarrow Mg^{2+}(aq) + 2e^-$ $E^\circ_{ox} = 2.357$ V

     Adding E°_{red} to E°_{ox} gives a positive E°, thus HCl **will** oxidize Mg.

c. $2H^+(aq) + 2e^- \rightarrow 2H_2(g)$ $E^\circ_{red} = 0.0$ V

     $Cu(s) \rightarrow Cu^{2+}(aq) + 2e^-$ $E^\circ_{ox} = -0.339$ V

     $Cu(s) \rightarrow Cu^+(aq) + e^-$ $E^\circ_{ox} = -0.518$ V

     Adding E°_{red} to either E°_{ox} gives a negative E°, thus HCl will **not** oxidize Cu.

d. $2H^+(aq) + 2e^- \rightarrow 2H_2(g)$ $E^\circ_{red} = 0.0$ V

     $2F^-(aq) \rightarrow F_2(g) + 2e^-$ $E^\circ_{ox} = -2.889$ V

     Adding E°_{red} to either E°_{ox} gives a negative E°, thus HCl will **not** oxidize F^-.

---

32. Refer to Sections 18.2 and 18.3 and Problem 26c (above).

Consider the possible half-reactions and associated potentials. Consider both cations and anions. If the potentials of a pair of redox half-reactions give a positive E°, then a reaction will occur.

$S(s) + 2H^+(aq) + 2e^- \rightarrow H_2S(aq)$ $E^\circ_{red} = 0.144$ V

a. $2Br^-(aq) \rightarrow Br_2(l) + 2e^-$ $E^\circ_{ox} = -0.769$ V

     Mg^{2+} is already oxidized.

     Adding E°_{ox} to E°_{red} gives a negative E°, thus no reaction will occur.

b. $Sn^{2+}(aq) \rightarrow Sn^{4+}(aq) + 2e^-$ $E^\circ_{ox} = -0.154$ V

     NO_3^- is already oxidized.

     Adding E°_{ox} to E°_{red} gives a negative E°, thus no reaction will occur.

c. Adding E°_{ox} (ClO_3^-) (see Problem 26(c)) to E°_{red} (S) gives a negative E°, thus S will not oxidize ClO_3^-.

     $2[Cr^{2+}(aq) \rightarrow Cr^{3+}(aq) + e^-]$ $E^\circ_{ox} = 0.408$ V

     $\underline{S(s) + 2H^+(aq) + 2e^- \rightarrow H_2S(aq) \qquad\qquad\qquad}$ $\underline{E^\circ_{red} = 0.144}$ V

     $2Cr^{2+}(aq) + S(s) + 2H^+(aq) + 2e^- \rightarrow 2Cr^{3+}(aq) + 2e^- + H_2S(aq)$

     $2Cr^{2+}(aq) + S(s) + 2H^+(aq) \rightarrow 2Cr^{3+}(aq) + H_2S(aq)$ $E^\circ = 0.552$ V

	ΔG°	E°	K
a.	**16 kJ**	**-0.041 V**	1.6×10^{-3}
b.	**-45.2 kJ**	0.117 V	**8.1×10^{7}**
c.	-5.8 kJ	**0.015 V**	**10**

a. $E^o = \dfrac{0.0257 \text{ V}}{n} \cdot \ln K = \dfrac{0.0257 \text{ V}}{4} \cdot \ln (1.6 \times 10^{-3})$

$E^\circ = -0.041$ V

$\Delta G^\circ = -nFE^\circ$
$\Delta G^\circ = -(4 \text{ mol.})(9.648 \times 10^4 \text{ J/mol.} \cdot \text{V})(-0.041 \text{ V})$
$\Delta G^\circ = 1.6 \times 10^4 \text{ J} = 16 \text{ kJ}$

b. $\Delta G^\circ = -nFE^\circ$
$\Delta G^\circ = -(4 \text{ mol.})(9.648 \times 10^4 \text{ J/mol.} \cdot \text{V})(0.117 \text{ V})$
$\Delta G^\circ = -4.52 \times 10^4 \text{ J} = -45.2 \text{ kJ}$

$E^o = \dfrac{0.0257 \text{ V}}{n} \cdot \ln K \quad \Rightarrow \quad 0.117 \text{ V} = \dfrac{0.0257 \text{ V}}{4} \cdot \ln K$

$\ln K = 18.2$
$K = e^{18.2} = 8.1 \times 10^7$

c. $\Delta G^\circ = -nFE^\circ$
$-5800 \text{ J} = -(4 \text{ mol.})(9.648 \times 10^4 \text{ J/mol.} \cdot \text{V})(E^\circ)$
$\Delta E^\circ = 0.015$ V

$E^o = \dfrac{0.0257 \text{ V}}{n} \cdot \ln K \quad \Rightarrow \quad 0.015 \text{ V} = \dfrac{0.0257 \text{ V}}{4} \cdot \ln K$

$\ln K = 2.3$
$K = e^{2.3} = 10$

a. $\Delta G^\circ = -nFE^\circ$
$\Delta G^\circ = -(1 \text{ mol.})(9.648 \times 10^4 \text{ J/mol.} \cdot \text{V})(1.20 \text{ V})$
$\Delta G^\circ = -1.16 \times 10^5 \text{ J} = -1.16 \times 10^2 \text{ kJ}$

b. $\Delta G^\circ = -nFE^\circ$
$\Delta G^\circ = -(2 \text{ mol.})(9.648 \times 10^4 \text{ J/mol.} \cdot \text{V})(1.20 \text{ V})$
$\Delta G^\circ = -2.32 \times 10^5 \text{ J} = -2.32 \times 10^2 \text{ kJ}$

c. $\Delta G^\circ = -nFE^\circ$
$\Delta G^\circ = -(3 \text{ mol.})(9.648 \times 10^4 \text{ J/mol.} \cdot \text{V})(1.20 \text{ V})$
$\Delta G^\circ = -3.47 \times 10^5 \text{ J} = -3.47 \times 10^2 \text{ kJ}$

The number of electrons exchanged has no effect on the spontaneity of the reaction since changes in the number of electrons does not change the sign of $\Delta G°$.

---

38. Refer to Section 18.3, Table 18.1, and Example 18.5.

Note that in the reaction given, OH⁻ is a product, thus the solution is basic.

a. $3[S^{2-}(aq) \rightarrow S(s) + 2e^-]$ $\qquad\qquad\qquad\qquad$ $E°_{ox} = 0.445$ V

$2[NO_3^-(aq) + 2H_2O(l) + 3e^- \rightarrow NO(g) + 4OH^-(aq)]$ $\qquad$ $E°_{red} = -0.140$ V

$3S^{2-}(aq) + 2NO_3^-(aq) + 4H_2O(l) + 6e^- \rightarrow 3S(s) + 6e^- + 2NO(g) + 8OH^-(aq)$

$3S^{2-}(aq) + 2NO_3^-(aq) + 4H_2O(l) \rightarrow 3S(s) + 2NO(g) + 8OH^-(aq)$ $\quad E° = 0.305$ V

b. $\Delta G° = -nFE°$
$\Delta G° = -(6 \text{ mol.})(9.648 \times 10^4 \text{ J/mol.·V})(0.305 \text{ V})$
$\Delta G° = -1.77 \times 10^5 \text{ J} = -1.77 \times 10^2 \text{ kJ}$

c. $E° = \dfrac{0.0257 \text{ V}}{n} \ln K \Rightarrow 0.305 \text{ V} = \dfrac{0.0257 \text{ V}}{6} \ln K$

$\ln K = 71.2$
$K = e^{71.2} = 8.4 \times 10^{30}$
Note: the answer is different if you use the equation
$\Delta G° = -RT\ln K$:
$-1.77 \times 10^5 \text{ J} = -(8.31 \text{ J/ mol.·K})(298\text{K})\ln K$
$71.4 = \ln K$
$K = e^{71.4} = 1.0 \times 10^{31}$

---

40. Refer to Section 18.3, Example 18.5 and Problem 16 (above).

a. $E° = 0.926$ V, $n = 2e^-$
$\Delta G° = -nFE°$
$\Delta G° = -(2 \text{ mol.})(9.648 \times 10^4 \text{ J/mol.·V})(0.926 \text{ V})$
$\Delta G° = -1.79 \times 10^5 \text{ J} = -1.79 \times 10^2 \text{ kJ}$

b. $E° = 0.460$ V, $n = 4e^-$
$\Delta G° = -nFE°$
$\Delta G° = -(4 \text{ mol.})(9.648 \times 10^4 \text{ J/mol.·V})(0.460 \text{ V})$
$\Delta G° = -1.78 \times 10^5 \text{ J} = -1.78 \times 10^2 \text{ kJ}$

c. $E° = 0.360$ V, $n = 2e^-$
$\Delta G° = -nFE°$
$\Delta G° = -(2 \text{ mol.})(9.648 \times 10^4 \text{ J/mol.·V})(0.360 \text{ V})$
$\Delta G° = -6.95 \times 10^4 \text{ J} = -69.5 \text{ kJ}$

42. Refer to Section 18.3, Example 18.5 and Problem 18 (above).

$$Sn^{4+}(aq) + 2Cr^{2+}(aq) \rightarrow 2Cr^{3+}(aq) + Sn^{2+}(aq) \qquad n = 2 \quad E° = 0.562 \text{ V}$$
$$Mn^{2+}(aq) + H_2O_2(aq) \rightarrow MnO_2(s) + 2H^+(aq) \qquad n = 2 \quad E° = 0.534 \text{ V}$$

a. $E° = \dfrac{RT}{nF} \cdot \ln K = \dfrac{0.0257}{n} \cdot \ln K \Rightarrow 0.562 \text{ V} = \dfrac{0.0257 \text{ V}}{2} \cdot \ln K$

$\ln K = 43.7$

$K = e^{43.7} = 9.5 \times 10^{18}$

b. $E° = \dfrac{RT}{nF} \cdot \ln K = \dfrac{0.0257}{n} \cdot \ln K \Rightarrow 0.534 \text{ V} = \dfrac{0.0257 \text{ V}}{2} \cdot \ln K$

$\ln K = 41.6$

$K = e^{41.6} = 1.1 \times 10^{18}$

---

44. Refer to Sections 18.2 and 18.4 and Example 18.6.

a. $4[NO(g) + 2H_2O(l) \rightarrow NO_3^-(aq) + 4H^+(aq) + 3e^-] \qquad E_{ox}° = -0.964 \text{ V}$

$\underline{3[O_2(g) + 4H^+(aq) + 4e^- \rightarrow 2H_2O(l)] \qquad\qquad\qquad E_{red}° = 1.229 \text{ V}}$

$4NO(g) + 8H_2O(l) + 3O_2(g) + 12H^+(aq) + 12e^-$
$\qquad\qquad \rightarrow 4NO_3^-(aq) + 16H^+(aq) + 12e^- + 6H_2O(l)$

$4NO(g) + 2H_2O(l) + 3O_2(g) \rightarrow 4NO_3^-(aq) + 4H^+(aq) \qquad E° = 0.265 \text{ V}$

b. $E = E° - \dfrac{RT}{nF} \cdot \ln Q = 0.265 \text{ V} - \dfrac{0.0257 \text{ V}}{12} \cdot \ln \dfrac{[H^+]^4[NO_3^-]^4}{(P_{O_2})^3(P_{NO})^4}$

c. $[H^+] = 10^{-2.85} = 1.4 \times 10^{-3}$

$$E = 0.265 \text{ V} - \dfrac{0.0257 \text{ V}}{12} \cdot \ln \dfrac{(1.4\times10^{-3})^4(0.750)^4}{(0.515)^3(0.993)^4}$$

$E = 0.265 \text{ V} - (2.14 \times 10^{-3}\text{V})\cdot\ln(9.2 \times 10^{-12}) = 0.319 \text{ V}$

---

46. Refer to Sections 18.2 and 18.4 and Example 18.6.

a. $2[NO_3^-(aq) + 2H_2O(l) + 3e^- \rightarrow NO(g) + 4OH^-(aq)] \qquad E_{red}° = -0.140 \text{ V}$

$\underline{3[H_2(g) + 2OH^-(aq) \rightarrow 2H_2O(l) + 2e^-] \qquad\qquad\qquad E_{ox}° = 0.828 \text{ V}}$

$2NO_3^-(aq) + 4H_2O(l) + 6e^- + 3H_2(g) + 6OH^-(aq) \rightarrow 2NO(g) + 8OH^-(aq) + 6H_2O(l) + 6e^-$

$2NO_3^-(aq) + 3H_2(g) \rightarrow 2NO(g) + 2OH^-(aq) + 2H_2O(l) \qquad E° = 0.688 \text{ V}$

b. $E = E° - \dfrac{RT}{nF} \cdot \ln Q = 0.688 \text{ V} - \dfrac{0.0257 \text{ V}}{6} \cdot \ln \dfrac{(P_{NO})^2[OH^-]^2}{(P_{H_2})^3[NO_3^-]^2}$

c. $E = 0.688 \text{ V} - \dfrac{0.0257 \text{ V}}{6} \cdot \ln \dfrac{(0.922)^2 (3.2 \times 10^{-3})^2}{(0.437)^3 (0.0315)^2}$

$E = 0.688 \text{ V} - (4.28 \times 10^{-3} \text{ V}) \cdot \ln (0.105) = 0.698 \text{ V}$

---

48. Refer to Sections 18.1, 18.2 and 18.4.

Write the half-reactions, calculate $E°$ and then calculate E using the Nernst equation.

a.　$Zn(s) \rightarrow Zn^{2+}(aq) + 2e^-$ 　　　　　　　　　　　$E°_{ox} = 0.762 \text{ V}$

$\dfrac{Cd^{2+}(aq) + 2e^- \rightarrow Cd(s)}{Cd^{2+}(aq) + Zn(s) \rightarrow Zn^{2+}(aq) + Cd(s)}$ 　　　$\dfrac{E°_{red} = -0.402 \text{ V}}{E° = 0.360 \text{ V}}$

$E = E° - \dfrac{RT}{nF} \cdot \ln Q = 0.360 \text{ V} - \dfrac{0.0257 \text{ V}}{2} \cdot \ln \dfrac{[Zn^{2+}]}{[Cd^{2+}]}$

$E = 0.360 \text{ V} - 0.0129 \cdot \ln \dfrac{0.50}{0.020} = 0.318 \text{ V}$

b.　$Cu(s) \rightarrow Cu^{2+}(aq) + 2e^-$ 　　　　　　　　　　$E°_{ox} = -0.339 \text{ V}$

$\dfrac{2H^+(aq) + 2e^- \rightarrow H_2(g)}{2H^+(aq) + Cu(s) \rightarrow Cu^{2+}(aq) + H_2(g)}$ 　　　$\dfrac{E°_{red} = 0.00 \text{ V}}{E° = -0.339 \text{ V}}$

$E = E° - \dfrac{RT}{nF} \cdot \ln Q = -0.339 \text{ V} - \dfrac{0.0257 \text{ V}}{2} \cdot \ln \dfrac{[Cu^{2+}](P_{H_2})}{[H^+]^2}$

$E = -0.339 \text{ V} - 0.0129 \cdot \ln \dfrac{(0.0010)(1.00)}{(0.010)^2} = -0.369 \text{ V}$

---

50. Refer to Section 18.4 and Example 18.7.

Write the half-reactions and calculate $E°$. Then solve the Nernst equation for $[H^+]$, then calculate pH.

$2[\ Ag(s) + Br^-(aq) \rightarrow AgBr(s) + e^-\ ]$ 　　　　　$E°_{ox} = -0.073 \text{ V}$

$S(s) + 2H^+(aq) + 2e^- \rightarrow H_2S(aq)$ 　　　　　　　$E°_{red} = 0.144 \text{ V}$

$2Ag(s) + 2Br^-(aq) + S(s) + 2H^+(aq) + 2e^- \rightarrow 2AgBr(s) + H_2S(aq) + 2e^-$

$2Ag(s) + 2Br^-(aq) + S(s) + 2H^+(aq) \rightarrow 2AgBr(s) + H_2S(aq)$ 　　$E° = 0.071 \text{ V}$

$E = E° - \dfrac{RT}{nF} \cdot \ln Q = 0.071 \text{ V} - \dfrac{0.0257 \text{ V}}{2} \cdot \ln \dfrac{[H_2S]}{[H^+]^2[Br^-]^2}$

$0 \text{ V} = 0.071 \text{ V} - \dfrac{0.0257 \text{ V}}{2} \cdot \ln \dfrac{(1.00)}{[H^+]^2(1.00)^2} \quad \Rightarrow \quad -0.071 \text{ V} = \dfrac{0.0257 \text{ V}}{2} \cdot (2)\ln[H^+]$

$-2.8 = \ln [H^+]$

$[H^+] = e^{-2.8} = 0.063\ M$

$pH = -\log(0.063) = 1.20$

---

52. Refer to Section 18.4 and Example 18.8.

Write the half-reactions and calculate $E°$. Then solve the Nernst equation for $[H^+]$.

$Zn(s) \rightarrow Zn^{2+}(aq) + 2e^-$ $\qquad\qquad\qquad\qquad E°_{ox} = 0.762\ V$

$2H^+(aq) + 2e^- \rightarrow H_2(g)$ $\qquad\qquad\qquad\qquad E°_{red} = 0.00\ V$

$\overline{Zn(s) + 2H^+(aq) + 2e^- \rightarrow H_2(g) + Zn^{2+}(aq) + 2e^-}$

$Zn(s) + 2H^+(aq) \rightarrow H_2(g) + Zn^{2+}(aq)$ $\qquad\qquad E° = 0.762\ V$

$$E = E° - \frac{RT}{nF} \cdot \ln Q \Rightarrow 0.40\ V = 0.762\ V - \frac{0.0257\ V}{2} \cdot \ln \frac{(P_{H_2})[Zn^{2+}]}{[H^+]^2}$$

$$-0.362\ V \left(-\frac{2}{0.0257\ V} \right) = \ln \frac{(1.00)(1.00)}{[H^+]^2} \Rightarrow 28.2 = \ln(1) - \ln[H^+]^2$$

$$28.2\ V = 0 - 2\ln[H^+] \Rightarrow \ln[H^+] = -14.1$$

$$[H^+] = e^{-14.1} = 8 \times 10^{-7}\ M$$

---

54. Refer to Sections 18.3 and 18.4.

---

a. As seen from the redox equation below, $E°$ is positive, thus, the reaction is spontaneous.

$2[Cr(s) \rightarrow Cr^{3+}(aq) + 3e^-]$ $\qquad\qquad\qquad\qquad E°_{ox} = +0.744\ V$

$3[SO_4^{2-}(aq) + 4H^+(aq) + 2e^- \rightarrow SO_2(g) + 2H_2O(l)\ ]$ $\qquad E°_{red} = 0.155\ V$

$\overline{2Cr(s) + 3SO_4^{2-}(aq) + 12H^+(aq) + 6e^- \rightarrow 2Cr^{3+}(aq) + 3SO_2(g) + 6H_2O(l) + 6e^-}$

$2Cr(s) + 3SO_4^{2-}(aq) + 12H^+(aq) \rightarrow 2Cr^{3+}(aq) + 3SO_2(g) + 6H_2O(l)$ $\quad E° = 0.899\ V$

b. $[H^+] = 10^{-pH} = 10^{-3.00} = 1.0 \times 10^{-3}\ M$

$$E = E° - \frac{RT}{nF} \cdot \ln Q = 0.899\ V - \frac{0.0257\ V}{6} \cdot \ln \frac{(P_{SO_2})^3[Cr^{3+}]^2}{[SO_4^{2-}]^3[H^+]^{12}}$$

$$E = 0.899\ V - \frac{0.0257\ V}{6} \cdot \ln \frac{(1.00)^3(0.100)^2}{(0.100)^3(1.0 \times 10^{-3})^{12}}$$

$$E = 0.899\ V - (0.00428\ V) \cdot \ln(1.0 \times 10^{37})$$

$$E = 0.534\ V$$

E is positive, so the reaction is spontaneous.

c. $[H^+] = 10^{-pH} = 10^{-8.00} = 1.0 \times 10^{-8}\ M$

$$E = 0.899\ V - \frac{0.0257\ V}{6} \cdot \ln \frac{(1.00)^3 (0.100)^2}{(0.100)^3 (1.0 \times 10^{-8})^{12}}$$

$E = 0.899\ V - (0.00428\ V) \cdot \ln (1.0 \times 10^{97})$

$E = -0.058\ V$

E is negative, so the reaction is not spontaneous.

d. At equilibrium, $E = 0$.

$$E^\circ = \frac{RT}{nF} \cdot \ln K \Rightarrow 0.899\ V = \frac{0.0257\ V}{6} \cdot \ln \frac{(P_{SO_2})^3 [Cr^{3+}]^2}{[SO_4^{2-}]^3 [H^+]^{12}}$$

$$0.899\ V = \frac{0.0257\ V}{6} \cdot \ln \frac{(1.00)^3 (0.100)^2}{(0.100)^3 [H^+]^{12}}$$

$$0.899\ V \cdot \frac{6}{0.0257\ V} = -\ln(0.100) - \ln[H^+]^{12}$$

$$0.899\ V \cdot \frac{6}{0.0257\ V} - 2.303 = -12\ln[H^+]$$

$[H^+] = e^{-17.3} = 3.07 \times 10^{-8}\ M$

$pH = \log(3.07 \times 10^{-8}) = 7.51$

---

56. Refer to Section 18.4.

a. | $Pb(s) \rightarrow Pb^{2+}(aq) + 2e^-$ | $E^\circ_{ox} = 0.127\ V$ |
    | --- | --- |
    | $2H^+(aq) + 2e^- \rightarrow H_2(g)$ | $E^\circ_{red} = 0.0\ V$ |
    | $Pb(s) + 2H^+(aq) \rightarrow Pb^{2+}(aq) + H_2(g)$ | $E^\circ = 0.127\ V$ |

b. $$E = E^\circ - \frac{RT}{nF} \cdot \ln Q \Rightarrow 0.210\ V = 0.127\ V - \frac{0.0257\ V}{2} \cdot \ln \frac{(P_{H_2})[Pb^{2+}]}{[H^+]^2}$$

$$-6.46\ V = \ln \frac{(1.00)[Pb^{2+}]}{(1.00)^2} \Rightarrow -6.46 = \ln[Pb^{2+}]$$

$[Pb^{2+}] = e^{-6.46} = 1.6 \times 10^{-3}\ M$

c. $K_{sp} = [Pb^{2+}][Cl^-]^2 = (1.6 \times 10^{-3})(0.10)^2$

$K_{sp} = 1.6 \times 10^{-5}$

58. Refer to Section 18.5.

$2H_2O(l) + 2e^- \rightarrow H_2(g) + 2OH^-(aq)$

$2Cl^-(aq) \rightarrow Cl_2(g) + 2e^-$

a. $0.288 \, mol. \, e^- \times \dfrac{6.022 \times 10^{23} \, e^-}{1 \, mol.} = 1.37 \times 10^{23} \, e^-$

b. $0.288 \, mol. \, e^- \times \dfrac{9.648 \times 10^4 \, C}{1 \, mol. \, e^-} = 2.20 \times 10^4 \, C$

c. $0.288 \, mol. \, e^- \times \dfrac{1 \, mol. \, H_2}{2 \, mol. \, e^-} \times \dfrac{2.016 \, g}{1 \, mol. \, H_2} = 0.230 \, g \, H_2$

$\quad 0.288 \, mol. \, e^- \times \dfrac{1 \, mol. \, Cl_2}{2 \, mol. \, e^-} \times \dfrac{70.90 \, g}{1 \, mol. \, Cl_2} = 8.08 \, g \, Cl_2$

60. Refer to Section 18.5.

a. Calculate the volume and mass of gold to be plated out. Bear in mind that the gold will be plated on both sides of the thin sheet.

$2(1.5 \, in \times 8.5 \, in \times 0.0020 \, in) = 0.051 \, in^3$

$0.051 \, in^3 \times \dfrac{(2.54 \, cm)^3}{(1 \, in)^3} \times \dfrac{19.3 \, g \, Au}{1 \, cm^3 \, Au} = 16 \, g \, Au$

b. $AuCN(s) + e^- \rightarrow Au(s) + CN^-(aq)$

$16 \, g \, Au \times \dfrac{1 \, mol. \, Au}{197.0 \, g \, Au} \times \dfrac{1 \, mol. \, e^-}{1 \, mol. \, Au} \times \dfrac{9.648 \times 10^4 \, C}{1 \, mol. \, e^-} = 7.8 \times 10^3 \, C$

$7.8 \times 10^3 \, C \times \dfrac{1 \, A \cdot s}{1 \, C} \times \dfrac{1}{7.00 \, A} = 1.1 \times 10^3 \, s$

$1.1 \times 10^3 \, s \times \dfrac{1 \, min}{60 \, s} = 19 \, min$

62. Refer to Sections 18.5, 18.6, and Example 18.8.

Use the equation on page 538 for the reaction of a lead storage battery to determine the moles of e^- per mole of Pb.

a. $12.0 \, lb \, Ca \times \dfrac{453.6 \, g}{1 \, lb} \times \dfrac{1 \, mol.}{40.078 \, g \, Ca} \times \dfrac{2 \, mol. \, e^-}{1 \, mol. \, Ca} = 272 \, mol. \, e^-$

$$272 \, mol. \, e^- \times \frac{9.648 \times 10^4 \, C}{1 \, mol. \, e^-} = 2.62 \times 10^7 \, C$$

$$3.2 \, V = \frac{J}{2.62 \times 10^7 \, C}$$

$$J = (3.2 \, V)(2.62 \times 10^7 \, C) = 8.4 \times 10^7 \, J$$

b. $\$0.09 = 1 \, kWh = 3.600 \times 10^6 \, J$

$$8.4 \times 10^7 \, J \times \frac{1 \, kWh}{3.600 \times 10^6 \, J} \times \frac{\$0.09}{1 \, kWh} = \$2.1$$

---

64. Refer to Section 18.5 and Chapter 5.

Use the ideal gas equation to calculate the moles of $H_2(g)$ needed. Write the half-reactions needed to derive the equation for the electrolysis of water. Then use the moles of $H_2(g)$ and moles of e^- to determine the current and then the time needed for the electrolysis.

$$n = \frac{PV}{RT} = \frac{(0.924 \, atm)(10.00 \, L)}{(0.0821 \, L \cdot atm/mol. \cdot K)(295 \, K)} = 0.382 \, mol. \, H_2$$

$2[\, 2H_2O(l) + 2e^- \rightarrow H_2(g) + 2OH^-(aq) \,]$
$2H_2O(l) \rightarrow O_2(g) + 4H^+(aq) + 4e^-$
$\underline{4[\, H^+(aq) + OH^-(aq) \rightarrow H_2O(l) \,]}$
$2H_2O(l) \rightarrow O_2(g) + 2H_2(g)$ (thus 4 e^- are passed for each 2 moles of $H_2(g)$ produced)

$$0.382 \, mol. \, H_2 \times \frac{4 \, mol. \, e^-}{2 \, mol. \, H_2} \times \frac{9.648 \times 10^4 \, C}{1 \, mol. \, e^-} = 7.36 \times 10^4 \, C$$

$$7.36 \times 10^4 \, C \times \frac{1 \, A \cdot s}{1 \, C} \times \frac{1}{12.0 \, A} = 6.13 \times 10^3 \, s$$

$$6.13 \times 10^3 \, s \times \frac{1 \, min}{60 \, s} \times \frac{1 \, hr}{60 \, min} = 1.70 \, hr$$

---

66. Refer to Section 18.5.

$$28.8 \, g \, Au \times \frac{1 \, mol.}{197 \, g \, Au} \times \frac{1 \, mol. \, e^-}{1 \, mol. \, Au} = 0.146 \, mol. \, e^-$$

$2.00 \, A = 2.00 \, C/s$

$$2h \times \frac{60 \, min}{1 \, hr} \times \frac{60 \, s}{1 \, min} \times \frac{2.00 \, C}{s} = 1.44 \times 10^4 \, C$$

$0.146 \, mol. \, e^- = 1.44 \times 10^4 \, C$

$$1\,\text{mol.}\,e^- \times \frac{1.44 \times 10^4 \text{ C}}{0.146 \text{ mol.}\,e^-} = 9.86 \times 10^4 \text{ C}$$

---

68. *Refer to Section 18.3 and Chapter 17.*

Calculate $\Delta H°$, $\Delta S°$ and $\Delta G°$ using the thermodynamic data from Appendix 1. Then calculate $E°$ from $\Delta G°$.

$\Delta H° = \Sigma\ \Delta H_f°_{\text{(products)}} - \Sigma\ \Delta H_f°_{\text{(reactants)}}$
$\Delta H° = [(1 \text{ mol.})(-393.5 \text{ kJ/mol.})] - [(1 \text{ mol.})(-110.5 \text{ kJ/mol.}) + (\frac{1}{2} \text{ mol.})(0.0 \text{ kJ/mol.})]$
$\Delta H° = -283.0 \text{ kJ}$

$\Delta S° = \Sigma\ S°_{\text{(products)}} - \Sigma\ S°_{\text{(reactants)}}$
$\Delta S° = [(1 \text{ mol.})(0.2136 \text{ kJ/mol.·K})]$
$\qquad\ - [(1 \text{ mol.})(0.1976 \text{ kJ/mol.·K}) + (\frac{1}{2} \text{ mol.})(0.2050 \text{ kJ/mol.·K})]$
$\Delta S° = -0.0865 \text{ kJ/K}$

$\Delta G° = \Delta H° - T\Delta S°$
$\Delta G° = -283.0 \text{ kJ} - (1275 \text{ K})(-0.0865 \text{ kJ/K})$
$\Delta G° = -172.9 \text{ kJ}$

$$CO(g) + \tfrac{1}{2}O_2(g) \rightarrow CO_2(g)$$
$$O^0 + 2e^- \rightarrow O^{2-}$$
$$C^{2+} \rightarrow C^{4+} + 2e^-$$

$\Delta G° = -nFE°$
$(-172.9 \text{ kJ})(1000 \text{ J/kJ}) = -(2 \text{ mol.})(9.648 \times 10^4 \text{ J/mol.·V})E°$
$E° = 0.896 \text{ V}$

---

70. *Refer to Sections 14.1 and 18.4.*

$2[\ 2Br^-(aq) \rightarrow Br_2(l) + 2e^-\ ]$	$E°_{\text{ox}} = -1.077 \text{ V}$
$O_2(g) + 4H^+(aq) + 4e^- \rightarrow 2H_2O(l)$	$E°_{\text{red}} = 1.229 \text{ V}$

$4Br^-(aq) + O_2(g) + 4H^+(aq) + 4e^- \rightarrow 2H_2O(l) + 2Br_2(l) + 4e^-$
$4Br^-(aq) + O_2(g) + 4H^+(aq) \rightarrow 2H_2O(l) + 2Br_2(l)$ $\qquad E° = 0.152 \text{ V}$

$$[H^+] = K_a \times \frac{[HB]}{[B^-]} = 1.8 \times 10^{-5} \times \frac{0.1}{0.1} = 1.8 \times 10^{-5}\ M$$

$$E = E° - \frac{RT}{nF} \cdot \ln Q = 0.152 \text{ V} - \frac{0.0257 \text{ V}}{4} \cdot \ln \frac{1}{(P_{O_2})[H^+]^4[Br^-]^4}$$

$$E = 0.152 \text{ V} - 0.00643 \text{ V} \cdot \ln \frac{1}{(1)(1.8 \times 10^{-5})^4(0.100)^4} = -0.188 \text{ V}$$

Water is electrolyzed to form H_2 and O_2 (see p 496). Look for the box containing two squares together (H_2) and two circles together (O_2).
Box (c) represents the electrolysis of water.

74. Refer to Section 18.4.

Write the equation for the cell reaction and evaluate Q.

$$Co(s) \rightarrow Co^{2+}(aq) + 2e^-$$
$$2H^+(aq) + 2e^- \rightarrow H_2(g)$$
$$\overline{Co(s) + 2H^+(aq) \rightarrow H_2(g) + Co^{2+}(aq)}$$

Use the Nernst equation to determine the effect on E.

$$E = E° - \frac{RT}{nF} \cdot \ln Q = E° - \frac{0.0257 \, V}{2} \cdot \ln \frac{[Co^{2+}](P_{H_2})}{[H^+]^2}$$

a. The voltage is dependent on the concentration of Co^{2+}, not the volume. So changing the volume will have no effect on the voltage of the cell.

b. The voltage of the cell is dependent on the concentration of H^+. Increasing $[H^+]$ will decrease Q, which will decrease $\ln Q$. Consequently, the value being subtracted from $E°$ will be smaller, and E will be larger.

c. The voltage of the cell is dependent on the pressure of H_2. Increasing $P(H_2)$ will increase Q, which will increase $\ln Q$. Consequently, the value being subtracted from $E°$ will be larger, and E will be smaller.

d. The voltage of the cell does not depend on the amount of $Co(s)$ present. Changing this value will have no effect on the voltage of the cell.

e. The voltage of the cell does not depend on the surface area of the electrode. Changing this value will have no effect on the voltage of the cell.

Thus, only (b) will increase the voltage of the cell (E).

76. Refer to Sections 18.2 and 18.3 and Chapter 17.

a. $2[Tl(s) \rightarrow Tl^+(aq) + e^-]$ $\qquad\qquad$ $E°_{ox} = 0.34 \, V$

$\quad\;\; Tl^{3+}(aq) + 2e^- \rightarrow Tl^+(aq)$ $\qquad\qquad$ $E°_{red} = 1.28 \, V$

$\quad\overline{2Tl(s) + Tl^{3+}(aq) + 2e^- \rightarrow Tl^+(aq) + 2Tl^+(aq) + 2e^-}$

$\quad\; 2Tl(s) + Tl^{3+}(aq) \rightarrow 3Tl^+(aq)$ $\qquad\qquad$ $E° = 1.62 \, V$

$$2[\text{Tl}(s) \rightarrow \text{Tl}^{3+}(aq) + 3e^-] \qquad\qquad E^\circ_{ox} = -0.74 \text{ V}$$

$$\underline{3[\text{Tl}^{3+}(aq) + 2e^- \rightarrow \text{Tl}^+(aq)]} \qquad\qquad E^\circ_{red} = 1.28 \text{ V}$$

$$2\text{Tl}(s) + 3\text{Tl}^{3+}(aq) + 6e^- \rightarrow 2\text{Tl}^{3+}(aq) + 3\text{Tl}^+(aq) + 6e^-$$
$$2\text{Tl}(s) + \text{Tl}^{3+}(aq) \rightarrow 3\text{Tl}^+(aq) \qquad\qquad E^\circ = 0.54 \text{ V}$$

$$3[\text{Tl}(s) \rightarrow \text{Tl}^+(aq) + e^-] \qquad\qquad E^\circ_{ox} = 0.34 \text{ V}$$

$$\underline{\text{Tl}^{3+}(aq) + 3e^- \rightarrow \text{Tl}(s)} \qquad\qquad E^\circ_{red} = 0.74 \text{ V}$$

$$3\text{Tl}(s) + \text{Tl}^{3+}(aq) + 3e^- \rightarrow \text{Tl}(s) + 3\text{Tl}^+(aq) + 3e^-$$
$$2\text{Tl}(s) + \text{Tl}^{3+}(aq) \rightarrow 3\text{Tl}^+(aq) \qquad\qquad E^\circ = 1.08 \text{ V}$$

b. (1) $E^\circ = 1.62$ V

   (2) $E^\circ = 0.54$ V

   (3) $E^\circ = 1.08$ V

c. (1) $\Delta G^\circ = -nFE^\circ = -(2 \text{ mol.})(9.648 \times 10^4 \text{ J/mol.·V})(1.62 \text{ V}) = -3.13 \times 10^5 \text{ J} = -313 \text{ kJ}$
   (2) $\Delta G^\circ = -nFE^\circ = -(6 \text{ mol.})(9.648 \times 10^4 \text{ J/mol.·V})(0.54 \text{ V}) = -3.13 \times 10^5 \text{ J} = -313 \text{ kJ}$
   (3) $\Delta G^\circ = -nFE^\circ = -(3 \text{ mol.})(9.648 \times 10^4 \text{ J/mol.·V})(1.08 \text{ V}) = -3.13 \times 10^5 \text{ J} = -313 \text{ kJ}$

d. ΔG° is a state property as seen in part (c). Chemical reactions (1) - (3) are the same, the paths to them are different. ΔG° will be the same because state properties are independent of path, but E° values will differ because E° depends on the path.

---

78. Refer to Sections 18.4 and 18.6 and Chapter 13.

Calculate E°, [H^+] in the acid solution, and from K_a, the [SO_4^{2-}]. Substitute these values into the Nernst equation and calculate E.

$$\text{Pb}(s) + \text{SO}_4^{2-}(aq) \rightarrow \text{PbSO}_4(s) + 2e^- \qquad\qquad E^\circ_{ox} = 0.356 \text{ V}$$

$$\underline{\text{PbO}_2(s) + \text{SO}_4^{2-}(aq) + 4\text{H}^+(aq) + 2e^- \rightarrow \text{PbSO}_4(s) + 2\text{H}_2\text{O}(l)} \qquad E^\circ_{red} = 1.687 \text{ V}$$

$$\text{Pb}(s) + \text{PbO}_2(s) + 2\text{SO}_4^{2-}(aq) + 4\text{H}^+(aq) + 2e^- \rightarrow 2\text{PbSO}_4(s) + 2\text{H}_2\text{O}(l) + 2e^-$$
$$\text{Pb}(s) + \text{PbO}_2(s) + 2\text{SO}_4^{2-}(aq) + 4\text{H}^+(aq) \rightarrow 2\text{PbSO}_4(s) + 2\text{H}_2\text{O}(l) \qquad E^\circ = 2.043 \text{ V}$$

$$\frac{1.286 \text{ g solution}}{1 \text{ cm}^3} \times \frac{1 \text{ cm}^3}{1 \text{ mL}} \times \frac{1000 \text{ mL}}{1 \text{ L}} \times \frac{38 \text{ g H}_2\text{SO}_4}{100 \text{ g sol'n}} \times \frac{1 \text{ mol.}}{98.1 \text{ g}} = 4.98 \text{ M H}_2\text{SO}_4$$

$$[\text{H}^+] = [\text{HSO}_4^-] = [\text{H}_2\text{SO}_4] = 4.98 \ M$$

$$K_a = \frac{[\text{H}^+][\text{SO}_4^{2-}]}{[\text{HSO}_4^-]} \quad \Rightarrow \quad 1.0 \times 10^{-2} = \frac{(4.98 + x)(x)}{(4.98 - x)}$$

$$K_a/a = 1.0 \times 10^{-2} / 4.98 = 2.0 \times 10^{-3}$$
$$K_a/a = 2.0 \times 10^{-3} < 2.5 \times 10^{-3}$$

(The 5% rule applies. See the explanation accompanying Problem 38 in Chapter 13.)

$$1.0 \times 10^{-2} = \frac{(4.98)(x)}{4.98}$$

$$x = [SO_4^{2-}] = 0.0100 \ M$$

$$E = E^\circ - \frac{0.0257}{n} \cdot \ln \frac{1}{[H^+]^4 [SO_4^{2-}]^2} = 2.043 - \frac{0.0257}{2} \cdot \ln \frac{1}{(4.98)^4 (0.0100)^2} = 2.01 \ V$$

---

79. Refer to Sections 18.2 and 18.4.

a.
$$Zn(s) \rightarrow Zn^{2+}(aq) + 2e^- \qquad\qquad E^\circ_{ox} = 0.762 \ V$$
$$\underline{Sn^{2+}(aq) + 2e^- \rightarrow Sn(s) \qquad\qquad E^\circ_{red} = -0.141 \ V}$$
$$Sn^{2+}(aq) + Zn(s) \rightarrow Zn^{2+}(aq) + Sn(s) \qquad E^\circ = 0.621 \ V$$

b. E° is positive, so the reaction is spontaneous in the forward direction. As the cell operates, $[Zn^{2+}]$ increases and $[Sn^{2+}]$ decreases.

c. $E = E^\circ - \dfrac{0.0257}{n} \cdot \ln \dfrac{[Zn^{2+}]}{[Sn^{2+}]} \quad \Rightarrow \quad 0 = 0.621 - \dfrac{0.0257}{2} \cdot \ln \dfrac{[Zn^{2+}]}{[Sn^{2+}]}$

$$\ln \frac{[Zn^{2+}]}{[Sn^{2+}]} = 48.3 \quad \Rightarrow \quad \frac{[Zn^{2+}]}{[Sn^{2+}]} = 1 \times 10^{21}$$

d. $[Zn^{2+}] = [Zn^{2+}]_i + x$ and $[Sn^{2+}] = [Sn^{2+}]_i - x$

$[Zn^{2+}] = 1.0 \ M + x$ and $[Sn^{2+}] = 1.0 \ M - x$

$$1 \times 10^{21} = \frac{[Zn^{2+}]}{[Sn^{2+}]} = \frac{1.0 \ M + x}{1.0 \ M - x}$$

$1 \times 10^{21} - 1 \times 10^{21}x = 1 + x$

$1 \times 10^{21} = 1 \times 10^{21}x$

$x = 1.00$

$[Zn^{2+}] = 1.0 \ M + x = 2.0 \ M$

$[Sn^{2+}] = 1.0 \ M - x = 0.0 \ M$ **or**

$[Sn^{2+}] = [Zn^{2+}] / 1 \times 10^{21} = 2.0 \ M / 1 \times 10^{21} = 2 \times 10^{-21} \ M \ (\cong 0.0 \ M)$

Since K is so large, the reaction lies almost completely to the right at equilibrium (when $E = 0$). Thus we can assume that essentially all the Sn^{2+} reacts to form Zn^{2+}. Consequently, $[Zn^{2+}] = 1.0 \ M + 1.0 \ M = 2.0 \ M$. The residual $[Sn^{2+}]$ can be calculated from:

$$K = 1 \times 10^{21} = \frac{[Zn^{2+}]}{[Sn^{2+}]}$$

80. Refer to Section 18.3.

a. For step 1:

$\Delta G^{\circ\prime} = -nFE^{\circ\prime}$

$\Delta G^{\circ\prime} = -(2 \text{ mol.})(9.648 \times 10^4 \text{ J/mol.}\cdot\text{V})(-0.581 \text{ V})$

$\Delta G^{\circ\prime} = 1.12 \times 10^5 \text{ J}$

For step 2:

$\Delta G^{\circ\prime} = -nFE^{\circ\prime}$

$\Delta G^{\circ\prime} = -(2 \text{ mol.})(9.648 \times 10^4 \text{ J/mol.}\cdot\text{V})(-0.197 \text{ V})$

$\Delta G^{\circ\prime} = 3.80 \times 10^4 \text{ J}$

Overall:

$\Delta G^{\circ\prime} = 1.12 \times 10^5 \text{ J} + 3.80 \times 10^4 \text{ J}$

$\Delta G^{\circ\prime} = 1.50 \times 10^5 \text{ J}$

b. In the overall process, 4 electrons are transferred.

$\Delta G^{\circ\prime} = -nFE^{\circ\prime}$

$1.50 \times 10^5 \text{ J} = -(4 \text{ mol.})(9.648 \times 10^4 \text{ J/mol.}\cdot\text{V})(E^{\circ\prime})$

$E^{\circ\prime} = -0.389 \text{ V}$

81. Refer to Sections 18.2 and 18.4.

$H_2(g) \rightarrow 2H^+(aq) + 2e^-$ $\qquad\qquad\qquad\qquad E^{\circ}_{ox} = 0.0 \text{ V}$

$\underline{2H^+(aq) + 2e^- \rightarrow H_2(g)} \qquad\qquad\qquad\qquad E^{\circ}_{red} = 0.0 \text{ V}$

$H_{2\,(anode)}(g) + 2H^+{}_{(cathode)}(aq) \rightarrow 2H^+{}_{(anode)}(aq) + H_{2\,(cathode)}(g) \qquad E^{\circ} = 0.00 \text{ V}$

$[H^+]_{(anode)} = 10^{-pH} = 10^{-7.0} = 1.0 \times 10^{-7} \text{ M}$

$[H^+]_{(cathode)} = 10^{-pH} = 10^{0.0} = 1.0 \text{ M}$

$$E = E^{\circ} - \frac{0.0257}{n} \cdot \ln \frac{(P_{H_2})_{cathode}[H^+]_{anode}{}^2}{(P_{H_2})_{anode}[H^+]_{cathode}{}^2} = 0.00 - \frac{0.0257}{2} \cdot \ln \frac{(1.0)(1.0 \times 10^{-7})^2}{(1.0)(1.0)^2}$$

$E = 0.414 \text{ V}$

Chapter 19: Nuclear Reactions

---

2.   Refer to Section 19.1 and Example 19.1.

$^{210}_{82}\text{Pb} \rightarrow {}^{210}_{83}\text{Bi} + {}^{A}_{Z}?$

$A = 210 - 210 = 0$
$Z = 82 - 83 = -1$

$^{210}_{82}\text{Pb} \rightarrow {}^{210}_{83}\text{Bi} + {}^{0}_{-1}e$ ^{210}Pb undergoes beta particle emission.

---

4.   Refer to Section 19.1 and Example 19.1.

a. $^{230}_{90}\text{Th} \rightarrow {}^{4}_{2}\text{He} + {}^{A}_{Z}?$

$A = 230 - 4 = 226$
$Z = 90 - 2 = 88$

$^{230}_{90}\text{Th} \rightarrow {}^{4}_{2}\text{He} + {}^{226}_{88}\textbf{Ra}$

b. $^{210}_{82}\text{Pb} \rightarrow {}^{0}_{-1}e + {}^{A}_{Z}?$

$A = 210 - 0 = 210$
$Z = 82 - (-1) = 83$

$^{210}_{82}\text{Pb} \rightarrow {}^{0}_{-1}e + {}^{210}_{83}\textbf{Bi}$

c.   Fission indicates that a neutron reacted with the nucleus. To produce an *excess* of 2 neutrons, a total of 3 must be produced (the initial one plus 2 more).

$^{235}_{92}\text{U} + {}^{1}_{0}n \rightarrow {}^{140}_{56}\text{Ba} + 3{}^{1}_{0}n + {}^{A}_{Z}?$

$A = (235 + 1) - (140 + 3(1)) = 93$
$Z = (92 + 0) - (56 + 3(0)) = 36$

$^{235}_{92}\text{U} + {}^{1}_{0}n \rightarrow {}^{140}_{56}\text{Ba} + 3{}^{1}_{0}n + {}^{93}_{36}\textbf{Kr}$

d. $^{37}_{18}\text{Ar} + {}^{0}_{-1}e \rightarrow {}^{A}_{Z}?$

$A = 37 + 0 = 37$
$Z = 18 + (-1) = 17$

$^{37}_{18}\text{Ar} + {}^{0}_{-1}e \rightarrow {}^{37}_{17}\textbf{Cl}$

6. *Refer to Section 19.1.*

The sum of the mass numbers and the atomic numbers on either side of the arrow must be equal to get the missing species. Calculate A and Z for the missing species and determine the element from the atomic number.

a. $^{209}_{83}\text{Bi} + ^{64}_{28}\text{Ni} \rightarrow ^{1}_{0}n + ^{A}_{Z}\text{X}$

  $A = 209 + 64 - 1 = 272$
  $Z = 83 + 28 - (0) = 111$

  $^{209}_{83}\text{Bi} + ^{64}_{28}\text{Ni} \rightarrow ^{1}_{0}n + ^{272}_{111}\textbf{Uuu}$

b. $^{272}_{111}\text{X} \rightarrow 3\,^{4}_{2}\text{He} + ^{A}_{Z}\text{X}$

  $A = 272 - 3(4) = 260$
  $Z = 111 - 3(2) = 105$

  $^{272}_{111}\text{X} \rightarrow 3\,^{4}_{2}\text{He} + ^{260}_{105}\textbf{Db}$

8. *Refer to Section 19.1 and Example 19.1.*

$^{282}_{115}\text{X} + ^{0}_{-1}e \rightarrow ^{A}_{Z}\text{Y}$

$A = 282 + 0 = 282$
$Z = 115 + (-1) = 114$

$^{282}_{115}\text{X} + ^{0}_{-1}e \rightarrow ^{282}_{114}\text{Y}$

$^{282}_{115}\text{X} \rightarrow ^{A}_{Z}\text{Y} + ^{0}_{1}e$

$A = 282 - 0 = 282$
$Z = 115 - 1 = 114$

$^{282}_{115}\text{X} \rightarrow ^{282}_{114}\text{Y} + ^{0}_{1}e$

The product nuclide is the same for both reactions.

10. *Refer to Section 19.1 and Example 19.1.*

a. $^{54}_{26}\text{Fe} + ^{4}_{2}\text{He} \rightarrow 2\,^{1}_{1}\text{H} + ^{A}_{Z}?$

  $A = 54 + 4 - 2(1) = 56$
  $Z = 26 + 2 - 2(1) = 26$

  $^{54}_{26}\text{Fe} + ^{4}_{2}\text{He} \rightarrow 2\,^{1}_{1}\text{H} + ^{56}_{26}\textbf{Fe}$

b. $^{96}_{42}\text{Mo} + ^{2}_{1}\text{H} \rightarrow ^{1}_{0}n + ^{A}_{Z}?$

  $A = 96 + 2 - 1 = 97$
  $Z = 42 + 1 - 0 = 43$

  $^{96}_{42}\text{Mo} + ^{2}_{1}\text{H} \rightarrow ^{1}_{0}n + ^{97}_{43}\textbf{Tc}$

c. $^{40}_{18}\text{Ar} + ^{A}_{Z}? \rightarrow ^{43}_{19}\text{K} + ^{1}_{1}\text{H}$

  $A = (43 + 1) - 40 = 4$
  $Z = (19 + 1) - 18 = 2$

  $^{40}_{18}\text{Ar} + ^{4}_{2}\textbf{He} \rightarrow ^{43}_{19}\text{K} + ^{1}_{1}\text{H}$

d. $^{A}_{Z}? + ^{1}_{0}n \rightarrow ^{1}_{1}\text{H} + ^{31}_{15}\text{P}$

  $A = 31 + 1 - 1 = 31$
  $Z = 15 + 1 - 0 = 16$

  $^{31}_{16}\textbf{S} + ^{1}_{0}n \rightarrow ^{1}_{1}\text{H} + ^{31}_{15}\text{P}$

12. Refer to Section 19.1 and Example 19.1.

a. $^{249}_{99}\text{Es} + ^{1}_{0}n \rightarrow ^{161}_{64}\text{Gd} + 2^{1}_{0}n + ^{A}_{Z}?$

$A = 249 + 1 - 161 - 2(1) = 87$
$Z = 99 + 0 - 64 - 2(0) = 35$

$^{249}_{99}\text{Es} + ^{1}_{0}n \rightarrow ^{161}_{64}\text{Gd} + 2^{1}_{0}n + ^{87}_{35}\textbf{Br}$

b. $^{A}_{Z}? \rightarrow ^{0}_{-1}e + ^{59}_{27}\text{Co}$

$A = 0 + 59 = 59$
$Z = -1 + 27 = 26$

$^{59}_{26}\textbf{Fe} \rightarrow ^{0}_{-1}e + ^{59}_{27}\text{Co}$

c. $4^{1}_{1}\text{H} \rightarrow 2^{0}_{1}e + ^{A}_{Z}?$

$A = 4(1) - 2(0) = 4$
$Z = 4(1) - 2(1) = 2$

$4^{1}_{1}\text{H} \rightarrow 2^{0}_{1}e + ^{4}_{2}\textbf{He}$

d. $^{24}_{12}\text{Mg} + ^{1}_{0}n \rightarrow ^{1}_{1}\text{H} + ^{A}_{Z}?$

$A = 24 + 1 - 1 = 24$
$Z = 12 + 0 - 1 = 11$

$^{24}_{12}\text{Mg} + ^{1}_{0}n \rightarrow ^{1}_{1}\text{H} + ^{24}_{11}\textbf{Na}$

14. Refer to Section 19.2.

$1 \text{ Ci} = 3.700 \times 10^{10} \text{ atoms/s}$

$12.0 \text{ mCi} \times \dfrac{1 \text{ Ci}}{1000 \text{ mCi}} \times \dfrac{3.700 \times 10^{10} \text{ atoms/s}}{1 \text{ Ci}} = 4.44 \times 10^{8} \text{ atoms/s}$

$\dfrac{4.44 \times 10^{8} \text{ atoms}}{1 \text{ s}} \times \dfrac{60 \text{ s}}{1 \text{ min}} \times \dfrac{60 \text{ min}}{1 \text{ h}} = 1.60 \times 10^{12} \text{ atoms / h}$

$3.17 \times 10^{17} \text{ atoms} \times \dfrac{1 \text{ h}}{1.60 \times 10^{12} \text{ atoms}} = 1.98 \times 10^{5} \text{ h}$

16. Refer to Section 19.2.

$\dfrac{1.94 \times 10^{3} \text{ counts}}{1 \text{ min}} \times \dfrac{100 \text{ particles}}{0.070 \text{ counts}} \times \dfrac{1 \text{ min}}{60 \text{ s}} \times \dfrac{1 \text{ Ci}}{3.700 \times 10^{10} \text{ atoms/s}} = 1.25 \times 10^{-5} \text{ Ci}$

18. Refer to Section 19.2 and Example 19.2.

Calculate N for the sample, then use $A = k \cdot N$ to calculate the activity.

$2.00 \text{ mg} \times \dfrac{1 \text{ g}}{1000 \text{ mg}} \times \dfrac{1 \text{ mol. Kr-87}}{87.0 \text{ g Kr-87}} \times \dfrac{6.022 \times 10^{23} \text{ atoms}}{1 \text{ mol.}} = 1.38 \times 10^{19} \text{ atoms}$

$A = (1.5 \times 10^{-4} /\text{sec})(1.38 \times 10^{19} \text{ atoms}) = 2.1 \times 10^{15} \text{ atoms/sec}$

$$2.1 \times 10^{15} \text{ atoms/sec} \times \frac{1 \text{ Ci}}{3.700 \times 10^{10} \text{ atoms/sec}} = 5.6 \times 10^{4} \text{ Ci}$$

20. *Refer to Section 19.2 and Example 19.2.*

Convert mass Cu to atoms Cu. Convert $t_{\frac{1}{2}}$ to seconds and then calculate k. Then calculate the activity (in atoms/s) and convert that to Ci.

$$1.50 \text{ mg Cu-64} \times \frac{1 \text{ g}}{1000 \text{ mg}} \times \frac{1 \text{ mol.}}{64.0 \text{ g}} \times \frac{6.022 \times 10^{23} \text{ atoms}}{1 \text{ mol.}} = 1.41 \times 10^{19} \text{ atoms}$$

$$12.8 \text{ h} \times \frac{60 \text{ min}}{1 \text{ h}} \times \frac{60 \text{ s}}{1 \text{ min}} = 4.61 \times 10^{4} \text{ s}$$

$$k = \frac{0.693}{t_{1/2}} = \frac{0.693}{4.61 \times 10^{4} \text{ s}} = 1.50 \times 10^{-5} \text{ /s}$$

$$A = kN = (1.50 \times 10^{-5} \text{ /s})(1.41 \times 10^{19} \text{ atoms}) = 2.12 \times 10^{14} \text{ atoms/s}$$

$$2.12 \times 10^{14} \text{ atoms/s} \times \frac{1 \text{ Ci}}{3.700 \times 10^{10} \text{ atoms/s}} = 5.72 \times 10^{3} \text{ Ci}$$

22. *Refer to Section 19.2.*

Calculate k, then N. Then convert the number of atoms to mass of Br-82.

$$36 \text{ h} \times \frac{60 \text{ min}}{1 \text{ h}} = 2.2 \times 10^{3} \text{ min}$$

$$k = \frac{0.693}{t_{1/2}} = \frac{0.693}{2.2 \times 10^{3} \text{ min}} = 3.2 \times 10^{-4} \text{ /min}$$

$$\frac{1.2 \times 10^{5} \text{ disintegrations}}{1 \text{ min}} \times \frac{1 \text{ atom}}{1 \text{ disintegration}} = 1.2 \times 10^{5} \text{ atoms/min}$$

$$A = kN \implies 1.2 \times 10^{5} \text{ atoms/min} = (3.2 \times 10^{-4} \text{ /min})(N)$$

$$N = 3.8 \times 10^{8} \text{ atoms}$$

$$3.8 \times 10^{8} \text{ atoms} \times \frac{1 \text{ mol.}}{6.022 \times 10^{23} \text{ atoms}} \times \frac{82 \text{ g}}{1 \text{ mol.}} = 5.2 \times 10^{-14} \text{ g}$$

$$^{36}_{17}Cl \rightarrow ^{0}_{-1}e + ^{36}_{18}\textbf{Ar} \quad \text{(Note there is 1 } \beta\text{-particle per 1 } ^{36}_{17}Cl \text{ decay.)}$$

$$\frac{2.3 \times 10^{-6}}{yr} \times \frac{1\,yr}{365\,d} \times \frac{1\,d}{24\,hr} \times \frac{1\,hr}{60\,min} = 4.4 \times 10^{-12} \text{ /min}$$

$$1.00\,mg\,Cl\text{-}36 \times \frac{1\,g}{1000\,mg} \times \frac{1\,mol.\,Cl\text{-}36}{36.0\,g} = 2.78 \times 10^{-5}\,mol.\,Cl\text{-}36$$

$$2.78 \times 10^{-5}\,mol.\,Cl\text{-}36 \times \frac{1\,mol.\,\beta\text{-particles}}{1\,mol.\,^{36}Cl} \times \frac{6.022 \times 10^{23}}{1\,mol.} = 1.67 \times 10^{19}\,\beta\text{-particles}$$

$$A = kN = (4.4 \times 10^{-12}\text{ /min})(1.67 \times 10^{19}\,\beta\text{-particles}) = 7.3 \times 10^{7} \text{ particles/min}$$

$$\frac{7.3 \times 10^{7} \text{ particles}}{1\,min} \times \frac{1\,min}{60\,s} \times \frac{1\,Ci}{3.700 \times 10^{10} \text{ particles/s}} = 3.3 \times 10^{-5}\,Ci$$

26. Refer to Section 19.2 and Example 19.3.

Calculate k from the half-life, then calculate X.

$$k = \frac{0.693}{5730\,yr} = 1.21 \times 10^{-4} \text{ /yr}$$

$$\ln\left(\frac{X_0}{X}\right) = kt \implies \ln\left(\frac{15.3}{X}\right) = (1.21 \times 10^{-4} \text{ /yr})(3.9 \times 10^{2}\,yr)$$

$\ln(15.3) - \ln X = 0.047$
$-\ln X = -2.68$
$X = e^{2.68} = 14.6$
Activity = 14.6 disintegrations/min/g

28. Refer to Section 19.2 and Example 19.3.

Calculate k from the half-life, then calculate time.

$$k = \frac{0.693}{5730\,yr} = 1.21 \times 10^{-4} \text{ /yr}$$

$$\ln\left(\frac{X_0}{X}\right) = kt \implies \ln\left(\frac{15.3}{5.0}\right) = (1.21 \times 10^{-4} \text{ /yr}) \cdot t \implies t = 9.2 \times 10^{3}\,yr$$

Calculate k from the half-life, then calculate time.

$$k = \frac{0.693}{12.3 \text{ yr}} = 0.0563 \text{ /yr}$$

If the tritium content is ⅗ the original, then $X = \tfrac{3}{5}X_0$. If $X_0 = 1$, then $X = 0.600$.

$$\ln\left(\frac{X_0}{X}\right) = kt \quad \Rightarrow \quad \ln\left(\frac{1}{0.600}\right) = (0.0563 \text{ /yr}) \cdot t$$

$t = 9.07$ yr

---

32. *Refer to Section 19.3 and Example 19.4.*

a. $^{90}_{38}\text{Sr} \rightarrow 2^{0}_{-1}e + {}^{90}_{40}\text{Zr}$ $\qquad$ (See problem 4a)

b. $\Delta m = 2(0.00055 \text{ g/mol.}) + 89.8824 \text{ g/mol.} - 89.8869 \text{ g/mol.} = -0.0034 \text{ g/mol.}$

c. Calculate ΔE for one mole, then convert to kJ/g.

$\Delta E = 9.00 \times 10^{10} \text{ kJ/g} \times \Delta m = (9.00 \times 10^{10} \text{ kJ/g})(-0.0034 \text{ g/mol.}) = -3.1 \times 10^{8} \text{ kJ/mol.}$

$$6.50 \text{ mg} \times \frac{1 \text{ g}}{1000 \text{ mg}} \times \frac{1 \text{ mol. Sr}}{89.8869 \text{ g}} \times \frac{-3.1 \times 10^{8} \text{ kJ}}{1 \text{ mol.}} = -2.2 \times 10^{4} \text{ kJ}$$

---

34. *Refer to Section 19.3, Table 19.3, and Example 19.5.*

a. Δm = [mass of neutrons + mass of protons] – mass of nucleus
Δm = [8(1.00867 g) + 6(1.00728 g)] – 13.99995 g = 0.11309 g

b. $\Delta E = 9.00 \times 10^{10}$ kJ/g $\times \Delta m$
$\Delta E = (9.00 \times 10^{10}$ kJ/g$)(0.11309$ g$) = 1.02 \times 10^{10}$ kJ

---

36. *Refer to Section 19.3, Example 19.5 and Table 19.3.*

Calculate the mass defect for each isotope. Then calculate and compare the ΔE's.

$^{26}_{12}\text{Mg} \rightarrow 12^{1}_{1}\text{H} + 14^{1}_{0}n$

$^{26}_{13}\text{Al} \rightarrow 13^{1}_{1}\text{H} + 13^{1}_{0}n$

Mg-26: $12(1.00728 \text{ g}) + 14(1.00867 \text{ g}) - 25.97600 \text{ g} = 0.2327 \text{ g}$
Al-26: $13(1.00728 \text{ g}) + 13(1.00867 \text{ g}) - 25.97977 \text{ g} = 0.2276 \text{ g}$

Since Mg-26 has the larger mass defect, it will have the greater binding energy.

$\Delta E_{\text{Mg-26}} = 9.00 \times 10^{10} \text{ kJ/g} \times \Delta m = 9.00 \times 10^{10} \text{ kJ/g} \times (0.2327 \text{ g}) = 2.09 \times 10^{10} \text{ kJ}$

$\Delta E_{\text{Al-26}} = 9.00 \times 10^{10} \text{ kJ/g} \times \Delta m = 9.00 \times 10^{10} \text{ kJ/g} \times (0.2276 \text{ g}) = 2.05 \times 10^{10} \text{ kJ}$

---

38. Refer to Sections 19.3 and 19.5, Examples 19.4 and 19.6 and Table 19.3.

Calculate the mass defect for the nuclear reaction, adjust for 1.00 g U and 1.00 g H, then calculate ΔE.

$\Delta m = 139.8734 \text{ g} + 93.8841 + 6(0.00055 \text{ g}) + 2(1.00867 \text{ g}) - 234.9934 \text{ g} - 1.00867 \text{ g}$
$\qquad\qquad = -0.2239 \text{ g (per 1 mole U)}$

$$1.00 \text{ g U - 235} \times \frac{1 \text{ mol. U - 235}}{234.9934 \text{ g}} \times \frac{-0.2239 \text{ g}}{1 \text{ mol. U - 235}} = -9.53 \times 10^{-4} \text{ g}$$

$\Delta E = 9.00 \times 10^{10} \text{ kJ/g} \times \Delta m = 9.00 \times 10^{10} \text{ kJ/g} \times (-9.53 \times 10^{-4} \text{ g}) = -8.57 \times 10^{-7} \text{ kJ}$

$\Delta m = 3.01550 \text{ g} + 1.0078 \text{ g} - 2(2.01355 \text{ g}) = -4.32 \times 10^{-3} \text{ g (per 1 mole H)}$

$$1.00 \text{ g H - 2} \times \frac{1 \text{ mol. H - 2}}{2.01355 \text{ g}} \times \frac{-4.32 \times 10^{-3} \text{ g}}{1 \text{ mol. H - 2}} = -0.00214 \text{ g}$$

$\Delta E = 9.00 \times 10^{10} \text{ kJ/g} \times \Delta m = 9.00 \times 10^{10} \text{ kJ/g} \times (-0.00214 \text{ g}) = -1.93 \times 10^{8} \text{ kJ}$
Thus, the fusion of deuterium produces more energy.

---

40. Refer to Sections 19.3 and 19.5 and Example 19.6.

Write the nuclear reaction for the fission of U-235 and calculate the mass defect. Calculate the energy associated with this Δm. Then convert to kJ per 1 mg. Finally, calculate the mass of NH_4NO_3 required to produce an equal amount of energy.

$$^{235}_{92}\text{U} + ^{1}_{0}n \rightarrow ^{144}_{58}\text{Ce} + ^{89}_{37}\text{Rb} + 3\,^{0}_{-1}e + 3\,^{1}_{0}n$$

$\Delta m = [143.8817 + 88.8913 + 3(0.00055) + 3(1.00867)] - [234.9934 + 1.00867]$
$\Delta m = -0.2014 \text{ g (per mole of U-235)}$

$\Delta E = 9.00 \times 10^{10} \text{ kJ/g} \times \Delta m = (9.00 \times 10^{10} \text{ kJ/g})(-0.2014 \text{ g/mol.}) = -1.81 \times 10^{10} \text{ kJ/mol.}$

Energy from one milligram of U-235:

$$1.00 \text{ mg} \times \frac{1 \text{ g}}{1000 \text{ mg}} \times \frac{1 \text{ mol.}}{234.99 \text{ g}} \times \frac{-1.81 \times 10^{10} \text{ kJ}}{1 \text{ mol.}} = -7.70 \times 10^{4} \text{ kJ}$$

Thus 7.70×10^{4} kJ of energy is released.

Mass of NH_4NO_3 need to produce an equivalent amount of energy:

$$7.70 \times 10^4 \text{ kJ} \times \frac{1 \text{ mol. } NH_4NO_3}{37.0 \text{ kJ}} \times \frac{80.04 \text{ g } NH_4NO_3}{1 \text{ mol. } NH_4NO_3} \times \frac{1 \text{ kg}}{1000 \text{ g}} = 167 \text{ kg}$$

42. Refer to Section 19.2.

a. $$k = \frac{0.693}{8.1 \text{ d}} = 0.086 \text{ /d}$$

$$\ln\left(\frac{X_0}{X}\right) = kt = (0.086 \text{ /d})(2.0 \text{ d}) = 0.17$$

$$\frac{X_0}{X} = 1.19 \implies X_0 = 1.19 X$$

If X_0 is set to 100%, then $X = 84.4\%$, thus 84.4% remains after 2.0 days, meaning: **15.6% has disintegrated**.

b. The goal is to provide, with this partially decayed I-131, the same activity the patient would
receive from fresh I-131. This can be expressed mathematically as:

$(15.0 \text{ mg})(100\%) = (x \text{ mg})(84.4\%)$
$x = 17.8 \text{ mg}$

44. Refer to Section 19.2 and Example 19.2.

Calculate k, then N. Then convert the number of atoms to mass of Am-241.

$$1 \text{ yr} \times \frac{365 \text{ d}}{1 \text{ yr}} \times \frac{24 \text{ hr}}{1 \text{ d}} \times \frac{3600 \text{ s}}{1 \text{ hr}} = 3.15 \times 10^7 \text{ s}$$

$$k = \frac{1.51 \times 10^{-3}}{3.15 \times 10^7 \text{ s}} = 4.79 \times 10^{-11} \text{ /s}$$

$$\frac{10 \text{ disintegrations}}{1 \text{ s}} \times \frac{1 \text{ atom}}{1 \text{ disintegration}} = 10 \text{ atoms/s}$$

$A = kN \implies 10 \text{ atoms/s} = (4.79 \times 10^{-11} \text{ /min})(N)$

$N = 2.1 \times 10^{11}$ atoms

$$2.1 \times 10^{11} \text{ atoms} \times \frac{1 \text{ mol.}}{6.022 \times 10^{23} \text{ atoms}} \times \frac{241 \text{ g}}{1 \text{ mol.}} = 8.4 \times 10^{-11} \text{ g}$$

46. Refer to Chapter 16.

The activity of the solution is proportional to the concentration of I^-. Use the activity to set up a ratio for the two solutions and solve for the concentration of I^- in the filtrate.

$$\frac{[I^-]_{0.050\,M}}{A_{0.050\,M}} = \frac{[I^-]_{filtrate}}{A_{filtrate}} \Rightarrow \frac{0.050\,M}{1.25 \times 10^{10}} = \frac{[I^-]_{filtrate}}{2.50 \times 10^3}$$

$[I^-]_{filtrate} = 1.0 \times 10^{-8}\,M$

$K_{sp} = [Ag^+][I^-] = [I^-]^2$ (since $[Ag^+] = [I^-]$)

$K_{sp} = (1.0 \times 10^{-8})^2 = 1.0 \times 10^{-16}$

48. Refer to Section 19.2.

Calculate k (in s^{-1}) from the half-life. Then calculate moles of $^3H_2$ using the ideal gas law. Finally, calculate N and A.

$$k = \frac{0.693}{12.3\,yr} \times \frac{1\,yr}{365\,d} \times \frac{1\,d}{24\,hr} \times \frac{1\,hr}{3600\,s} = 1.79 \times 10^{-9}\,/s$$

$$n = \frac{PV}{RT} = \frac{(1\,atm)(0.00100\,L)}{(0.0821\,L \cdot atm/mol. \cdot K)(273\,K)} = 4.46 \times 10^{-5}\,mol.\,^3H_2$$

$$N = 4.46 \times 10^{-5}\,mol.\,^3H_2 \times \frac{6.022 \times 10^{23}\,molecules\,^3H_2}{1\,mol.} \times \frac{2\,atoms\,^3H}{1\,^3H_2} = 5.37 \times 10^{19}\,atoms\,^3H$$

$$A = 5.37 \times 10^{19}\,atoms \times 1.79 \times 10^{-9}\,/s \times \frac{1\,Ci}{3.700 \times 10^{10}\,atoms/s} = 2.60\,Ci$$

50. Refer to Section 19.2.

The A_0/A ratio represents a "dilution factor" that can be used to calculate the total volume (volume of blood).

$$5.0\,mL \times \frac{1.7 \times 10^5\,cps}{1.3 \times 10^3\,cps} = 6.5 \times 10^2\,mL$$

Convert cpm/g to cpm/mol. Use this to convert cpm to mol. Determine the ratio of $C_2O_4^{2-}$ to Cr^{3+} by dividing moles of each by the smallest number of moles (as with empirical formula).

$$\frac{765\ cpm}{g\ Na_2CrO_4} \times \frac{162.0}{1mol.\ Na_2CrO_4} \times \frac{1mol.\ Na_2CrO_4}{1mol.\ Cr^{3+}} = \frac{1.24\times10^5\ cpm}{1mol.\ Cr^{3+}}$$

$$314\ cpm \times \frac{1mol.\ Cr^{3+}}{1.24\times10^5\ cpm} = 2.53\times10^{-3}\ mol.\ Cr^{3+}$$

$$\frac{512\ cpm}{g\ H_2C_2O_4} \times \frac{90.04\ g}{1mol.\ H_2C_2O_4} \times \frac{1mol.\ H_2C_2O_4}{1mol.\ C_2O_4^{2-}} = \frac{4.61\times10^4\ cpm}{1mol.\ C_2O_4^{2-}}$$

$$235\ cpm \times \frac{1mol.\ C_2O_4^{2-}}{4.61\times10^4\ cpm} = 5.10\times10^{-3}\ mol.\ C_2O_4^{2-}$$

2.53 x 10^{-3} mol. Cr^{3+}/2.53 x 10^{-3} mol. Cr^{3+} = 1
5.10 x 10^{-3} mol. $C_2O_4^{2-}$/2.53 x 10^{-3} mol. Cr^{3+} = 2
There are 2 $C_2O_4^{2-}$ ions bound to one Cr^{3+} ion.

Calculate k using $t_{1/2}$. Then convert the mass of Po-210 to moles to get X_0. Then solve for X to determine the amount of Po-210 remaining. Calculate the amount of Po-210 that decayed (and thus the amount of He that is formed). Finally, apply the ideal gas equation to determine the volume of He produced.

$$^{210}_{84}Po \rightarrow\ ^4_2He +\ ^{206}_{82}Pb$$

$$k = \frac{0.693}{138\ d} \times \frac{1\ d}{24\ hr} = 2.09\times10^{-4}\ /hr$$

$$25.00\ g\ Po\text{-}210 \times \frac{1\ mol.\ Po\text{-}210}{209.94\ g\ Po\text{-}210} = 0.1191\ mol.\ Po\text{-}210$$

$$\ln\left(\frac{X_0}{X}\right) = kt \implies \ln\left(\frac{0.1191\ mol.}{X}\right) = (2.09\times10^{-4}\ /hr)(75\ hr)$$

ln (0.1191 mol.) - ln (X) = 0.016
ln (X) = -2.144
$X = e^{-2.144} = 0.117$ (mol. of Po-210 remaining)

$X_0 - X$ = 0.1191 - 0.117 = 0.002   (mol. Po-210 that decayed = mol. He produced)

$$V = \frac{nRT}{P} = \frac{(0.002 \text{ mol.})(0.0821 \text{ L} \cdot \text{atm/mol.} \cdot \text{K})(298 \text{ K})}{1.20 \text{ atm}} = 0.04 \text{ L He} = 40 \text{ mL}$$

56. Refer to Section 19.3 and Chapter 6.

mass = 0.00055 g + 0.00055 g = 0.0011 g

$$\frac{0.0011 \text{ g}}{2 \text{ mol. photons}} \times \frac{2 \text{ mol. photons}}{6.022 \times 10^{23} \text{ photons}} = 9.13 \times 10^{-28} \text{ g/photon}$$

$$E = mc^2 = (9.13 \times 10^{-28} \text{ g/photon})(9.00 \times 10^{10} \text{ kJ/g}) = 8.22 \times 10^{-17} \text{ kJ} = 8.22 \times 10^{-14} \text{ J}$$

$$E = hc/\lambda \Rightarrow \lambda = hc/E = \frac{(6.626 \times 10^{-34} \text{ J} \cdot \text{s})(3.00 \times 10^8 \text{ m/s})}{8.22 \times 10^{-14} \text{ J}} = 2.42 \times 10^{-12} \text{ m} = 2.42 \times 10^{-3} \text{ nm}$$

58. Refer to Sections 19.1 and 19.4.

a. Alpha rays (He nuclei) are attracted to the negative pole while beta rays (electrons) are attracted to the positive pole.

b. Glucose labeled with C-11 is given to the patient. C-11 is a positron emitter. The brain is scanned using Positron Emission Tomography (PET).

c. Fission produces neutrons. The neutrons bombard more uranium producing more neutrons. This cycle continues on its own, constituting a chain reaction.

60. Refer to Sections 11.3 and 19.2.

Plot ln(A) versus time (in hours). The slope of the line is -k, which is used to calculate $t_{1/2}$.

Time (h)	0.00	0.50	1.00	1.50	2.00	2.50
A (disintegrations/h)	14,472	13,095	11,731	10,615	9,605	8,504
ln (A)	9.580	9.480	9.370	9.270	9.170	9.048

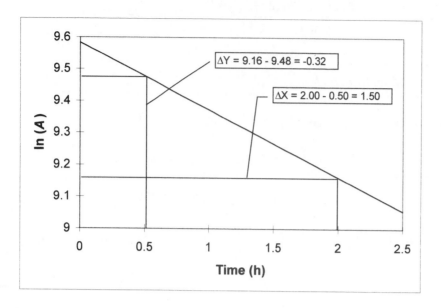

$$-k = \text{slope} = \frac{\Delta Y}{\Delta X} = \frac{-0.32}{1.5} = -0.21$$

$$k = 0.21$$

$$k = \frac{0.693}{t_{1/2}} \implies 0.21 = \frac{0.693}{t_{1/2}}$$

$$t_{1/2} = 3.3 \text{ hr}$$

---

62. Refer to Sections 19.1 and 19.2.

Calculate the percent Am-241 that remains after one year, setting X_0 to 100%.

$$\ln\left(\frac{X_0}{X}\right) = kt \implies \ln\left(\frac{100}{X}\right) = (1.51 \times 10^{-3} \text{ /yr})(1 \text{ yr})$$

$$\ln(100) - \ln(X) = 1.51 \times 10^{-3}$$

$$\ln(X) = 4.60$$

$$X = e^{4.60} = 99.8 \qquad\qquad \text{(percent Am-241 remaining)}$$

Thus 99.8% of the Am-241 remains. At that rate, the Am-241 will last a lifetime (only 14% decays in 100 years).

64. Refer to Section 19.2.

From $t_{1/2}$, calculate k and from k and A, calculate N.

$$k = \frac{0.693}{3.82\,d} \times \frac{1\,d}{24\,hr} \times \frac{1\,hr}{3600\,s} = 2.10 \times 10^{-6}\,/s$$

$$N = \frac{20 \times 10^{-12}\,Ci}{2.10 \times 10^{-6}\,/s} \times \frac{3.700 \times 10^{10}\,atoms/s}{1\,Ci} = 3.5 \times 10^{5}\,atoms$$

This is a concentration of 3.5×10^{-5} atoms/L. Convert to moles per liter.

$$\frac{3.5 \times 10^{-5}\,atoms}{1\,L} \times \frac{1\,mol.}{6.022 \times 10^{23}\,atoms} = 5.8 \times 10^{-19}\,mol./L$$

65. Refer to Sections 19.2 and 19.3 and "Chemistry Beyond the Classroom."

a. $1.00\,g \times 5.5 \times 10^{-11}\,/min. \times 45\,min. = 2.5 \times 10^{-9}\,g$

b. $\Delta m = 234.9934 + 4.00150 - 239.0006 = -0.0057\,g$

$$\Delta m_{(1g)} = 2.5 \times 10^{-9}\,g \times \frac{1\,mol.}{239.0\,g} \times \frac{-0.0057\,g}{1\,mol.} = -6.0 \times 10^{-14}\,g$$

$\Delta E = 9.00 \times 10^{10}\,kJ/g \times (-6.0 \times 10^{-14})\,g = -5.4 \times 10^{-3}\,kJ = -5.4\,J$

Thus 5.4 J of energy is given off.

c. 1 rad = 10^{-2} J/kg
    rems = n(rads) $n = 10$ for α rays

$$\frac{5.4\,J}{75\,kg} \times \frac{1\,rad}{10^{-2}\,J/kg} = 7.2\,rads$$

rems = 7.2 rads $\times$ 10 rems/rad = 72 rems

66. Refer to the equations stated in the problem.

a. $E = 8.99 \times 10^{9} q_1 q_2/r = (8.99 \times 10^{9})(1.60 \times 10^{-19}\,C)^2 / (2 \times 10^{-15}\,m) = 1 \times 10^{-13}\,J$

b. Convert the mass of the deuteron to kilograms.

$$2\,deuterons \times \frac{2.01355\,g}{1\,mol.} \times \frac{1\,mol.}{6.022 \times 10^{23}} \times \frac{1\,kg}{1000\,g} = 6.687 \times 10^{-27}\,kg$$

$E = mv^2/2 \quad = \quad 2E\,/\,m = v^2$

$$v = [(2)(1 \times 10^{-13} \text{ J}) / (6.687 \times 10^{-27} \text{ kg})]^{\frac{1}{2}} = 5 \times 10^{6} \text{ m/s}$$

67. Refer to Section 19.3 and Example 19.4.

a. $\Delta m = 4.00150 - 2(2.01355) = -0.02560$ g (per 2 mol. H-2)

$$\Delta m_{(1g)} = 1.0000 \text{ g} \times \frac{1 \text{ mol.}}{2.01355 \text{ g}} \times \frac{-0.02560 \text{ g}}{2 \text{ mol.}} = -0.006357 \text{ g}$$

$\Delta E = 9.00 \times 10^{10}$ kJ/g $\times$ (-0.006357 g) = -5.72 $\times 10^{8}$ kJ (per gram of deuterium fused)

b. $1.3 \times 10^{24} \text{ g} \times \dfrac{0.0017 \text{ g}}{100 \text{ g}} = 2.2 \times 10^{19} \text{ g H-2}$

2.2×10^{19} g $\times$ -5.72 $\times 10^{8}$ kJ/g = -1.3 $\times 10^{28}$ kJ

c. $\dfrac{2.3 \times 10^{17} \text{ kJ}}{1.3 \times 10^{28} \text{ kJ}} = 1.8 \times 10^{-11}$

Chapter 20: Chemistry of the Metals

2. Refer to Section 20.1 and Chapter 5.

$2Al_2O_3(l) \rightarrow 4Al(l) + 3O_2(g)$

$$n = \frac{PV}{RT} = \frac{(0.988\,atm)(2.00\,L)}{(0.0821\,L\cdot atm/mol\cdot K)(298\,K)} = 0.0808\,mol.\,O_2$$

$$0.0808\,mol.\,O_2 \times \frac{4\,mol.\,Al}{3\,mol.\,O_2} \times \frac{26.98\,g\,Al}{1\,mol.\,Al} = 2.91\,g\,Al$$

4. Refer to Section 20.1.

$Cu_2S(s) + O_2(g) \rightarrow 2Cu(s) + SO_2(g)$

6. Refer to Example 20.1 and Chapter 17.

$\Delta H° = \Sigma\,\Delta H_f°_{(products)} - \Sigma\,\Delta H_f°_{(reactants)}$
$\Delta H° = [(2\,mol.)(0.0\,kJ/mol.) + (1\,mol.)(-296.8\,kJ/mol.)]$
$\qquad - [(1\,mol.)(-79.5\,kJ/mol.) + (1\,mol.)(0.0\,kJ/mol.)]$
$\Delta H° = -217.3\,kJ$

$\Delta S° = \Sigma\,S°_{(products)} - \Sigma\,S°_{(reactants)}$
$\Delta S° = [(2\,mol.)(0.0332\,kJ/mol.\cdot K) + (1\,mol.)(0.2481\,kJ/mol.\cdot K)]$
$\qquad - [(1\,mol.)(0.1209\,kJ/mol.\cdot K) + (1\,mol.)(0.2050\,kJ/mol.\cdot K)]$
$\Delta S° = -0.0114\,kJ/K$

$\Delta G° = \Delta H° - T\Delta S°$
$\Delta G° = -217.3\,kJ - (473\,K)(-0.0114\,kJ/K) = -211.9\,kJ$

8. Refer to Section 20.1.

a. $Fe_2O_3(s) + 3CO(g) \rightarrow 2Fe(l) + 3CO_2(g)$

b. $C(s) + O_2(g) \rightarrow CO_2(g)$

10. Refer to Section 20.1 and Chapter 18.

one metric ton $= 1 \times 10^3$ kg

$$1.000 \times 10^3 \text{ kg} \times \frac{1000 \text{ g}}{1 \text{ kg}} \times \frac{1 \text{ mol. Zn}}{65.39 \text{ g}} = 1.529 \times 10^4 \text{ mol. Zn}$$

$$1.529 \times 10^4 \text{ mol. Zn} \times \frac{2 \text{ mol. } e^-}{1 \text{ mol. Zn}} \times \frac{9.648 \times 10^4 \text{ C}}{1 \text{ mol. } e^-} = 2.950 \times 10^9 \text{ C}$$

$$J = VC = (3.0 \text{ V})(2.950 \times 10^9 \text{ C}) = 8.9 \times 10^9 \text{ J}$$

$$8.9 \times 10^9 \text{ J} \times \frac{1 \text{ kWh}}{3.600 \times 10^6 \text{ J}} = 2.5 \times 10^3 \text{ kWh}$$

12. Refer to Section 20.1.

Chalcopyrite, $CuFeS_2$, has a ratio of Cu:S of 1:2; thus for each mole of Cu, two moles of SO_2 are formed. Chalcopyrite ore is not pure chalcopyrite, but contains $CuFeS_2$ and other minerals. The ore contains 0.75% Cu, meaning 100 g ore has 0.75 g Cu.

$$4.00 \times 10^3 \text{ ft}^3 \times \left(\frac{12 \text{ in}}{1 \text{ ft}}\right)^3 \times \left(\frac{2.54 \text{ cm}}{1 \text{ in}}\right)^3 \times \frac{2.6 \text{ g ore}}{1 \text{ cm}^3} = 2.9 \times 10^8 \text{ g ore}$$

$$2.9 \times 10^8 \text{ g ore} \times \frac{0.75 \text{ g Cu}}{100 \text{ g ore}} \times \frac{1 \text{ mol. Cu}}{63.55 \text{ g Cu}} = 3.4 \times 10^4 \text{ mol. Cu}$$

$$3.4 \times 10^4 \text{ mol. Cu} \times \frac{2 \text{ mol. } SO_2}{1 \text{ mol. Cu}} = 6.8 \times 10^4 \text{ mol. } SO_2$$

$$V = \frac{nRT}{P} = \frac{(6.8 \times 10^4 \text{ mol.})(0.0821 \text{ L} \cdot \text{atm/mol.} \cdot \text{K})(298 \text{ K})}{1 \text{ atm}} = 1.7 \times 10^6 \text{ L } SO_2$$

14. Refer to Section 20.2 and Table 20.1.

a. potassium nitride K_3N

b. potassium iodide   KI

c. potassium hydroxide   KOH

d. potassium hydride   KH

e. potassium sulfide K_2S

a. $Na_2O_2(s) + 2H_2O(l) \rightarrow 2Na^+(aq) + 2OH^-(aq) + H_2O_2(aq)$
sodium ion, hydroxide ion, hydrogen peroxide.

b. $2Ca(s) + O_2(g) \rightarrow 2CaO(s)$
calcium oxide.

c. $Rb(s) + O_2(g) \rightarrow RbO_2(s)$
rubidium superoxide.

d. $SrH_2(s) + 2H_2O(l) \rightarrow Sr^{2+}(aq) + 2OH^-(aq) + 2H_2(g)$
strontium ion, hydroxide ion, hydrogen gas.

18. Refer to Section 20.2.

$$n = \frac{PV}{RT} = \frac{(1.00\,\text{atm})(1.00\,\text{L})}{(0.0821\,\text{L}\cdot\text{atm/mol}\cdot\text{K})(310\,\text{K})} = 0.0393\,\text{mol. air}$$

$0.0393\,\text{mol. air} \times \dfrac{5.00\,\text{mol. CO}_2}{100\,\text{mol. air}} = 0.00196\,\text{mol. CO}_2$

$0.00196\,\text{mol. CO}_2 \times 0.900 = 0.00177\,\text{mol. CO}_2$ (to be removed).

$4KO_2(s) + 2H_2O(g) \rightarrow 3O_2(g) + 4KOH(s)$
$KOH(s) + CO_2(g) \rightarrow KHCO_3(g)$

$0.00177\,\text{mol. CO}_2 \times \dfrac{1\,\text{mol. KOH}}{1\,\text{mol. CO}_2} \times \dfrac{4\,\text{mol. KO}_2}{4\,\text{mol. KOH}} \times \dfrac{71.10\,\text{g}}{1\,\text{mol. KO}_2} = 0.126\,\text{g KO}_2$

20. Refer to Section 20.3 and Example 20.6.

a.
$Co^0 \rightarrow Co^{2+} + 2e^-$ oxidation half-reaction
$\underline{2H^+ + 2e^- \rightarrow H_2}$ reduction half-reaction
$Co(s) + 2H^+(aq) \rightarrow Co^{2+}(aq) + H_2(g)$

b. $Cu^0 \rightarrow Cu^{2+} + 2e^-$ oxidation half-reaction
$4H^+ + NO_3^- + 3e^- \rightarrow NO + 2H_2O$ reduction half-reaction

     balancing the electrons gives:
$3Cu^0 \rightarrow 3Cu^{2+} + 6e^-$
$\underline{8H^+ + 2NO_3^- + 6e^- \rightarrow 2NO + 4H_2O}$
$3Cu(s) + 8H^+(aq) + 2NO_3^-(aq) \rightarrow 3Cu^{2+}(aq) + 2NO(g) + 4H_2O(l)$

c. $14H^+(aq) + Cr_2O_7^{2-}(aq) + 6e^- \rightarrow 2Cr^{3+}(aq) + 7H_2O(l)$
    (see Chapter 4, Problem 54.)

22. ***Refer to Section 20.3 and Chapter 4.***

$Cd(s) + H^+(aq) + Cl^-(aq) + NO_3^-(aq) \rightarrow CdCl_4^{2-}(aq) + NO(g) + H_2O(l)$

$Cd^0 \rightarrow Cd^{2+} + 2e^-$ Oxidation half-reaction
$Cd(s) + 4Cl^-(aq) \rightarrow CdCl_4^{2-}(aq) + 2e^-$

$N^{5+} + 3e^- \rightarrow N^{2+}$ Reduction half-reaction
$NO_3^-(aq) + 3e^- \rightarrow NO(g)$
$4H^+(aq) + NO_3^-(aq) + 3e^- \rightarrow NO(g) + 2H_2O(l)$

    Balancing electrons gives:
$3Cd(s) + 12Cl^-(aq) + 8H^+(aq) + 2NO_3^-(aq) \rightarrow 2NO(g) + 4H_2O(l) + 3CdCl_4^{2-}(aq)$

24. ***Refer to Section 20.3 and Chapter 4.***

a. $Fe^0 \rightarrow Fe^{3+} + 3e^-$ Oxidation half-reaction

    $N^{5+} + e^- \rightarrow N^{4+}$ Reduction half-reaction
    $2H^+(aq) + NO_3^-(aq) + e^- \rightarrow NO_2(g) + H_2O(l)$

    Balancing electrons gives:
    $6H^+(aq) + Fe(s) + 3NO_3^-(aq) \rightarrow 3NO_2(g) + Fe^{3+}(aq) + 3H_2O(l)$

b. $Cr^{3+} \rightarrow Cr^{6+} + 3e^-$ Oxidation half-reaction
    $Cr(OH)_3(s) \rightarrow CrO_4^{2-}(aq) + 3e^-$
    $5\ OH^-(aq) + Cr(OH)_3(s) \rightarrow CrO_4^{2-}(aq) + 3e^-$
    $5\ OH^-(aq) + Cr(OH)_3(s) \rightarrow CrO_4^{2-}(aq) + 3e^- + 4H_2O(l)$

    $O^0 + 2e^- \rightarrow O^{2-}$ Reduction half-reaction
    $O_2(g) + 4e^- \rightarrow 2H_2O(l)$
    $O_2(g) + 4e^- \rightarrow 2H_2O(l) + 4\ OH^-(aq)$
    $4H_2O(l) + O_2(g) + 4e^- \rightarrow 2H_2O(l) + 4OH^-(aq)$
    $2H_2O(l) + O_2(g) + 4e^- \rightarrow 4\ OH^-(aq)$

    Balancing electrons gives:
    $20\ OH^-(aq) + 4Cr(OH)_3(s) + 6H_2O(l) + 3O_2(g) + 12e^-$
                       $\rightarrow 12\ OH^-(aq) + 4CrO_4^{2-}(aq) + 12e^- + 16H_2O(l)$
    $8\ OH^-(aq) + 4Cr(OH)_3(s) + 3O_2(g) \rightarrow 4CrO_4^{2-}(aq) + 10H_2O(l)$

Calculate $E°$ for the reaction. If $E°$ is positive, then the reaction will occur.

$NO_3^-(aq) \rightarrow NO(g)$ $E°_{red} = +0.964$ V

a. $Cd \rightarrow Cd^{2+}$ $E°_{ox} = 0.402$ V

$E° = E°_{red} + E°_{ox} = 0.964 + 0.402 = 1.366$ V
$E°$ is positive, so the reaction will occur.

b. $Cr \rightarrow Cr^{2+}$ $E°_{ox} = 0.912$ V

$E° = E°_{red} + E°_{ox} = 0.964 + 0.912 = 1.876$ V
$E°$ is positive, so the reaction will occur.

$Cr \rightarrow Cr^{3+}$ $E°_{ox} = 0.744$ V

$E° = E°_{red} + E°_{ox} = 0.964 + 0.744 = 1.708$ V
$E°$ is positive, so the reaction will occur.

c. $Co \rightarrow Co^{2+}$ $E°_{ox} = 0.282$ V

$E° = E°_{red} + E°_{ox} = 0.964 + 0.282 = 1.246$ V
$E°$ is positive, so the reaction will occur.

d. $Ag \rightarrow Ag^+$ $E°_{ox} = -0.799$ V

$E° = E°_{red} + E°_{ox} = 0.964 - 0.799 = 0.165$ V
$E°$ is positive, so the reaction will occur.

e. $Au \rightarrow Au^{3+}$ $E°_{ox} = -1.498$ V

$E° = E°_{red} + E°_{ox} = 0.964 - 1.498 = -0.534$ V
$E°$ is negative, so the reaction will **not** occur.

28. Refer to Section 20.3 and Chapter 18.

a. $2Co^{3+}(aq) + 2e^- \rightarrow 2Co^{2+}(aq)$ $E°_{red} = 1.953$ V
$H_2O \rightarrow \frac{1}{2}O_2(g) + 2H^+(aq) + 2e^-$ $E°_{ox} = -1.229$ V

$E° = E°_{red} + E°_{ox} = 1.953 - 1.229 = 0.724$ V

b. $I_2(s) + 2e^- \rightarrow 2I^-(aq)$ $\qquad\qquad\qquad E^\circ_{red} = 0.534$ V

$2Cr^{2+}(aq) \rightarrow 2Cr^{3+}(aq) + 2e^-$ $\qquad\quad E^\circ_{ox} = 0.408$ V

$E^\circ = E^\circ_{red} + E^\circ_{ox} = 0.534 + 0.408 = 0.942$ V

---

30. Refer to Section 20.3 and Chapter 18.

a. $Au^+ + e^- \rightarrow Au^0$ $\qquad\qquad\qquad\qquad E^\circ_{red} = 1.695$ V

$\quad Au^+ \rightarrow Au^{3+} + 2e^-$ $\qquad\qquad\qquad E^\circ_{red} = -1.400$ V

$\qquad\qquad$ Balancing electrons:

$3Au^+ \rightarrow Au^{3+} + 2Au^0$ $\qquad\qquad\quad E^\circ = 0.295$ V

$$E^\circ = \frac{0.0257}{n} \ln K \quad \Rightarrow \quad 0.295 = \frac{0.0257}{2} \ln K$$

$\ln K = 23.0$

$K = 9.74 \times 10^9$

b. $\quad K = \dfrac{[Au^{3+}]}{[Au^+]^3} \quad \Rightarrow \quad 9.74 \times 10^9 = \dfrac{0.10\,M}{[Au^+]^3}$

$[Au^+]^3 = 1.03 \times 10^{-11}$

$[Au^+] = 2.2 \times 10^{-4}$

---

32. Refer to Section 20.2, Problem 18 (above) and Chapter 5.

Calculate the volume of water vapor in the exhaled air, the moles of water and then the amount of KO_2 that reacts, and finally the amount of KO_2 remaining.

$$116 \text{ L air} \times \frac{6.2 \text{ L H}_2\text{O}}{100 \text{ L air}} = 7.2 \text{ L H}_2\text{O}(g)$$

$$n = \frac{PV}{RT} = \frac{(0.984 \text{ atm})(7.2 \text{ L})}{(0.0821 \text{ L} \cdot \text{atm/mol} \cdot \text{K})(310 \text{ K})} = 0.28 \text{ mol. H}_2\text{O}$$

$$0.28 \text{ mol. H}_2\text{O} \times \frac{4 \text{ mol. KO}_2}{2 \text{ mol. H}_2\text{O}} \times \frac{71.10 \text{ g KO}_2}{1 \text{ mol. KO}_2} = 40 \text{ g KO}_2 \text{ (reacts)}$$

248 g - 40 g = 208 g (remains)

Calculate the [H$^+$] for the given concentrations of chromate and dichromate, then calculate pH from the result.

$$K_a = \frac{[Cr_2O_7^{2-}]}{[CrO_4^{2-}]^2[H^+]^2} \quad \Rightarrow \quad 3 \times 10^{14} = \frac{0.10\,M}{(0.10\,M)^2[H^+]^2}$$

$[H^+]^2 = 3 \times 10^{-14}$
$[H^+] = 2 \times 10^{-7}\,M$

pH = -log[H$^+$] = - log(2 × 10^{-7}) = 6.7

Calculate the total pressure of H$_2$(g) and then the moles of H$_2$. From the result calculate the mass of Zn and the mass percent of zinc in the alloy.

$$P_{H_2} = P_{total} - P_{H_2O} = 755 - 26.74 = 728\,mm\,Hg = 0.958\,atm$$

$$n = \frac{PV}{RT} = \frac{(0.958\,atm)(0.1057\,L)}{(0.0821\,L \cdot atm/mol. \cdot K)(300\,K)} = 0.00411\,mol.\,H_2$$

$$Zn(s) + 2H^+(aq) \rightarrow Zn^{2+}(aq) + H_2(g)$$

$$0.00411\,mol.\,H_2 \times \frac{1\,mol.\,Zn}{1\,mol.\,H_2} \times \frac{65.39\,g\,Zn}{1\,mol.\,Zn} = 0.269\,g\,Zn$$

$$mass\,\%\,Zn = \frac{0.269\,g\,Zn}{0.500\,g\,alloy} \times 100\% = 53.8\,\%\,Zn$$

100% - 53.8% Zn = 46.2% Cu

Write a balanced redox equation for the process, assuming basic conditions (since gold is extracted similarly, under basic conditions). To simplify the process, consider the oxygen to be reduced from O$_2$(g) to hydroxide ion, and ignore the spectator ions (silver and cyanide).

$Ag_2S(s) + CN^-(aq) + O_2(g) \rightarrow SO_2(g) + Ag(CN)_2^-(aq)$

$O^0 + 2e^- \rightarrow O^{2-}$
$O_2(g) + 4e^- \rightarrow 2OH^-(aq)$
$O_2(g) + 4e^- \rightarrow 2OH^-(aq) + 2OH^-(aq)$
$2H_2O(l) + O_2(g) + 4e^- \rightarrow 4OH^-(aq)$

$S^{2-} \rightarrow S^{4+} + 6e^-$

$S^{2-}(aq) \rightarrow SO_2(g) + 6e^-$

$4OH^-(aq) + S^{2-}(aq) \rightarrow SO_2(g) + 6e^-$

$4OH^-(aq) + S^{2-}(aq) \rightarrow SO_2(g) + 6e^- + 2H_2O(l)$

Balancing electrons gives:

$6H_2O(l) + 3O_2(g) + 12e^- + 8OH^-(aq) + 2S^{2-}(aq) \rightarrow 2SO_2(g) + 12e^- + 4H_2O(l) + 12OH^-(aq)$

$2H_2O(l) + 3O_2(g) + 2S^{2-}(aq) \rightarrow 2SO_2(g) + 4OH^-(aq)$

Now add in the spectator ions, keeping the proper ratio.

$2H_2O(l) + 3O_2(g) + 2Ag_2S(s) \rightarrow 2SO_2(g) + 4OH^-(aq) + 4Ag^+(aq)$

$2H_2O(l) + 3O_2(g) + 2Ag_2S(s) + 8CN^-(aq) \rightarrow 2SO_2(g) + 4OH^-(aq) + 4Ag(CN)_2^-(aq)$

This equation describes the conversion of insoluble Ag_2S to the soluble $Ag(CN)_2^-$ complex. The final step is the reduction of the silver cation to silver metal.

$Zn^0 \rightarrow Zn^{2+} + 2e^-$

$Ag^+ + e^- \rightarrow Ag^0$

Balancing electrons gives:

$Zn(s) + 2Ag^+(aq) + 2e^- \rightarrow 2Ag(s) + Zn^{2+}(aq) + 2e^-$

$Zn(s) + 2Ag^+(aq) \rightarrow 2Ag(s) + Zn^{2+}(aq)$

Adding back the cyanide spectator ion:

$Zn(s) + 2Ag(CN)_2^-(aq) \rightarrow 2Ag(s) + Zn(CN)_4^{2-}(aq)$

---

40. Refer to Section 20.3 and Chapter 18.

---

a. The strongest reducing agent is the one most easily oxidized. Of the choices, **Cr^{2+}** is the only one with a positive E_{ox}° (+0.408 V) and is thus the strongest reducing agent.

b. The strongest oxidizing agent is the one most easily reduced. Of the choices, **Au$^+$** has the highest E_{red}° (+1.695 V), and is thus the strongest oxidizing agent.

c. The weakest reducing agent is the one least readily oxidized. Of the choices, **Co^{2+}** has the lowest E_{ox}° (-1.953 V), and is thus the weakest reducing agent.

d. The weakest oxidizing agent is the one least readily reduced. Of the choices, **Mn^{2+}** has the lowest E_{red}° (-1.182 V), and is thus the weakest oxidizing agent.

The only reactants in the two reactions are barium and oxygen. Calculate the amount of oxygen that reacts and then the mole ratio of barium to oxygen. The excess moles of oxygen must be used to form BaO_2, and thus indicate the amount of BaO_2 formed.

22.38 g total - 20.00 g Ba = 2.38 g oxygen

$$20.00 \text{ g} \times \frac{1 \text{ mol. Ba}}{137.3 \text{ g}} = 0.1457 \text{ mol. Ba}$$

$$2.38 \text{ g} \times \frac{1 \text{ mol. O}}{16.00 \text{ g}} = 0.149 \text{ mol. O}$$

$$\text{mole ratio} = \frac{0.149 \text{ mol. O}}{0.1457 \text{ mol. Ba}} = 1.02 \text{ mol. O per mol. Ba} \quad \text{(in the product mixture)}$$

This indicates that there are 0.02 moles 'excess' oxygen. Since barium oxide (BaO) has a 1:1 Ba:O ratio, the 'excess' oxygens must belong to barium peroxide (BaO_2); giving 0.02 mol. BaO_2, with the remainder (0.98 mol.) BaO.

This can be shown mathematically by setting x = mol. Ba in BaO, and y = mol. Ba in BaO_2. Consequently, x = mol. O in Ba, and $2y$ = mol. O in BaO_2. Then:

Ba:	$x + y = 1$	total moles Ba
O:	$x + 2y = 1.02$	total moles O
	$y = 0.02$	This is moles of BaO_2 (and $x = 0.98$ mol. BaO)

$$\frac{0.98 \text{ mol. BaO}}{1 \text{ mol. total}} \times 100\% = 98\% \text{ BaO}$$

$$\frac{0.02 \text{ mol. BaO}_2}{1 \text{ mol. total}} \times 100\% = 2\% \text{ BaO}_2$$

a. The oxalic acid is a chelating ligand, and Fe^{3+} has 6 coordination sites, suggesting that a complex with 3 oxalates forms. Recall that oxalic acid is a diprotic acid (has 2 H^+).

$$Fe(OH)_3(s) + 3H_2C_2O_4(aq) \rightarrow [Fe(C_2O_4)_3]^{3-}(aq) + 3H_2O(l) + 3H^+(aq)$$

b. $1.0 \text{ g Fe(OH)}_3 \times \dfrac{1 \text{ mol. Fe(OH)}_3}{106.9 \text{ g}} \times \dfrac{3 \text{ mol. H}_2\text{C}_2\text{O}_4}{1 \text{ mol. Fe(OH)}_3} \times \dfrac{1 \text{ L solution}}{0.10 \text{ mol. H}_2\text{C}_2\text{O}_4} = 0.28 \text{ L}$

Write a balanced redox equation for the dichromate-ferric ion couple. Then calculate the initial amount of Fe^{2+} used, and the amount in excess. The difference gives the amount that reacted with MnO_4^-. Write a balanced redox equation for the reduction of permanganate with ferrous ion to get a mole ratio of iron to manganese. From the mole ratio, calculate the amount of permanganate and then the percent manganese in the sample.

$$Fe^{2+} \rightarrow Fe^{3+} + e^- \qquad\qquad \text{oxidation half-reaction}$$

$$Cr^{6+} + 3e^- \rightarrow Cr^{3+} \qquad\qquad \text{reduction half-reaction}$$
$$14H^+(aq) + Cr_2O_7^{2-}(aq) + 6e^- \rightarrow 2Cr^{3+}(aq) + 7H_2O(l) \quad \textit{(see problem 20c)}$$

balancing electrons and adding the equations gives:
$$14H^+(aq) + Cr_2O_7^{2-}(aq) + 6Fe^{2+}(aq) \rightarrow 6Fe^{3+}(aq) + 2Cr^{3+}(aq) + 7H_2O(l)$$

$$0.07500 \text{ L sol'n} \times \frac{0.125 \text{ mol. FeSO}_4}{1 \text{ L sol'n}} \times \frac{1 \text{ mol. Fe}^{2+}}{1 \text{ mol. FeSO}_4} = 0.00938 \text{ mol. Fe}^{2+} \quad \text{(initial)}$$

$$0.01350 \text{ L sol'n} \times \frac{0.100 \text{ mol. Cr}_2O_7^{2-}}{1 \text{ L sol'n}} \times \frac{6 \text{ mol. Fe}^{2+}}{1 \text{ mol. Cr}_2O_7^{2-}} = 0.00810 \text{ mol. Fe}^{2+} \quad \text{(excess)}$$

$$0.00938 \text{ mol.} - 0.00810 = 0.00128 \text{ mol. Fe}^{2+} \text{ (reacted with MnO}_4^-)$$

$$Fe^{2+} \rightarrow Fe^{3+} + e^- \qquad\qquad \text{oxidation half-reaction}$$

$$Mn^{7+} + 5e^- \rightarrow Mn^{2+} \qquad\qquad \text{reduction half-reaction}$$
$$8H^+(aq) + MnO_4^-(aq) + 5e^- \rightarrow Mn^{2+}(aq) + 4H_2O(l) \qquad \textit{(see Chapter 4, problem 50a)}$$

balancing electrons and adding the equations gives:
$$8H^+(aq) + MnO_4^-(aq) + 5Fe^{2+}(aq) \rightarrow 5Fe^{3+}(aq) + Mn^{2+}(aq) + 4H_2O(l)$$

$$0.00128 \text{ mol. Fe}^{2+} \times \frac{1 \text{ mol. MnO}_4^-}{5 \text{ mol. Fe}^{2+}} \times \frac{1 \text{ mol. Mn}}{1 \text{ mol. MnO}_4^-} \times \frac{54.94 \text{ g}}{1 \text{ mol. Mn}} = 0.0141 \text{ g Mn}$$

$$\text{mass \% Mn} = \frac{\text{mass Mn}}{\text{mass sample}} \times 100\% = \frac{0.0141 \text{ g}}{0.500 \text{ g}} \times 100\% = 2.82 \%$$

$\Delta G° = \Delta H° - T\Delta S°$ and,
$\Delta G° = -RT \ln K,$ but $\ln(1.00) = 0,$ thus,
$\Delta G° = 0 = \Delta H° - T\Delta S°$

Calculate $\Delta H°$ and $\Delta S°$, and then solve for T.

$\Delta H_{(reaction)} = \Sigma \; \Delta H_f^\circ \;_{(products)} - \Sigma \; \Delta H_f^\circ \;_{(reactants)}$

$\Delta H_{(reaction)} = [(1 \; mol.)(0 \; kJ/mol.) + (1 \; mol.)(0 \; kJ/mol.)] - [(1 \; mol.)(-520.0 \; kJ/mol.)]$

$\Delta H_{(reaction)} = +520.0 \; kJ$

$\Delta S_{(reaction)} = \Sigma \; S_f^\circ \;_{(products)} - \Sigma \; S_f^\circ \;_{(reactants)}$

$\Delta S_{(reaction)} = [(1 \; mol.)(0.0320 \; kJ/mol.\cdot K) + (1 \; mol.)(0.2050 \; kJ/mol.\cdot K)]$
$\qquad\qquad\; - [(1 \; mol.)(0.0530 \; kJ/mol.\cdot K)]$

$\Delta S_{(reaction)} = 0.184 \; kJ/K$

$\Delta G^\circ = 0 = 520.0 \; kJ - (T)(0.184 \; kJ/K)$
$T = 2.83 \times 10^3 \; K = 2.56 \times 10^3 \; ^\circ C$

---

45. Refer to Section 20.3, Figure 20.11 and Chapters 4 and 15.

$Cr_2O_7^{2-}(aq) + 2OH^-(aq) \; \rightarrow \; 2CrO_4^{2-}(aq) + H_2O(l)$

$CrO_4^{2-}(aq) + 2Ag^+(aq) \; \rightarrow \; Ag_2CrO_4(s)$

$Ag_2CrO_4(s) + 4NH_3(aq) \; \rightarrow \; 2[Ag(NH_3)_2]^+(aq) + CrO_4^{2-}(aq)$

$2[Ag(NH_3)_2]^+(aq) + CrO_4^{2-}(aq) + 4H^+(aq) \; \rightarrow \; Ag_2CrO_4(s) + 4NH_4^+(aq)$

Chapter 21: Chemistry of the Nonmetals

2. *Refer to Chapter 2.*

    a. bromic acid

    b. potassium hypoiodate

    c. sodium chlorite

    d. sodium perbromate

4. *Refer to Chapter 2.*

    a. $KBrO_2$

    b. $CaBr_2$

    c. $NaIO_4$

    d. $Mg(ClO)_2$

6. *Refer to Section 21.4.*

To act as a reducing agent, the anion itself must be oxidized. Thus an oxoanion in which the nonmetal is in its highest oxidation state cannot be further oxidized and cannot act as a reducing agent.

    a. NO_3^-

    b. SO_4^{2-}

    c. ClO_4^-

8. *Refer to Section 21.2 and Chapter 2.*

    a. $SO_2 + H_2O \rightarrow H_2SO_3$

    b. $Cl_2O + H_2O \rightarrow 2HClO$

    c. $P_4O_6 + 6H_2O \rightarrow 4H_3PO_3$

10. *Refer to Section 21.2 and Chapter 2.*

    a. NaN_3

    b. H_2SO_3

    c. N_2H_4

    d. NaH_2PO_4

12. *Refer to Section 21.2 and Chapter 2.*

    a. H_2S

    b. N_2H_4

    c. PH_3

14. Refer to Sections 21.2 and 21.4.

a. NH_3, N_2H_4

b. HNO_3

c. HNO_2

d. HNO_3

---

16. Refer to Section 21.1 and Chapter 4.

First write the half-reactions, then combine them and cancel common elements.

a. $I^- \rightarrow I^0 + e^-$

$2I^-(aq) \rightarrow I_2(s) + 2e^-$ ⟶ oxidation half-reaction

$S^{6+} + 2e^- \rightarrow S^{4+}$
$SO_4^{2-}(aq) + 2e^- \rightarrow SO_2(g)$
$4H^+(aq) + SO_4^{2-}(aq) + 2e^- \rightarrow SO_2(g)$
$4H^+(aq) + SO_4^{2-}(aq) + 2e^- \rightarrow SO_2(g) + 2H_2O(l)$ ⟶ reduction half-reaction

$4H^+(aq) + SO_4^{2-}(aq) + 2e^- + 2I^-(aq) \rightarrow I_2(s) + 2e^- + SO_2(g) + 2H_2O(l)$
$4H^+(aq) + SO_4^{2-}(aq) + 2I^-(aq) \rightarrow I_2(s) + SO_2(g) + 2H_2O(l)$

b. $I^- \rightarrow I^0 + e^-$

$2I^-(aq) \rightarrow I_2(s) + 2e^-$ ⟶ oxidation half-reaction

$Cl^0 + e^- \rightarrow Cl^-$
$Cl_2(g) + 2e^- \rightarrow 2Cl^-(aq)$ ⟶ reduction half-reaction

$Cl_2(g) + 2e^- + 2I^-(aq) \rightarrow I_2(s) + 2e^- + 2Cl^-(aq)$
$Cl_2(g) + 2I^-(aq) \rightarrow I_2(s) + 2Cl^-(aq)$

---

18. Refer to Sections 21.1 and 21.4 and Chapter 4.

First write the half-reactions, then combine them and cancel common elements.

a. In this reaction, one HClO molecule oxidizes a second HClO molecule.

$Cl^+ \rightarrow Cl^{3+} + 2e^-$
$HClO(aq) \rightarrow HClO_2(aq) + 2e^-$
$HClO(aq) \rightarrow HClO_2(aq) + 2e^- + 2H^+(aq)$
$H_2O(l) + HClO(aq) \rightarrow HClO_2(aq) + 2e^- + 2H^+(aq)$ ⟶ oxidation half-reaction

$Cl^+ + e^- \rightarrow Cl^0$
$2HClO(aq) + 2e^- \rightarrow Cl_2(g)$
$2H^+(aq) + 2HClO(aq) + 2e^- \rightarrow Cl_2(g)$
$2H^+(aq) + 2HClO(aq) + 2e^- \rightarrow Cl_2(g) + 2H_2O(l)$ ⟶ reduction half-reaction

$2H^+(aq) + 2HClO(aq) + 2e^- + H_2O(l) + HClO(aq)$
$$\rightarrow HClO_2(aq) + 2e^- + 2H^+(aq) + Cl_2(g) + 2H_2O(l)$$

$3HClO(aq) \rightarrow HClO_2(aq) + Cl_2(g) + H_2O(l)$

(Note: this equation could also be written with the anions ClO^- and ClO_2^- instead of the acids. However, since these are weak acids, they exist in solution mainly as the acid.)

b. $Cl^{5+} \rightarrow Cl^{7+} + 2e^-$
$ClO_3^-(aq) \rightarrow ClO_4^-(aq) + 2e^-$
$ClO_3^-(aq) \rightarrow ClO_4^-(aq) + 2e^- + 2H^+(aq)$
$H_2O(l) + ClO_3^-(aq) \rightarrow ClO_4^-(aq) + 2e^- + 2H^+(aq)$ oxidation half-reaction

$Cl^{5+} + 2e^- \rightarrow Cl^{3+}$
$ClO_3^-(aq) + 2e^- \rightarrow ClO_2^-(aq)$
$2H^+(aq) + ClO_3^-(aq) + 2e^- \rightarrow ClO_2^-(aq)$
$2H^+(aq) + ClO_3^-(aq) + 2e^- \rightarrow ClO_2^-(aq) + H_2O(l)$ reduction half-reaction

$2H^+(aq) + ClO_3^-(aq) + 2e^- + H_2O(l) + ClO_3^-(aq)$
$$\rightarrow ClO_4^-(aq) + 2e^- + 2H^+(aq) + ClO_2^-(aq) + H_2O(l)$$

$2ClO_3^-(aq) \rightarrow ClO_4^-(aq) + ClO_2^-(aq)$

---

20. Refer to Section 21.1 and Table 18.1.

---

a. $2Br^-(aq) \rightarrow Br_2(l) + 2e^-$ $E°_{ox} = -1.077$ V
$Cl_2(g) + 2e^- \rightarrow 2Cl^-(aq)$ $E°_{red} = +1.360$ V
$Cl_2(g) + 2Br^-(aq) \rightarrow 2Cl^-(aq) + Br_2(l)$ $E°$ is (+) thus reaction occurs.

b. $2Cl^-(aq) \rightarrow Cl_2(g) + 2e^-$ $E°_{ox} = -1.360$ V
$I_2(s) + 2e^- \rightarrow 2I^-(aq)$ $E°_{red} = +0.534$ V
$I_2(s) + Cl^-(aq) \rightarrow$ **N.R.** $E°$ is (-) thus no reaction occurs.

c. $2Br^-(aq) \rightarrow Br_2(l) + 2e^-$ $E°_{ox} = -1.077$ V
$I_2(s) + 2e^- \rightarrow 2I^-(aq)$ $E°_{red} = +0.534$ V
$I_2(s) + Br^-(aq) \rightarrow$ **N.R.** $E°$ is (-) thus no reaction occurs.

d. $2Cl^-(aq) \rightarrow Cl_2(g) + 2e^-$ $E°_{ox} = -1.360$ V
$Br_2(l) + 2e^- \rightarrow 2Br^-(aq)$ $E°_{red} = +1.077$ V
$Br_2(l) + Cl^-(aq) \rightarrow$ **N.R.** $E°$ is (-) thus no reaction occurs.

a. $Pb(N_3)_2(s) \rightarrow Pb(s) + 3N_2(g)$

b. $2O_3(g) \rightarrow 3O_2(g)$

c. $2H_2S(g) + O_2(g) \rightarrow 2H_2O(l) + 2S(s)$

a. $Cd^{2+}(aq) + H_2S(aq) \rightarrow 2H^+(aq) + CdS(s)$ (metathesis reaction)

b. $OH^-(aq) + H_2S(aq) \rightarrow H_2O(l) + HS^-(aq)$ (acid-base reaction)

c. $O_2(g) + 2H_2S(aq) \rightarrow 2H_2O(l) + 2S(s)$ (redox reaction)

a. $CaCO_3(s) + H_2SO_4(aq) \rightarrow CaSO_4(aq) + H_2O(l) + CO_2(g)$
Writing the soluble species as ions gives:
$$CaCO_3(s) + 2H^+(aq) + SO_4^{2-}(aq) \rightarrow Ca^{2+}(aq) + SO_4^{2-}(aq) + H_2O(l) + CO_2(g)$$
Net ionic: $CaCO_3(s) + 2H^+(aq) \rightarrow Ca^{2+}(aq) + H_2O(l) + CO_2(g)$

b. $2NaOH(aq) + H_2SO_4(aq) \rightarrow Na_2SO_4(aq) + 2H_2O(l)$
Writing the soluble species as ions gives:
$$2Na^+(aq) + 2OH^-(aq) + 2H^+(aq) + SO_4^{2-}(aq) \rightarrow 2Na^+(aq) + SO_4^{2-}(aq) + 2H_2O(l)$$
Net ionic: $H^+(aq) + OH^-(aq) \rightarrow H_2O(l)$

c. This is a redox reacton and best approached from that standpoint.

$Cu^0 \rightarrow Cu^{2+}(aq) + 2e^-$ oxidation half-reaction

$SO_4^{2-}(aq) \rightarrow SO_2(g)$
$S^{6+} + 2e^- \rightarrow S^{4+}$
$SO_4^{2-}(aq) + 2e^- \rightarrow SO_2(g)$
$4H^+(aq) + SO_4^{2-}(aq) + 2e^- \rightarrow SO_2(g)$
$4H^+(aq) + SO_4^{2-}(aq) + 2e^- \rightarrow SO_2(g) + 2H_2O(l)$ reduction half-reaction

$4H^+(aq) + 2e^- + SO_4^{2-}(aq) + Cu(s) \rightarrow Cu^{2+}(aq) + 2e^- + SO_2(g) + 2H_2O(l)$
$4H^+(aq) + SO_4^{2-}(aq) + Cu(s) \rightarrow Cu^{2+}(aq) + SO_2(g) + 2H_2O(l)$

28. *Refer to Section 21.3 and Chapter 7.*

a. 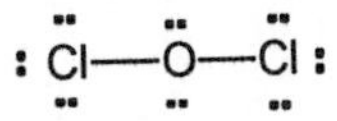 $:\overset{..}{\underset{..}{Cl}}-\overset{..}{\underset{..}{O}}-\overset{..}{\underset{..}{Cl}}:$

b. 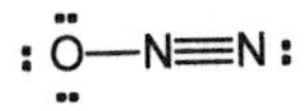 $:\overset{..}{\underset{..}{O}}-N\equiv N:$

c. 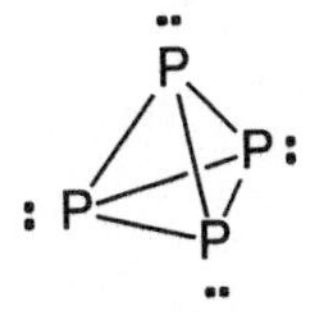 P_4 tetrahedral structure

d.  $:N\equiv N:$

30. *Refer to Chapter 7.*

Both **(a)** Cl_2O and **(b)** N_2O are polar. (Since Cl_2O is bent (AX_2E_2), the dipoles do not cancel.)

P_4 and N_2 both lack dipoles and are nonpolar.

32. *Refer to Section 21.4 and Chapters 7 and 13.*

a. NO_3^-

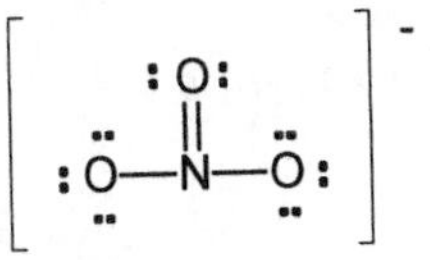

b. HSO_4^- 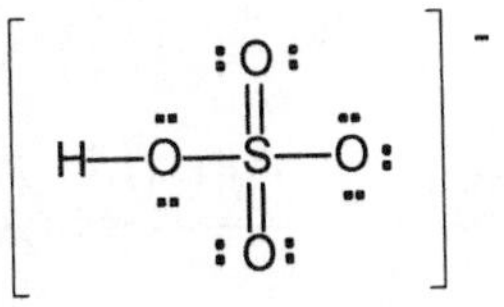   This Lewis structure has a -1 formal charge on the most electronegative atom (O).

c. $H_2PO_4^-$ 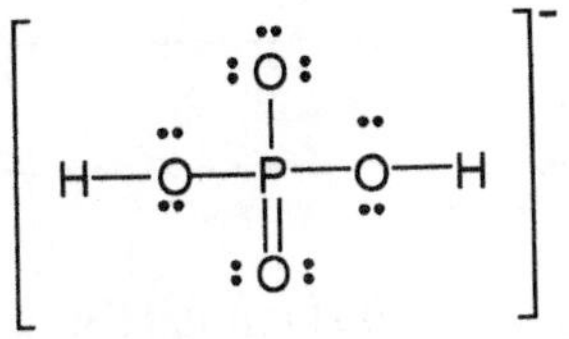   This Lewis structure has a -1 formal charge on the most electronegative atom (O).

34. Refer to Section 21.4 and Chapter 7.

a. N_2O_5

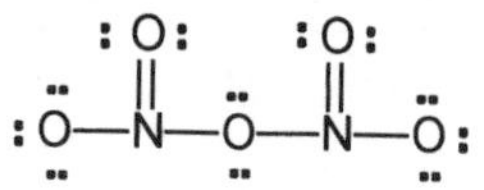

b. HNO_3

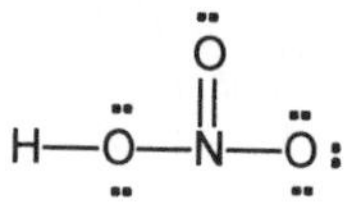

c. SO_4^{2-}

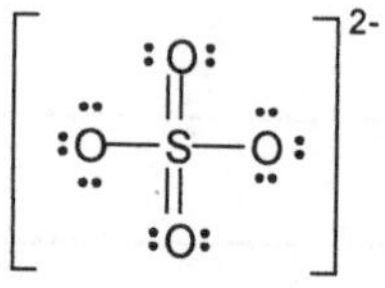

This Lewis structure has a -1 formal charge on two of the oxygens, which are the most electronegative atoms.

36. Refer to Section 21.1, Problem 16b (above) and Chapter 5.

Write the balanced equation and calculate the moles of Cl_2 needed for the oxidation. Then convert the moles to liters using the ideal gas law.

$$2NaI(aq) + Cl_2(g) \rightarrow I_2(s) + 2NaCl(aq)$$

$$175 \text{ g NaI} \times \frac{1 \text{ mol. NaI}}{149.89 \text{ g NaI}} \times \frac{1 \text{ mol. Cl}_2}{2 \text{ mol. NaI}} = 0.584 \text{ mol. Cl}_2$$

$$758 \text{ mm Hg} \times \frac{1 \text{ atm}}{760 \text{ mmHg}} = 0.997 \text{ atm}$$

$$V = \frac{nRT}{P} = \frac{(0.584 \text{ mol.})(0.0821 \text{ L} \cdot \text{atm/mol.} \cdot \text{K})(298 \text{ K})}{0.997 \text{ atm}} = 14.3 \text{ L}$$

38. Refer to Chapters 5 and 10.

$$n = \frac{PV}{RT} = \frac{(0.974 \text{ atm})(1.283 \text{ L})}{(0.0821 \text{ L} \cdot \text{atm/mol.} \cdot \text{K})(298 \text{ K})} = 0.0511 \text{ mol. HBr}$$

$$M = \frac{\text{mol.}}{\text{L}} = \frac{0.0511 \text{ mol.}}{0.250 \text{ L}} = 0.204 \, M \text{ HBr}$$

40. Refer to Chapters 1 and 5.

Calculate the amount of SO_2 produced, and from that, the moles of H_2S needed. Then calculate the volume of H_2S and the mass of S produced.

$$S(s) + O_2(g) \rightarrow SO_2(g)$$

$$1.00 \times 10^3 \text{ kg} \times \frac{1000 \text{ g}}{1 \text{ kg}} \times \frac{5.0 \text{ g S}}{100 \text{ g coal}} \times \frac{1 \text{ mol. S}}{32.066 \text{ g}} \times \frac{1 \text{ mol. SO}_2}{1 \text{ mol. S}} = 1.6 \times 10^3 \text{ mol. SO}_2$$

$$755 \text{ mm Hg} \times \frac{1 \text{ atm}}{760 \text{ mmHg}} = 0.993 \text{ atm}$$

$$2H_2S(g) + SO_2(g) \rightarrow 3S(s) + 2H_2O(l)$$

$$1.6 \times 10^3 \text{ mol. SO}_2 \times \frac{2 \text{ mol. H}_2S}{1 \text{ mol. SO}_2} = 3.2 \times 10^3 \text{ mol. H}_2S$$

$$V = \frac{(3.2 \times 10^3 \text{ mol.})(0.0821 \text{ L} \cdot \text{atm/mol.} \cdot \text{K})(300 \text{ K})}{0.993 \text{ atm}} = 7.9 \times 10^4 \text{ L}$$

$$1.6 \times 10^3 \text{ mol. SO}_2 \times \frac{3 \text{ mol. S}}{1 \text{ mol. SO}_2} \times \frac{32.066 \text{ g S}}{1 \text{ mol. S}} = 1.5 \times 10^5 \text{ g} = 0.15 \text{ metric tons S}$$

---

42. Refer to Chapters 1, 5, 10 and 13.

Convert gal. water to grams H_2S. Then calculate moles of Cl_2 required to react with that mass of H_2S.

$$1.00 \times 10^3 \text{ gal.} \times \frac{4 \text{ qt.}}{1 \text{ gal.}} \times \frac{1 \text{ L}}{1.057 \text{ qt.}} \times \frac{1000 \text{ mL}}{1 \text{ L}} = 3.78 \times 10^6 \text{ mL H}_2O$$

$$3.78 \times 10^6 \text{ mL H}_2O \times \frac{1.00 \text{ g}}{1 \text{ mL}} \times \frac{5 \text{ g H}_2S}{10^6 \text{ g H}_2O} = 19 \text{ g H}_2S$$

$$19 \text{ g H}_2S \times \frac{1 \text{ mol. H}_2S}{34.08 \text{ g H}_2S} \times \frac{1 \text{ mol. Cl}_2}{1 \text{ mol. H}_2S} = 0.56 \text{ mol. Cl}_2$$

$$V = \frac{(0.56 \text{ mol.})(0.0821 \text{ L} \cdot \text{atm/mol.} \cdot \text{K})(273 \text{ K})}{1 \text{ atm}} = 13 \text{ L}$$

To calculate the pH, calculate the moles of H^+, concentration of H^+, and then the pH.

$$19 \text{ g H}_2S \times \frac{1 \text{ mol. H}_2S}{34.08 \text{ g H}_2S} \times \frac{2 \text{ mol. H}^+}{1 \text{ mol. H}_2S} = 1.1 \text{ mol. H}^+$$

$$1.1 \text{ mol. H}^+ \times \frac{1}{3.78 \times 10^6 \text{ mL H}_2O} \times \frac{1000 \text{ mL}}{1 \text{ L}} = 2.9 \times 10^{-4} \ M$$

$$\text{pH} = -\log [H^+] = -\log (2.9 \times 10^{-4}) = 3.54$$

$$HClO \rightleftharpoons H^+ + ClO^-$$

$$K_a = 2.8 \times 10^{-8} = \frac{[H^+][ClO^-]}{[HClO]}$$

If $[H^+] = x$, then $[ClO^-] = x$ and $[HClO] = 0.10 - x$.

$$K_a = 2.8 \times 10^{-8} = \frac{[x][x]}{[0.10 - x]} \Rightarrow 2.8 \times 10^{-9} - (2.8 \times 10^{-8})x = x^2$$

Using successive approximations: $x \cong \sqrt{2.8 \times 10^{-9}}$
$x = [H^+] = [ClO^-] = 5.3 \times 10^{-5} \, M$

Equilibrium concentration of $HClO = 0.10 - (5.3 \times 10^{-5}) = 0.10 \, M$

$$pH = -\log [H^+] = -\log (5.3 \times 10^{-5}) = 4.28$$

The equilibrium equation for the final reaction is:

$$K_{overall} = \frac{[H^+][HF_2^-]}{[HF]^2} \qquad \text{eq. 1}$$

Write the equilibrium equation for the second reaction, and solve for HF_2^-.

$$K_1 = 2.7 = \frac{[HF_2^-]}{[HF][F^-]} \Rightarrow [HF_2^-] = (2.7)[HF][F^-]$$

Substituting this equation into equation 1 gives:

$$K_{overall} = \frac{[H^+](2.7)[HF][F^-]}{[HF]^2} = \frac{[H^+](2.7)[F^-]}{[HF]} \qquad \text{eq. 2}$$

Write the equilibrium equation for the first reaction, and solve for HF.

$$K_a = 6.9 \times 10^{-4} = \frac{[H^+][F^-]}{[HF]} \Rightarrow [HF] = \frac{[H^+][F^-]}{6.9 \times 10^{-4}}$$

Substituting this equation into equation 2 gives:

$$K_{overall} = \frac{[H^+](2.7)[F^-]}{\dfrac{[H^+][F^-]}{6.9 \times 10^{-4}}} = (2.7)(6.9 \times 10^{-4}) = 1.9 \times 10^{-3}$$

Thus we see that the equilibrium constant for the overall process is the product of the individual constants.

---

48. Refer to Chapter 16.

$K_{sp} = [Ba^{2+}][F^-]^2$

If x = mol. BaF_2, then $[F^-] = 2x$, and $[Ba^{2+}] = x$

$K_{sp} = 1.8 \times 10^{-7} = (0.10 - x)(2x)^2$

$K_{sp} = 1.8 \times 10^{-7} = 0.40x^2 - 4x^3$

Using successive approximations: $\quad 1.8 \times 10^{-7} = 0.40x^2$

$\qquad\qquad\qquad\qquad\qquad\qquad\qquad x = 6.7 \times 10^{-4}\ M$

Thus 6.7×10^{-4} moles of BaF_2 would dissolve in 1 L of 0.10 M $BaCl_2$

$$100\ mL \times \frac{1\ L}{1000\ mL} \times \frac{6.7 \times 10^{-4}\ mol.}{1\ L} \times \frac{175.33\ g\ BaF_2}{1\ mol.} = 0.012\ g\ BaF_2\ (per\ 100\ mL)$$

---

50. Refer to Chapter 17.

Calculate the enthalpy and entropy and then calculate the Gibbs free energy.

$\Delta S° = \Sigma\ S°_{(products)} - \Sigma\ S°_{(reactants)}$
$\Delta S° = [(2\ mol.)(0.1431\ kJ/mol.·K) + (1\ mol.)(0.1522\ kJ/mol.·K)]$
$\qquad\quad - [(2\ mol.)(0.1492\ kJ/mol.·K) + (1\ mol.)(0.2230\ kJ/mol.·K)]$
$\Delta S° = -0.0830\ kJ/K$

$\Delta H° = \Sigma\ \Delta H_f°_{(products)} - \Sigma\ \Delta H_f°_{(reactants)}$
$\Delta H° = [(2\ mol.)(-397.7\ kJ/mol.) + (1\ mol.)(0.0\ kJ/mol.)]$
$\qquad\quad - [(2\ mol.)(-360.2\ kJ/mol.) + (1\ mol.)(0.0\ kJ/mol.)]$
$\Delta H° = -75.0\ kJ$

$\Delta G° = \Delta H° - T\ \Delta S°$
$\Delta G° = -75.0\ kJ - (298\ K)(-0.0830\ kJ/K)$
$\Delta G° = -50.2\ kJ$

ΔG is negative, therefore **the reaction is spontaneous**.

The temperature at which the reaction is no longer spontaneous is that at which $\Delta G° = 0$

$0 = -75.0\ kJ/mol. - (T)(0.0832\ kJ/mol.·K)$

$T = 901\ K.$ Thus the reaction is spontaneous **up** to 901 K, and the

lowest temperature at which the reaction is spontaneous is 0 K.

a. Calculate the enthalpy and entropy and then calculate the Gibbs free energy.

$\Delta S^\circ = \Sigma\, S^\circ_{(products)} - \Sigma\, S^\circ_{(reactants)}$

$\Delta S^\circ = [(2\ mol.)(0.2230\ kJ/mol.\cdot K) + (2\ mol.)(0.0699\ kJ/mol.\cdot K)]$
$\qquad - [(4\ mol.)(0.1868\ kJ/mol.\cdot K) + (1\ mol.)(0.2050\ kJ/mol.\cdot K)]$

$\Delta S^\circ = -0.3664\ kJ/K$

$\Delta H^\circ = \Sigma\, \Delta H_f^\circ{}_{(products)} - \Sigma\, \Delta H_f^\circ{}_{(reactants)}$

$\Delta H^\circ = [(2\ mol.)(0.0\ kJ/mol.) + (2\ mol.)(-285.8\ kJ/mol.)]$
$\qquad - [(4\ mol.)(-92.3\ kJ/mol.) + (1\ mol.)(0.0\ kJ/mol.)]$

$\Delta H^\circ = -202.4\ kJ$

$\Delta G^\circ = \Delta H^\circ - T\,\Delta S^\circ$

$\Delta G^\circ = -202.4\ kJ - (298\ K)(-0.3664\ kJ/K)$

$\Delta G^\circ = -93.2\ kJ$

ΔG is negative, therefore **the reaction is spontaneous**.

b. $\Delta G^\circ = -RT \ln K$

$$\ln K = \frac{-\Delta G}{RT} = \frac{93200\ J}{(8.314\ J/mol.\cdot K)(298\ K)} = 37.6$$

$$K = e^{37.6} = 2.14 \times 10^{16}$$

Sublimation is the phase change from solid to gas. Calculate T using $\Delta G^\circ = 0$, the point at which the phase change is becoming spontaneous.

$P_4(s)\ =\ P_4(g)$

$\Delta S^\circ = (1\ mol.)(0.2800\ kJ/mol.\cdot K) - (1\ mol.)(0.1644\ kJ/mol.\cdot K)$
$\Delta S^\circ = 0.1156\ kJ/\cdot K$

$\Delta H^\circ = (1\ mol.)(58.9\ kJ/mol.) - (1\ mol.)(0.0\ kJ/mol.)$
$\Delta H^\circ = 58.9\ kJ$

$\Delta G^\circ = \Delta H^\circ - T\,\Delta S^\circ$
$0 = 58.9\ kJ - (T)(0.1156\ kJ/K)$
$58.9\ kJ = (T)(0.1156\ kJ/K)$
$T = 510\ K = 237°C$

56. Refer to Chapter 18.

Calculate the moles of electrons produced by the current. Then calculate the mass of F_2 produced with the given efficiency.

$$7.00 \times 10^3 \text{ C/s} \times \frac{3600 \text{ s}}{1 \text{ hr}} \times \frac{24 \text{ hr}}{1 \text{ d}} \times 2 \text{ d} \times \frac{1 \text{ mol.}}{9.648 \times 10^4 \text{ C}} = 1.25 \times 10^4 \text{ mol. } e^-$$

$$2F^- \rightarrow F_2 + 2e^-$$

$$1.25 \times 10^4 \text{ mol. } e^- \times 0.95 \times \frac{1 \text{ mol. } F_2}{2 \text{ mol. } e^-} \times \frac{37.996 \text{ g } F_2}{1 \text{ mol. } F_2} = 2.26 \times 10^5 \text{ g } F_2$$

58. Refer to Chapter 18.

Calculate the moles of electrons produced by the current. Then calculate the mass of $NaClO_4$ produced.

$$Cl^{5+} \rightarrow Cl^{7+} + 2e^-$$
$$ClO_3^- \rightarrow ClO_4^- + 2e^-$$

$$1.50 \times 10^3 \text{ C/s} \times \frac{3600 \text{ s}}{1 \text{ hr}} \times 8 \text{ hr} \times \frac{1 \text{ mol.}}{9.648 \times 10^4 \text{ C}} = 448 \text{ mol. } e^-$$

$$448 \text{ mol. } e^- \times \frac{1 \text{ mol. } NaClO_4}{2 \text{ mol. } e^-} \times \frac{122.44 \text{ g}}{1 \text{ mol. } NaClO_4} = 2.74 \times 10^4 \text{ g} = 27.4 \text{ kg } NaClO_4$$

60. Refer to Chapter 18 and Table 18.1.

To determine if the species will be oxidized by H_2O_2, add E_{ox}° and E_{red}° (H_2O_2). If the sum is negative, then the species will not be oxidized.
    (Note: The values listed in Table 18.1 are *reduction* potentials.)

a. Co^{2+} 1.763 V + (-1.953 V) = -0.190 V   Will **not** be oxidized.

b. Cl^- 1.763 V + (-1.360 V) = 0.403 V   Will be oxidized.

c. Fe^{2+} 1.763 V + (-0.769 V) = 0.994 V   Will be oxidized.

d. Sn^{2+} 1.763 V + (-0.154 V) = 1.609 V   Will be oxidized.

The first steps are to calculate $E°$ and write the balanced redox reaction. Then apply the Nernst equation, solving for $[H^+]$. With $[H^+]$ in hand, calculate pH.

$$\begin{array}{ll} SO_2 \rightarrow SO_4^{2-} & \text{-0.115 V} \\ NO_3^- \rightarrow NO & \text{0.964 V} \\ \hline NO_3^- + SO_2 \rightarrow NO + SO_4^{2-} & \text{0.809 V} \end{array}$$

$S^{4+} \rightarrow S^{6+} + 2e^-$

$SO_2(g) \rightarrow SO_4^{2-}(aq) + 2e^-$

$SO_2(g) \rightarrow SO_4^{2-}(aq) + 2e^- + 4H^+(aq)$

$2H_2O(l) + SO_2(g) \rightarrow SO_4^{2-}(aq) + 2e^- + 4H^+(aq)$ Oxidation half-reaction

$N^{5+} + 3e^- \rightarrow N^{2+}$

$NO_3^-(aq) + 3e^- \rightarrow NO(g)$

$4H^+(aq) + NO_3^-(aq) + 3e^- \rightarrow NO(g)$

$4H^+(aq) + NO_3^-(aq) + 3e^- \rightarrow NO(g) + 2H_2O(l)$ Reduction half-reaction

To balance the electrons, multiply the oxidation half-reaction by 3 and the reduction half-reaction by 2, and then add the results and simplify.

$$6H_2O(l) + 3SO_2(g) + 8H^+(aq) + 2NO_3^-(aq) + 6e^-$$
$$\rightarrow 2NO(g) + 4H_2O(l) + 3SO_4^{2-}(aq) + 6e^- + 12H^+(aq)$$
$$2H_2O(l) + 3SO_2(g) + 2NO_3^-(aq) \rightarrow 2NO(g) + 3SO_4^{2-}(aq) + 4H^+(aq)$$

$$E = E° - \frac{0.0257}{n} \ln Q \Rightarrow 1.000\ V = 0.809\ V - \frac{0.0257}{6} \ln \left(\frac{(P_{NO})^2 [SO_4^{2-}]^3 [H^+]^4}{(P_{SO_2})^3 [NO_3^-]^2} \right)$$

$$0.191\ V = -0.00428 \ln \left(\frac{(1)^2 [0.100]^3 [H^+]^4}{(1)^3 [0.100]^2} \right)$$

$$-44.6 = \ln (0.100)[H^+]^4$$

$$(0.100)[H^+]^4 = e^{-44.6} = 4.3 \times 10^{-20}$$

$$[H^+]^4 = 4.3 \times 10^{-19}$$

$$[H^+] = 2.6 \times 10^{-5}\ M$$

$$pH = -\log (2.6 \times 10^{-5}) = 4.59$$

a. dispersion forces

b. dispersion and dipole forces

c. dispersion forces, dipole forces and hydrogen bonding

d. dispersion forces, dipole forces and hydrogen bonding

e. none, this is an ionic solid

---

66. Refer to Chapter 4.

a. NO_2^- O: -2
          N: +3 $(N + 2(-2) = -1)$

b. NO_2 O: -2
          N: +4 $(N + 2(-2) = 0)$

c. HNO_3 O: -2
          H: +1
          N: +5 $(N + 1 + 3(-2) = 0)$

d. NH_4^+ H: +1
          N: -3 $(N + 4(+1) = +1)$

---

68. Refer to Sections 21.1 and 21.2.

a. HClO

b. S, $KClO_3$

c. NH_3, NaClO

d. HF

---

70. Refer to Sections 21.3 and 21.4.

a. Increasing oxidation number corresponds to an increase in oxygens atoms around the central atom. These oxygen atoms stabilize the negative charge of the resulting conjugate base by i) delocalizing the negative charge (consider resonance structures) and ii) by the electronegative oxygens pulling electron density away from the atom bearing the charge.

b. NO_2 has an odd number of electrons and will thus have an unpaired electron.

c. In general, the reduction reactions of the oxoanions (oxoanions acting as oxidizing agents) involve H^+ as a reactant. Therefore higher $[H^+]$ (lower pH) causes the reaction to be more spontaneous.

d. The sugar is oxidized to carbon, which is black.

71. Refer to Chapter 1.

To determine the amount of sulfuric acid that could be produced, one would first need to know the mass of sulfur present. This would require that one know the **depth of the deposit** to calculate the volume of the deposit, the **density** to calculate the mass of the deposit, and the **percent by mass (or purity) of the sulfur**, in the deposit to calculate the mass of sulfur.

72. Refer to Chapters 1 and 3.

Use a given mass of quartz (i.e. 1000 g) to calculate the amount of gold present and the amount of HF solution needed. Then calculate the cost of each.

$$1000 \text{ g SiO}_2 \times \frac{1.0 \times 10^{-3} \text{ g Au}}{100 \text{ g SiO}_2} = 0.010 \text{ g Au}$$

$$1000 \text{ g SiO}_2 \times \frac{1 \text{ mol. SiO}_2}{60.08 \text{ g SiO}_2} \times \frac{4 \text{ mol. HF}}{1 \text{ mol. SiO}_2} \times \frac{20.01 \text{ g}}{1 \text{ mol. HF}} = 1330 \text{ g HF}$$

$$1330 \text{ g HF} \times \frac{100 \text{ g sol'n}}{50 \text{ g HF}} \times \frac{1 \text{ mL}}{1.17 \text{ g sol'n}} = 2270 \text{ mL solution}$$

$$2270 \text{ mL} \times \frac{1 \text{ L}}{1000 \text{ mL}} \times \frac{\$0.75}{1 \text{ L}} = \$1.70$$

$$0.010 \text{ g Au} \times \frac{1 \text{ troy oz}}{31.1 \text{ g}} \times \frac{\$425}{1 \text{ troy oz}} = \$0.14$$

Thus it would cost $1.70 of HF solution to recover $0.14 of gold. Definitely not economical.

73. Refer to Chapters 3 and 4.

Determine the mole ratios of the reductant and oxidant for each redox couple. Then use mass relationships of thiosulfate to iodine and iodine to NaClO to determine the mass of NaClO and from that, the mass percent of NaClO in the bleach solution.

$$S^{2+} \rightarrow S^{2.5+} + \tfrac{1}{2}e^-$$
$$2S_2O_3^{2-} \rightarrow S_4O_6^{2-} + 2e^-$$

$$I^0 + e^- \rightarrow I^-$$
$$I_2 + 2e^- \rightarrow 2I^-$$

$$2S_2O_3^{2-} + I_2 \rightarrow 2I^- + S_4O_6^{2-}$$

This gives the mole ratio of $S_2O_3^{2-}$ to I_2 for the calculation of moles of I_2.

$$25.00 \text{ mL} \times \frac{1 \text{ L}}{1000 \text{ mL}} \times \frac{0.0700 \text{ mol. S}_2\text{O}_3^{2-}}{1 \text{ L}} \times \frac{1 \text{ mol. I}_2}{2 \text{ mol. S}_2\text{O}_3^{2-}} = 8.75 \times 10^{-4} \text{ mol. I}_2$$

$$Cl^+(aq) + 2e^- \rightarrow Cl^-(aq)$$
$$2H^+(aq) + ClO^-(aq) + 2e^- \rightarrow Cl^-(aq) + H_2O(l)$$

$$2I^- \rightarrow I_2 + 2e^-$$

$$2H^+(aq) + ClO^-(aq) + 2I^-(aq) \rightarrow I_2(aq) + Cl^-(aq) + H_2O(l)$$

This gives the mole ratio of $ClO^-(aq)$ (or NaClO) to I_2 for the calculation of mass of NaClO.

$$8.75 \times 10^{-4} \text{ mol. I}_2 \times \frac{1 \text{ mol. NaClO}}{1 \text{ mol. I}_2} \times \frac{74.44 \text{ g NaClO}}{1 \text{ mol. NaClO}} = 0.0651 \text{ g NaClO}$$

$$\text{mass \%} = \frac{0.0651 \text{ g NaClO}}{5.00 \text{ g sol'n}} \times 100\% = 1.30\%$$

---

74. Refer to Section 21.1, Chapter 5 and "Chemistry Beyond the Classroom: Airbags" (p. 123).

Assume that none of the N_2 escapes from the bag. Then use the ideal gas law to calculate the moles of N_2 needed to fill 20.0 L, and the amount of NaN_3 to produce that N_2.

$$n = \frac{PV}{RT} = \frac{(1 \text{ atm})(20.0 \text{ L})}{(0.0821 \text{ L} \cdot \text{atm/mol} \cdot \text{K})(298 \text{ K})} = 0.817 \text{ mol. N}_2$$

$$2NaN_3(s) \rightarrow 2Na(s) + 3N_2(g)$$

$$0.817 \text{ mol. N}_2 \times \frac{2 \text{ mol. NaN}_3}{3 \text{ mol. N}_2} \times \frac{65.01 \text{ g NaN}_3}{1 \text{ mol. NaN}_3} = 35.4 \text{ g NaN}_3$$

Chapter 22: Organic Chemistry

---

2. *Refer to Sections 22.1 and 22.2 and Example 22.2.*

For a hydrocarbon with n carbons, if the number of hydrogens equals $2n+2$, then it is an alkane, if it equals $2n$, then it is an alkene, if it equals $2n-2$, then it is an alkyne. This assumes the hydrocarbons are **not** cyclic and contain at most, one double bond.

a. $C_{12}H_{24}$ $n=12$, number of hydrogens = 24 = $2n$, thus it is an **alkene**.

b. C_7H_{12} $n=7$, number of hydrogens = 12 = $2n-2$, thus it is an **alkyne**.

c. $C_{13}H_{28}$ $n=13$, number of hydrogens = 28 = $2n+2$, thus it is an **alkane**.

---

4. *Refer to Sections 22.1 and 22.2 and Example 22.2.*

For a hydrocarbon with n carbons, the number of hydrogens in an alkane equals $2n +2$, in alkene it equals $2n$, in an alkyne it equals $2n -2$. This assumes the hydrocarbons are **not** cyclic and contain at most, one double bond.

a.   For alkynes, the number of hydrogens = $2n -2 = 16$, thus $n = 9$. $\mathbf{C_9H_{16}}$

b.   For alkenes, the number of hydrogens = $2n = 44$, thus $n = 22$. $\mathbf{C_{22}H_{44}}$

c.   For alkanes, the number of hydrogens = $2n +2 = 2(10) + 2 = 22$. $\mathbf{C_{10}H_{22}}$

---

6.   *Refer to Section 22.4 and Example 22.3.*

a.   **Alcohol**. The -OH group is not attached to a C=O.

b.   **Ester**. The C=O is attached to a -O-, but not an -OH.

c.   **Carboxylic acid and Ester**. The central C=O (the third C from the left) is attached to an -O- which is not an -OH, thus that group is an ester. The terminal COOH is a C=O with and -OH attached, thus it is a carboxylic acid.

8. Refer to Section 22.4 and Example 22.3.

a.
$$OH$$
$$CH_3-CH-CH_3$$

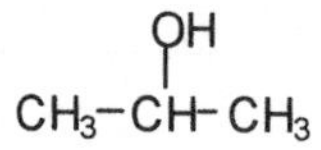

b.
$$CH_3$$
$$\backslash$$
$$CH-COOH$$
$$CH_3 \diagup$$

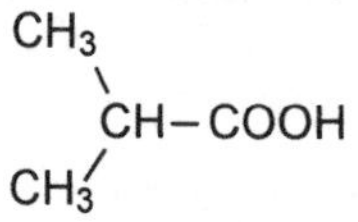

c.
$$CH_3-CH-CH_3$$
$$| $$
$$C=O$$
$$| $$
$$O$$
$$| $$
$$CH_3-CH-CH_3$$

From the acid

From the alcohol

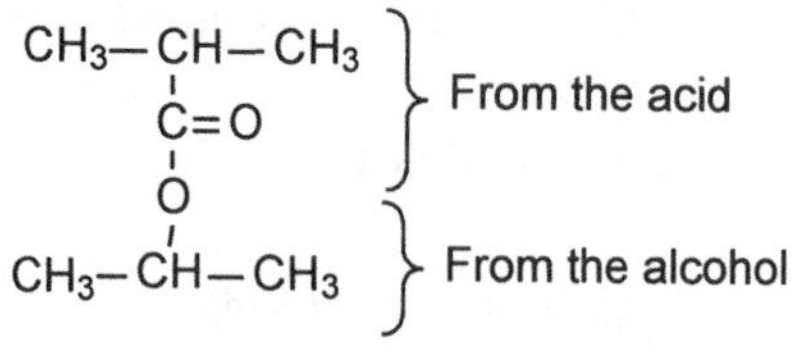

10. Refer to Sections 22.2 and 22.5 and Example 22.4.

Recall that placing the double bond between the first two carbons is identical to placing the double bond between the last two (as shown in the 1st structure). Thus there are only three unique isomers.

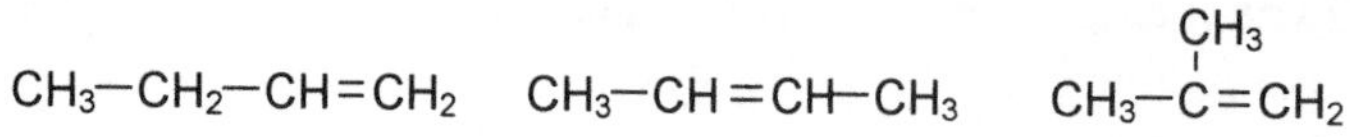

$$CH_3-CH_2-CH=CH_2 \qquad CH_3-CH=CH-CH_3 \qquad CH_3-\underset{|}{\overset{CH_3}{C}}=CH_2$$

Note that the last structure is identical to the last structure given in the answer in Appendix 6.

---

12. Refer to Section 22.5 and Example 22.4.

This type of problem is best approached by systematically moving the Cl atoms, first keeping two to a carbon, then moving them one at a time. Be aware that the isomer with both Cl's on the 1st carbon is identical to that with both on the last carbon.

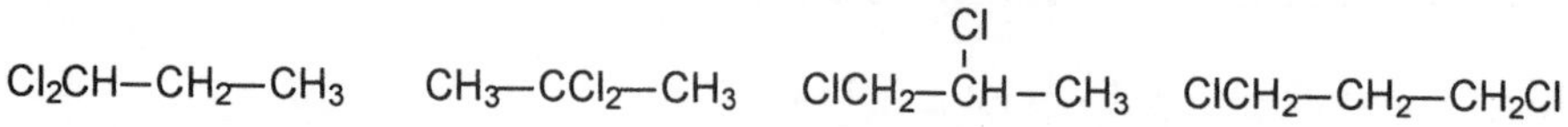

$$Cl_2CH-CH_2-CH_3 \qquad CH_3-CCl_2-CH_3 \qquad ClCH_2-\underset{|}{\overset{Cl}{CH}}-CH_3 \qquad ClCH_2-CH_2-CH_2Cl$$

---

14. Refer to Sections 22.3 and 22.5 and Example 22.4.

This type of problem is best approached by systematically moving the Cl atoms. Be aware that the isomer with Cl's on carbons 1, 2 and 4 is identical to that with Cl's on carbons 1, 2 and 5.

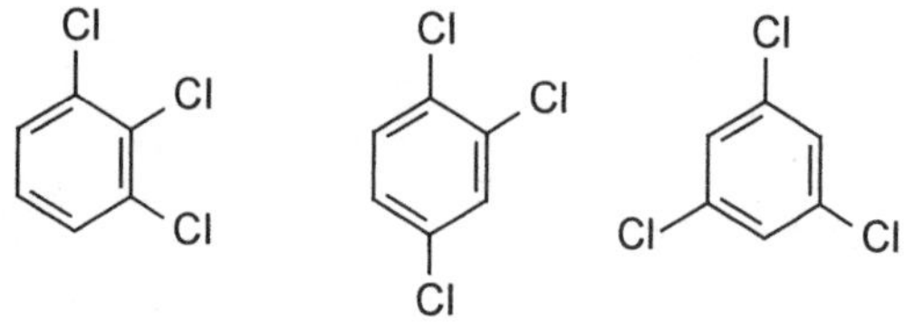

16. *Refer to Section 22.5.*

Refer to Problem 10 (above) for the three butene isomers, then systematically replace 2 H's of the double-bonded carbon(s), with Cl and Br.

$$CH_3-CH_2-\overset{\underset{|}{Cl}}{C}=\overset{\underset{|}{Br}}{CH} \qquad CH_3-CH_2-CH=\overset{\underset{|}{Br}}{C}-Cl \qquad CH_3-CH_2-\overset{\underset{|}{Br}}{C}=\overset{\underset{|}{Cl}}{CH}$$

$$CH_3-\overset{\underset{|}{Br}}{C}=\overset{\underset{|}{Cl}}{C}-CH_3 \qquad CH_3-\overset{\underset{|}{CH_3}}{C}=C\overset{Cl}{\underset{Br}{}}$$

18. *Refer to Sections 22.4 and 22.5.*

Since the carboxylic acid group can only be on an end of the hydrocarbon chain, the number of isomers is severely limited.

$$CH_3-CH_2-CH_2-COOH \qquad \overset{CH_3}{\underset{CH_3}{}}CH-COOH$$

20. *Refer to Section 22.5.*

All the isomers, except the last one ($(CH_3)_2C=CClBr$), have geometric isomers. The last does not have geometric isomers since it has two identical groups ($-CH_3$) on the same carbon. In assigning *cis / trans* to isomers, the terms refer to the relationship between the larger groups attached to each carbon. For example, in the first pair of isomers, Cl is larger than C, and Br is larger than H; thus *cis* refers to the isomer in which the Cl and Br are on the same side.

22. *Refer to Section 22.5.*

a. No. The two groups attached to any one of the double-bonded carbons are identical.

b. Yes. Refer to the pair of structures in problem 22(d) and replace the bromines with Cl; note that the two isomers are still unique after the substitution.

c. Yes. Refer to the pair of structures in problem 22(d).

---

24.   ***Refer to Section 22.5.***

To show optical isomerism, the molecule must have a carbon atom with 4 different groups attached.

a. H_2CCl_2. No, the carbon atom has two identical groups attached.

b. $ClCH_2\text{-}CH_2Cl$. No, each of the carbons has two identical groups (H's).

c. CBrClFH. Yes, the carbon has 4 different groups (H, F, Cl and Br).

d. $CH_3\text{-}CH(Br)OH$. Yes, the alcoholic carbon has 4 different groups (OH, H, Br and CH_3).

---

26. *Refer to Section 22.5 and Example 22.6.*

The chiral carbons (denoted with an *) are those that have four different groups attached.

a. $CH_3-\overset{\overset{\displaystyle H}{|}}{\underset{\underset{\displaystyle OH}{|}}{C}}{}^{*}-\overset{\overset{\displaystyle H}{|}}{\underset{\underset{\displaystyle OH}{|}}{C}}-H$ b. $H-\overset{\overset{\displaystyle}{|}}{\underset{\underset{\displaystyle H}{|}}{C}}=\overset{\overset{\displaystyle}{|}}{\underset{\underset{\displaystyle H}{|}}{C}}-CH_2\text{-}OH$ no chiral carbons   c. $CH_3-\overset{\overset{\displaystyle Cl}{|}}{\underset{\underset{\displaystyle Cl}{|}}{C}}-\overset{\overset{\displaystyle F}{|}}{\underset{\underset{\displaystyle H}{|}}{C}}{}^{*}-Cl$

---

28. *Refer to Section 22.6 and Example 22.7.*

---

a. Refer to Table 22.3 for the monomer. Replace double bonds in the monomer with single bonds connecting monomers together.

$$-\overset{\overset{\displaystyle F}{|}}{\underset{\underset{\displaystyle F}{|}}{C}}-\overset{\overset{\displaystyle F}{|}}{\underset{\underset{\displaystyle F}{|}}{C}}-\overset{\overset{\displaystyle F}{|}}{\underset{\underset{\displaystyle F}{|}}{C}}-\overset{\overset{\displaystyle F}{|}}{\underset{\underset{\displaystyle F}{|}}{C}}-$$

b. $\dfrac{5.0\times10^4 \text{ CF}_2 \text{ units}}{1\,\text{mol.}}\times\dfrac{50.01\,\text{g}}{1\,\text{CF}_2\,\text{unit}}=2.5\times10^6\,\text{g/mol.}$

c. $\dfrac{1\,\text{CF}_2\,\text{unit}}{50.01\,\text{g}}\times\dfrac{1\,\text{C}}{1\,\text{CF}_2\,\text{unit}}\times\dfrac{12.01\,\text{g C}}{1\,\text{C}}\times100\%=24.02\%$

% F = 100% - %C = $100-24.02=75.98\%$

b. Yes. Refer to the pair of structures in problem 22(d) and replace the bromines with Cl; note that the two isomers are still unique after the substitution.

c. Yes. Refer to the pair of structures in problem 22(d).

---

24. **_Refer to Section 22.5._**

To show optical isomerism, the molecule must have a carbon atom with 4 different groups attached.

a. H_2CCl_2. No, the carbon atom has two identical groups attached.

b. $ClCH_2$-CH_2Cl. No, each of the carbons has two identical groups (H's).

c. CBrClFH. Yes, the carbon has 4 different groups (H, F, Cl and Br).

d. CH_3-CH(Br)OH. Yes, the alcoholic carbon has 4 different groups (OH, H, Br and CH_3).

---

26. _Refer to Section 22.5 and Example 22.6._

The chiral carbons (denoted with an *) are those that have four different groups attached.

a.
$$CH_3\overset{*}{-}\overset{\overset{\displaystyle H}{|}}{\underset{\underset{\displaystyle OH}{|}}{C}}-\overset{\overset{\displaystyle H}{|}}{\underset{\underset{\displaystyle OH}{|}}{C}}-H$$

b.
$$H-\overset{\overset{\displaystyle H}{|}}{C}=\overset{\overset{\displaystyle H}{|}}{C}-CH_2\text{-}OH$$

no chiral carbons

c.
$$CH_3-\overset{\overset{\displaystyle Cl}{|}}{\underset{\underset{\displaystyle Cl}{|}}{C}}-\overset{\overset{\displaystyle F}{|}}{\underset{\underset{\displaystyle H}{|}}{\overset{*}{C}}}-Cl$$

---

28. _Refer to Section 22.6 and Example 22.7._

a. Refer to Table 22.3 for the monomer. Replace double bonds in the monomer with single bonds connecting monomers together.

$$-\overset{\overset{\displaystyle F}{|}}{\underset{\underset{\displaystyle F}{|}}{C}}-\overset{\overset{\displaystyle F}{|}}{\underset{\underset{\displaystyle F}{|}}{C}}-\overset{\overset{\displaystyle F}{|}}{\underset{\underset{\displaystyle F}{|}}{C}}-\overset{\overset{\displaystyle F}{|}}{\underset{\underset{\displaystyle F}{|}}{C}}-$$

b. $\dfrac{5.0\times10^4\ CF_2\ units}{1\ mol.}\times\dfrac{50.01\ g}{1\ CF_2\ unit}=2.5\times10^6\ g/mol.$

c. $\dfrac{1\ CF_2\ unit}{50.01\ g}\times\dfrac{1\ C}{1\ CF_2\ unit}\times\dfrac{12.01\ g\ C}{1\ C}\times100\%=24.02\%$

   % F = 100% - %C = $100-24.02=75.98\%$

Refer to Problem 10 (above) for the three butene isomers, then systematically replace 2 H's of the double-bonded carbon(s), with Cl and Br.

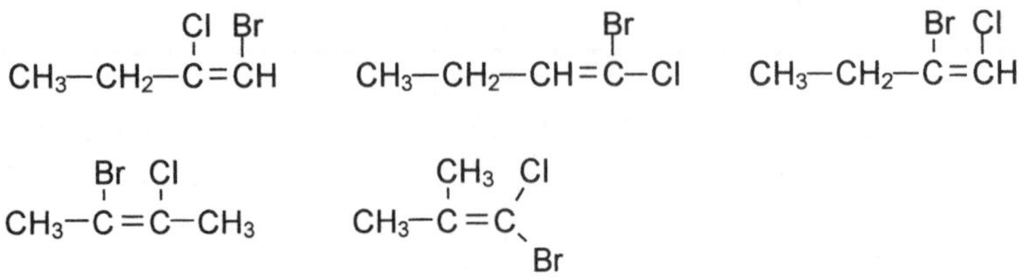

$$\begin{array}{ccc}
\overset{\displaystyle Cl\ \ Br}{CH_3-CH_2-C=CH} & \overset{\displaystyle Br}{CH_3-CH_2-CH=C-Cl} & \overset{\displaystyle Br\ \ Cl}{CH_3-CH_2-C=CH}
\end{array}$$

$$\begin{array}{cc}
\overset{\displaystyle Br\ \ Cl}{CH_3-C=C-CH_3} & \overset{\displaystyle CH_3\ \ Cl}{CH_3-C=C}\diagdown{Br}
\end{array}$$

Since the carboxylic acid group can only be on an end of the hydrocarbon chain, the number of isomers is severely limited.

$$\begin{array}{cc}
CH_3-CH_2-CH_2-COOH & \overset{\displaystyle CH_3}{\underset{\displaystyle CH_3}{\diagup}}CH-COOH
\end{array}$$

All the isomers, except the last one (($CH_3)_2C=CClBr$), have geometric isomers. The last does not have geometric isomers since it has two identical groups ($-CH_3$) on the same carbon. In assigning *cis* / *trans* to isomers, the terms refer to the relationship between the larger groups attached to each carbon. For example, in the first pair of isomers, Cl is larger than C, and Br is larger than H; thus *cis* refers to the isomer in which the Cl and Br are on the same side.

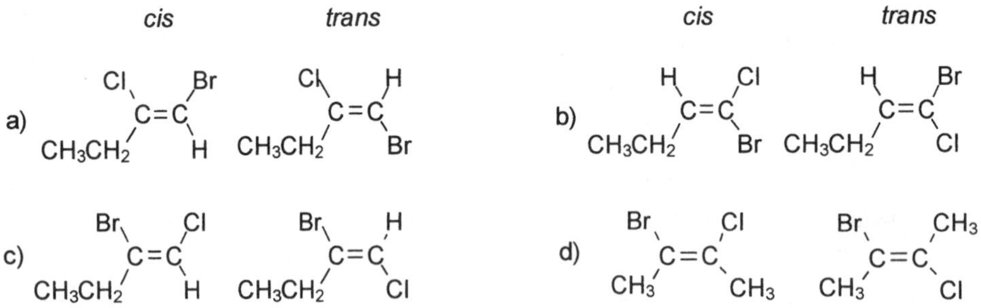

a)
cis   trans

b)
cis   trans

c)

d)

a.   No. The two groups attached to any one of the double-bonded carbons are identical.

30. Refer to Section 22.6 and Example 22.7.

Table 22.3 shows that addition polymers come from monomers with double bonds. Look for the smallest repeat unit; that will be the monomer skeleton structure. Place a double bond between the carbon atoms to get the structure of the monomer.

a. 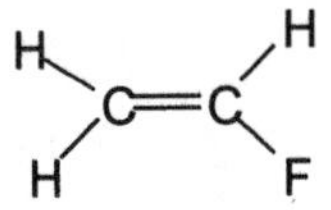   b.

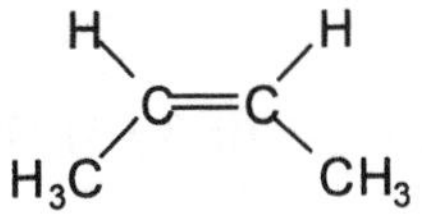

32. Refer to Section 22.6 and Example 22.8.

Condensation polymers are made by removing an OH group from one monomer and H from the other monomer. Remove OH from carbonic acid and H from the other monomer. Join the two monomers together where the groups have been removed.

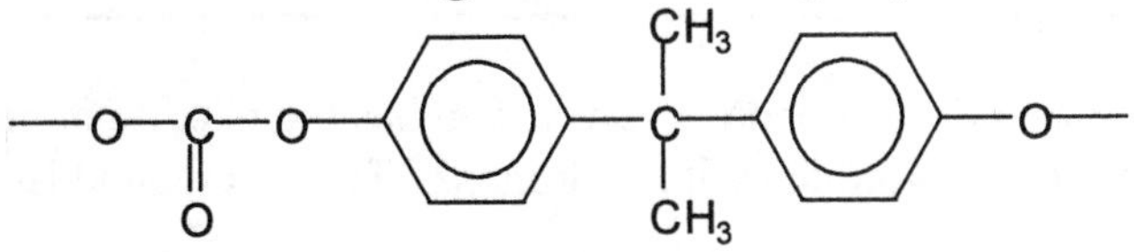

34. Refer to Section 22.6 and Example 22.8.

Condensation polymers are made by removing OH from one monomer and H from the other. To identify the monomer, locate the bond forming the polymer linkage and add OH to one end of the bond and H to the other.

a.   This one is similar to the polymer on p. 596.
     H_2N-CH_2-CH_2-NH_2 and HOOC-CH_2-COOH

b.   This one is similar to the polymer on p. 595.

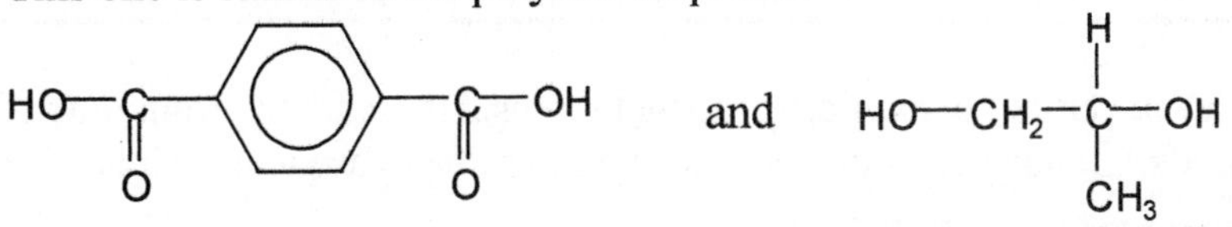

     and

36. Refer to Section 22.3.

The circle represents the six delocalized electrons of an aromatic ring.

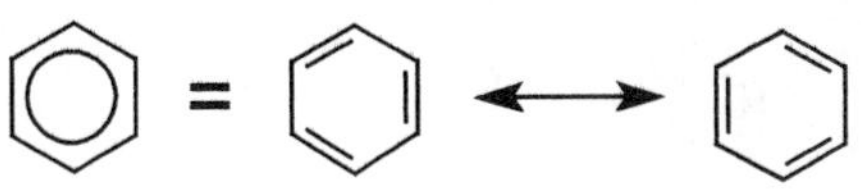

401

The first two reactions are given in the text. The text shows the reaction to form methyl acetate. The formation of ethyl acetate is identical except that ethanol is used instead of methanol.

a. $CO(g) + 2H_2(g) \rightarrow CH_3OH(g)$

b. $CH_3CH_2OH(aq) + O_2(g) \rightarrow CH_3CO_2H(aq) + H_2O(l)$

c. $CH_3CO_2H(l) + CH_3CH_2OH(l) \rightarrow CH_3CO_2CH_2CH_3(l) + H_2O(l)$

ethyl acetate =

a.   A saturated fat is one that has no multiple bonds. Each carbon has the maximum number of H's, thus each carbon is said to be saturated with hydrogens. This compound is an alkane.

b.   Soaps are sodium salts of "fatty acids" (long chain carboxylic acids).

c.   Proof refers to the concentration of an aqueous solution of ethanol. The proof is twice the concentration, expressed as a volume percent.

d.   Ethanol that has been poisoned to render it undrinkable. This is done so that alcohol intended for industrial (non-consumption) purposes, and thus exempt from federal taxes, cannot be bootlegged.

In the structure, each vertex not explicitly labeled represents a carbon atom. Each carbon atom forms 4 bonds. Bonds not shown are understood to be hydrogens. Thus:

counting the C's, H's and O gives: $C_{27}H_{46}O$.

This type of problem is best approached systematically. Start with the linear molecule and move the OH group. Then repeat with a branched molecule, then with a second branched molecule and so on, until all possibilities have been exhausted. Note that the solution below lists the same molecules in the same order as that in Appendix 6; many of the molecules have been "flipped," but they are identical to those in the appendix.

CH_3—CH_2—CH_2—CH_2—CH_2—CH_2OH H_3C—$\underset{\underset{OH}{|}}{CH}$—$CH_2$—$CH_2$—$CH_2$—$CH_3$ H_3C—CH_2—$\underset{\underset{OH}{|}}{CH}$—$CH_2$—$CH_2$—$CH_3$

$HOCH_2$—$\underset{\underset{CH_3}{|}}{CH}$—$CH_2$—$CH_2$—$CH_3$ CH_3—$\underset{\underset{CH_3}{|}}{\overset{\overset{OH}{|}}{C}}$—$CH_2$—$CH_2$—$CH_3$ CH_3—$\underset{\underset{CH_3}{|}}{CH}$—$\overset{\overset{OH}{|}}{CH}$—$CH_2$—$CH_3$

CH_3—$\underset{\underset{CH_3}{|}}{CH}$—$CH_2$—$\underset{\underset{OH}{|}}{CH}$—$CH_3$ CH_3—$\underset{\underset{CH_3}{|}}{CH}$—$CH_2$—$CH_2$—$CH_2OH$ CH_3—CH_2—$\underset{\underset{CH_3}{|}}{\overset{\overset{OH}{|}}{C}}$—$CH_2$—$CH_3$

CH_3—CH_2—$\underset{\underset{CH_2OH}{|}}{CH}$—$CH_2$—$CH_3$ CH_3—$\underset{\underset{OH}{|}}{CH}$—$\underset{\underset{CH_3}{|}}{CH}$—$CH_2$—$CH_3$ $HOCH_2$—CH_2—$\underset{\underset{CH_3}{|}}{CH}$—$CH_2$—$CH_3$

CH_3—$\underset{\underset{H_3C}{|}}{\overset{\overset{OH}{|}}{C}}$—$\underset{\underset{CH_3}{|}}{CH}$—$CH_3$ $HOCH_2$—$\underset{\underset{CH_3}{|}}{CH}$—$\underset{\underset{CH_3}{|}}{CH}$—$CH_3$ CH_3—CH_2—$\underset{\underset{CH_2OH}{|}}{\overset{\overset{H_3C}{|}}{C}}$—$CH_3$ H_3C—$\underset{\underset{OH}{|}}{CH}$—$\underset{\underset{CH_3}{|}}{\overset{\overset{H_3C}{|}}{C}}$—$CH_3$

$HOCH_2$—CH_2—$\underset{\underset{CH_3}{|}}{\overset{\overset{CH_3}{|}}{C}}$—$CH_3$

Start by calculating the amount of heat needed to raise the temperature of the water to boiling.

$q = mc\Delta t$

$m = 1\ qt \times \dfrac{1\ L}{1.057\ qt} \times \dfrac{1000\ mL}{1\ L} \times \dfrac{1\ g}{1\ mL} = 946\ g$

$\Delta t = 100°C - 25°C = 75°C$

$q = (946\ g)(4.184\ J/°C)(75°C) = 297000\ J = 297\ kJ$

Now calculate the amount of heat given off by the combustion of one mole of propane.

$C_3H_8(g) + 5O_2(g) \rightarrow 3CO_2(g) + 4H_2O(l)$

$\Delta H° = \Sigma \Delta H_f°$ products $- \Delta H_f°$ reactants

$\Delta H° = [3\Delta H_f°CO_2 + 4\Delta H_f°H_2O] - [\Delta H_f°C_3H_8 + 5\Delta H_f°O_2]$

$\Delta H° = [(3 \text{ mol.})(-393.2 \text{ kJ/mol.}) + (4 \text{ mol.})(-285.8 \text{ kJ/mol.})]$
$\qquad - [(1 \text{ mol.})(-103.8 \text{ kJ/mol.}) + (5 \text{ mol.})(0 \text{ kJ/mol.})]$

$\Delta H° = -2220 \text{ kJ/mol.}$

Finally, calculate the moles and grams of propane needed to raise the temperature of the water.

$$297 \text{ kJ} \times \frac{1 \text{ mol. } C_3H_8}{2220 \text{ kJ}} \times \frac{44.10 \text{ g } C_3H_8}{1 \text{ mol. } C_3H_8} = 5.9 \text{ g}$$

About 5.9 g propane is needed to raise the temperature of one quart of water from 25°C (roughly room temperature) to boiling. This neglects the specific heat of the pan and also assumes that all the heat given off is actually transferred to the water.